DISCARD

the
SCIENTIFIC
AMERICAN
book of
DINOSAURS

the
SCIENTIFIC AMERICAN
book of
DINOSAURS

GREGORY S. PAUL, EDITOR

A BYRON PREISS BOOK

ST. MARTIN'S PRESS
NEW YORK

A Byron Preiss Book.
Editorial Project Coordinator: Howard Zimmerman
Associate Editor: Dinah Dunn
Assistant Editor: Kelly Smith
Design by Tom Draper Design

The following essays first appeared in the pages of *Scientific American* magazine:
"Dinosaur Renaissance" by Robert Bakker, April 1975; "The Mass Extinctions of the Late Mesozoic" by
Dale Russell, January 1982; "A Volcanic Eruption" by Vincent Courtillot, October 1990; "An
Extraterrestrial Impact" by Walter Alvarez and Frank Asaro, October 1990; "How Dinosaurs Ran" by R.
McNeil Alexander, April 1991; "Australia's Polar Dinosaurs" by Patricia Vickers-Rich and Thomas Hewitt
Rich, July 1993 (updated by the authors and their co-workers for this book); "The Art of Charles R.
Knight" by Gregory S. Paul, June 1996; "Tracking a Dinosaur Attack" by David A. Thomas and James O.
Farlow, December 1997; "The Origin of Birds and Their Flight" by Kevin Padian and Luis M. Chiappe,
February 1998; "The Teeth of the Dinosaurs" by William L. Abler, September 1999; "Breathing Life into
Tyrannosaurus Rex" by Gregory M. Erickson, September 1999.
ISBN 0-312-26226-4

First Edition: November 2000

10 9 8 7 6 5 4 3 2 1

Books are available in quantity for promotional and premium use.
Write Director of Special Sales, St. Martin's Press, 175 Fifth Avenue, New York, NY 10010, for information
on discounts and terms, or call toll-free 1-(800)-221-7945. In New York, call (212)-674-5151 (ext. 645).

TABLE OF CONTENTS

INTRODUCTION

by

Gregory S. Paul

It may be difficult for younger dinosaur enthusiasts to know what it was like to be one back in the 1960s. True, dinosaurs were part of popular culture: the Sinclair Oil dinosaur logo could be seen adorning gas stations across much of the country; there were children's books on the subject, such as the *How and Why Wonder Book of Dinosaurs*; the comic strip *Alley Oop* was filled with the great beasts; and there were even dinosaur flicks such as *One Million Years BC* starring the lovely Raquel Welch. Ever since the discovery of dinosaurs in the early 1800s, and especially since the uncovering of complete skulls and skeletons of the often huge and weird monsters of the ancient world, dinosaurs have captured the public's imagination, but it was nothing compared to today's interest. I never saw a documentary on dinosaurs because no one had yet made one. Adult books featuring dinosaurs were scarce, so I pored over the dinosaur section of *The Fossil Book* trying to satisfy my need to learn about all things archosaurian. Even in science magazines, essays on dinosaurs were scarcer than a hen's tooth. For years, I looked for a skeletal restoration of my favorite dinosaur, *Brachiosaurus*. But none was to be found, even though it was the world's largest dinosaur! At the Smithsonian, which I often visited, the prehistoric mammal hall had vast and superb murals of Cenozoic marvels by Jay Matternes, but in the dinosaur hall, we had to make do with a few black-and-white sketches. Sigh.

Even the illustrations of the restored dinosaurs were different, as was the established view of what these creatures were and how they had lived. They were seen as reptiles, in the main, sluggish reptiles—ponderous, slow-moving, and barely able to carry their vast bulk across steamy, fern-filled Mesozoic landscapes. Indeed, the enormous sauropods appeared too big to drag their overwhelming bulk onto dry land, except when driven— by the instinct to reproduce—to lay their eggs. They, and the later duckbilled hadrosaurs, were thought to have fed on the soft aquatic plants that satisfied the low, reptilian rates of metabolism common to the scaly skinned dinosaurs. The water-loving dinosaurs had little to fear from the great predatory dinosaurs of the day, for the latter, it seemed, were as fearful of water as a person with rabies. On land, when *Triceratops* faced off with *Tyrannosaurus*, they thrust and parried at each other in dinosaurian slow motion. Oddly enough, dinosaurs never seemed to lift their feet off the ground. In almost every illustration, every massive foot was planted on terra firma. Walk into a dinosaur hall circa 1960, and all the skeletons had their two or four feet bearing their massive loads. How did they ever manage to get around?

Before the recent explosion in the popularity of dinosaurs, the Sinclair logo was one of the best-known dinosaur images. Oil is not, however, derived from the remains of dinosaurs.

And dinosaurs were regarded as dumb colossi driven by walnut-sized brains, without the wits to live in organized groups and raise and protect their young. And they were a dead end—completely extinct saurians that had left no descendants. Why they all went "belly-up" no one knew. Perhaps, some hypothesized, it had become too cold for their reptilian thermo-regulatory system. No wonder those big-brained, energetic mammals arose to take their place.

Evolutionary biologist Stephen J. Gould has characterized this view of dinosaurs as the "Postmodernist Consensus." And those of us with an affinity for dinosaurs at the time did not recognize that research in the field was pretty much moribund. The Great Depression and World War II had brought most fieldwork to a halt, and most postwar vertebrate paleontologists had turned to groups with living descendants—fish, amphibians, and mammals. Mammal teeth, in particular, were a popular area of study, as their complexity facilitated phylogenetic and functional studies. Scientists had better things to do than bother with a bunch of clunky, simple-toothed reptiles that were nothing more than an evolutionary sideline; they were kid stuff, below the notice of the serious professional. Consequently, technical papers on dinosaurs were few and far between. New dinosaur finds were even scarcer.

Then things started to stir. Sickle-clawed *Deinonychus* was revealed to be a big-brained, agile, sophisticated little predator whose form was more like that of birds than reptiles. A fellow named Jim "Dinosaur" Jensen out in western Colorado was digging up what he said were the world's largest dinosaurs. A new kid on the block, Robert Bakker, was saying that those sauropods spent their lives not wallowing in the water, but striding about conifer-dominated landscapes like super-elephants and giraffes. The duckbills, too, were being dragged, without too much fuss, out of the rivers and lakes onto dry land. *Tyrannosaurus* was re-restored with its tail up in the air and one of its feet swinging forward in a long stride. Horned dinosaurs were illustrated galloping—galloping!—like a herd of charging rhinos. One wondered: What next?

The answer came in a short blurb in a museum magazine revealing that Bakker was now arguing that dinosaurs were "warm-blooded" like birds and mammals. Though I was shocked, a lot of this new thinking on dinosaur biology was not making the big-time media. But for me, personally, it was a time both exciting and frustrating.

I'll never forget walking into the library at Northern Virginia Community College, Annandale Campus, and seeing displayed on the new periodical rack the April 1975 issue of *Scientific American*. Adorning its cover was a bright green image of the strange climbing reptile *Longisquama,* with featherlike scales sprouting from its back below the title, "Dinosaur Renaissance." Even before I reached for the magazine, I knew Bakker was the author of the essay. What wonders it contained: Illustrations showed furry, mammal-like reptiles facing off in ancient snows; furry pterosaurs; and *Archaeopteryx* using its feathered wings to snare an insect. Charts, graphs, and maps presented evidence that dinosaurs, and even their thecodont ancestors, had shared varying degrees of elevated metabolic rates. Plus, there were comments on John Ostrom's hypothesis that birds were the direct descendants of small predatory dinosaurs; and there was Bakker's notion that they should be grouped with thecodonts in a new class called Archosauria. So dinosaurs had not gone extinct after all; they were still among us in winged form! A revolutionary idea indeed.

But perhaps packing the most punch of all in that article—to a degree that is difficult to comprehend today—was the speculative sketch of a feathered *Syntarsus* dashing down the slope of a dune in hot pursuit of a gliding lizard, only one of its feet barely touching the sand. This had never been considered before: a feathered, not scaly, dinosaur, a dinosaur that was near avian, not reptilian.

"Dinosaur Renaissance" was the first major presentation of the postmodern revolution in dinosaurology. Its impact on both the scientific community and the public cannot be overemphasized. Dinosaurs were now a hot, and controversial, topic in both arenas. No longer did those mammal tooth cusps seem so intriguing. Growing numbers of researchers turned their attention to the ancient archosaurs, many hoping to refute parts of the radical revolution they profoundly disagreed with. The first dinosaur documentaries began appearing. In short, it was a whole new game.

In August 1978, I was with the remnants of the Johns Hopkins University field crew, packing up our summer's worth of work in Boulder, Colorado. One warm evening on the local campus, we heard news reports that our friend John Horner had found the eggs and even nests of duckbilled dinosaurs up in Montana. Horner's crews would, in the following years, dig up the evidence that these small-brained creatures were smart enough to build complex nests in which to incubate their eggs, and then may have taken care of the hatchlings until they were large enough to leave the nest. The first major popular presentation of this body of work was Horner's "The Nesting Behavior of Dinosaurs," published in *Scientific American* in 1984. Dinosaurs, it turned out, were not so dumb after all.

Horner's work also encouraged people to look for more dinosaur eggs. Now, what was once considered extremely rare has proven to be abundant, with some sites containing

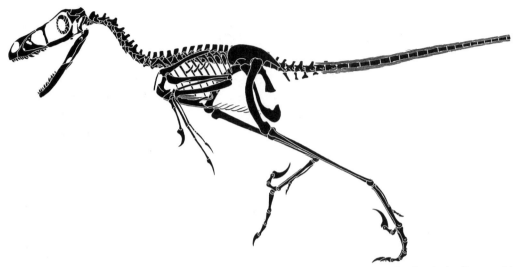

Little, bird-like *Bambiraptor*, only 3 feet long, is just one of a host of new dinosaurs that has been discovered in recent years.

tens of thousands of dinosaur eggs, and a few containing the remains of embryonic skeletons.

During World War II, Luis Alvarez designed the first radar guidance system for landing aircraft in dense fog; he was then called to Los Alamos, New Mexico, to help devise the explosive trigger for the Fat-Man plutonium bomb fired at Trinity and Nagasaki. After the war, Alvarez was involved in development of the H-bomb, and his work in astrophysics won him a Nobel Prize in science. But he didn't stop there. In the late 1970s, he and his son Walter were cooperating on a project to use the iridium deposited by cosmic dust to determine the rate at which marine sediments were laid down at the very end of the dinosaur era, 65 million years ago. To their surprise, they found the boundary layer loaded with far more iridium than could be accounted for by the normal inflow of the extraterrestrial heavy metal. They proposed that an enormous asteroid or comet, as big as Mt. Everest, had hit the Earth at that time. The resulting aftereffects of the explosion—millions of times more powerful than the H-bombs Luis had helped build—led to the extinction of dinosaurs and many other life-forms. If the debate over dinosaur energetics and bird origins had been intense, this was a topic as white hot as the fireball created by the extraterrestrial impact—one that even led to a fistfight in a conference hallway. The continuing debate, with more decorum, graced the pages of *Scientific American* in 1990.

In the late 1970s and early 1980s, when dinosaurology was still revving up, we few dinosaur researchers were grateful when the annual Society of Vertebrate Paleontology (SVP) meeting featured eight or even—praise the paleo gods—ten talks on dinosaurs. As the 1980s and 1990s progressed, the field ballooned. All over the world, paleontologists scoured dinosaur beds for the abundant remains they contained. To these finds, they applied new techniques and technologies unavailable to previous researchers. What had been a slow-moving area of research suddenly became fast-paced, with new dinosaurs named at a rate not charted for a hundred years. The new discoveries and ideas boosted popular interest as the public became increasingly fascinated with the notion that our feathered friends were actually living dinosaurs, whose energetics might not be very different from those of their ancestors—ancestors that had succumbed to the super-ecodisaster caused by a visitor from the heavens. The result was a growing body of adult and children's books, articles, and documentaries on the dinosaurs and related matters.

Just when things looked like they could not get hotter, Michael Crichton's *Jurassic Park* became a best-seller, soon translated by Steven Speilberg into a motion picture blockbuster as big as the dinosaurs it featured. A feedback system developed: Popular interest boosts scientific interest that generates new finds and innovative concepts, which in turn further spurs popular interest, and so on. Nowadays, dinosaur talks at SVP fill two days' worth of the schedule and pack the largest rooms. Bookstore shelves groan under the weight of dinosaur books, and there are so many dinosaur documentaries that one is hard-pressed to watch them all. Most successful of these documentaries has been the recent *Walking with Dinosaurs*, a spectacular (albeit filled with minor flaws) techno-opus that drew the largest cable-viewing audience ever for a nonsporting event. Within weeks of the dawning of the new millennium, 2000 became the year of the dinosaur when the tyrannosaur named Sue made its official debut; the finding of a dinosaur heart was announced; and Disney made a big splash with its anthropomorphized animated beasts in the movie *Dinosaur.*

So popular are dinosaurs and other fossils today that the topic has, in some regards, gotten out of hand. It used to be that researchers could knock on a rancher's door, ask if they could explore and dig on his or her property, and if they found a stegosaur or tyrannosaur skeleton, cart it back to the museum and put it up for all to wonder at. Things have become more complicated—and much more lucrative. Fossils small and large have become collectors' items, with price tags to match. The *Tyrannosaurus* skeleton named Sue became the focus of an FBI raid, and criminal charges were filed, followed by the sale of Sue at Sotheby's for a cool $8.5 million. Researchers and museums often find themselves unable to compete with the private market for the best dinosaur skulls and skeletons. On the other hand, the new value of fossils has encouraged new finds. In China, farmers to the northeast of Beijing realized that the feathered fossil birds they could dig up easily from local deposits would fetch a pretty penny on the fossil black market. Thus, professionals became aware of the extraordinarily productive 120-million-year-old Yixian lakebed fossil deposits that not only have produced more than a thousand bird skeletons, but also a set of feathered dinosaurs. These fossils have verified the once-radical concept. Alas, the same commercial pressures have caused the locals to occasionally "improve" the birds and dinosaurs, resulting in repair work that, for example, some have mistaken for crocodilian-like livers and lungs; in another case, tail transfers have been reported as new dino-birds.

As interest and research in dinosaurs has flourished like never before, *Scientific American* has kept up. In its pages, Kevin Padian and Luis Chiappe covered the bird origins debate from the dinosaurian perspective in "The Origins of the Birds and Their Flight." From down under, Thomas Hewitt Rich and Patricia Vickers-Rich, in "Australia's Polar Dinosaurs," presented evidence that during the balmy Mesozoic, dinosaurs living near the south pole were surviving long, dark winters of hard freezes and snow. Was nothing of the postmodernist consensus sacred? Apparently not. In his *Scientific American* essay "How Dinosaurs Ran," one of the originators of biomechanics, McNeil Alexander, demonstrated why illustrations of galloping *Triceratops* might not be so outrageous after all. In "Tracking a Dinosaur Attack," Jim Farlow and the late David Thomas took old data—the spectacular giant sauropod and theropod trackways uncovered in Depression-era Texas—and argued that it recorded a direct attack of a dinosaurian predator upon a great herbivore. Of course, the Tyrant King has not been forgotten. Gregory Erickson looked at how modern paleontology was reexamining this classic dinosaur in his *Scientific American* offering, "Breathing Life into Tyrannosaurus rex."

The science of dinosaurs is evolving so fast that it is hard even for the paleontologist to keep pace with all the new information and concepts, harder still for the interested layperson. That is why *The Scientific American Book of Dinosaurs* is so important: it gathers the classic pieces from the pages of *Scientific American* and combines them with new, cutting-edge essays by leading researchers in the field. Jim Farlow catches us up on the latest on dinosaur trackways. Per Christiansen, who has done innovative technical research into the biomechanics and locomotion of dinosaurs great and small, presents a popular account of his and others' work. Kristina Curry Rogers showed that the sauropods achieved their vast bulk by growing as fast as whales, and here explains how she did so. Reese Barrick has used bone isotopes to examine the thermo-regulation of dinosaurs, and here he discusses this still-controversial subject, including the fascinating discovery of what appears to be a fossil heart! Philip Currie catches us up on the equally extraordinary finding of feathered dinosaurs in China. As with all science, dinosaurologists often disagree with one another. This is reflected in the pages of this multiauthor volume. Compare the differing opinions of Barrick and David Norman on the metabolic condition of dinosaurs, for example.

Dinosaurs first appeared on the scene some 220 million years ago. They dominated the land for more than 150 million years, and they still rule the skies. Dinosaurs of the non-avian sort were real dragons that lived in a world alien in time, rather than place. They, too, are still with us. Having fascinated people since the early 1800s, they will continue to do so in the 2000s. Consider that vast stretches of territory remain unexplored for dinosaurs. Who knows what remains await discovery in parts of the Sahara or in Tibet, where barely a geologist has passed by. Or consider that the peculiar croclike dinosaur *Baryonyx* was discovered just a few years ago on the outskirts of London, from the same beds that produced the original *Iguanodon*. How these dinosaurs will be found and who will analyze them may be another matter. The days when human beings, baking in badland heat, swatting at pesky insects, slowly walking across barren grounds, eyeballing for the traces of fragmented bone that indicate the presence of a buried skeleton, may soon pass. Will new super-scientists soon take over the task, with minds cyber, rather than human, in form? Perhaps these cyber-sleuths will be able to use sophisticated remote sensing systems to probe deep into ancient sediments to find, and then examine, the bones of dinosaurs without ever bothering to dig them up. If so, then we humans may become as obsolete as the extinct dinosaurs we so enjoy. But don't fret, it's just evolution.

Chapter One
Discovering Dinosaurs

Introduction

Everything we know about dinosaurs has been compiled from less than two centuries' worth of scientific research, so the best place to start our journey into dinosaur paleontology is with a history of the effort. And because dinosaur research got its start in the British Isles, who better than Michael Benton, who hails from those lands, to handle the subject?

A Brief History of Dinosaur Paleontology

by Michael Benton

First Finds

The first dinosaur bones must have been found in ancient times, but no record was kept of these discoveries. In medieval times, philosophers commented on fossil shells and sharks' teeth they had seen, and they debated the origin of these strange artifacts. Were these stony petrifactions in any way related to modern shells and fishes, or were they simply odd pebbles that happened to look like the remains of plants and animals? A popular view was that fossils were "sports of nature" formed in the rocks by plastic forces. (The nature of these plastic forces was never made clear.)

The first dinosaur bone to be described was found while this debate raged, and the line of argument followed by its describer is revealing. Robert Plot, professor of "Chymistry" at the University of Oxford was known to be preparing a monumental book on the natural history of Oxfordshire, and local naturalists sent him unusual specimens to include. These included a weighty specimen that had been collected in a shallow limestone quarry at Cornwell in north Oxfordshire. Plot shared the specimen in a figure that also contained illustrations of numerous other strangely shaped stones, some of which he interpreted as preserved kidneys, hearts, and feet of humans. His interpretation of the rock from Cornwell was, however, quite different.

Plot saw that the specimen looked like a bone. It had a broken end, which was circular, and seemed to have a hollow core that was full of sand. The fractured surface round the core showed clear porous patterns exactly as in bone, and the shape of the opposite end, with its two large rounded processes, was just like the knee end of a thigh bone. Unlike the mystical interpretations that Plot gave to many of the other stones illustrated in his book, he seemed to have no doubts about this giant bone. Plot elaborated on the kind of creature that could have produced such a monster bone. He stated that it had come from an animal that was larger than an ox or horse; and he considered the possibility that it might have come from an elephant brought to Britain by the Romans, though he ruled out that possibility since the bone was even bigger than that of an elephant.

Plot's final decision was that the Cornwell bone came from a giant man or woman. He referred to mythical, historical, and biblical authorities in support of this interpretation ("there *were* giants in those days"). After a promising discussion, Plot's conclusion, to identify the bone as human, might seem perverse to us, but recall that no one at the time had an inkling of the former existence of dinosaurs and other extraordinary animals in the history of the Earth. Indeed, there was no acceptance of the idea that some plants and animals might have become extinct, since that would imply that God had made a mistake; nor was there any idea that the Earth was very old.

In fact, Plot's bone was later identified from his illustration as the lower end of the thigh bone of *Megalosaurus*, a dinosaur that is now relatively well known from the middle Jurassic of Oxfordshire. Plot's specimen is now lost, and there is a twist in the tale: It was illustrated again in 1763 by R. Brookes, and he named it *Scrotum humanum* in honor of its appearance. This is the first named dinosaur, although unfortunately the name has never been used seriously.

The Idea of Extinction

Extinction seems a very obvious aspect of nature to us now, but in the seventeenth and eighteenth centuries, this was not the case. By 1700, most naturalists accepted that fossils truly represented the remains of ancient organisms that had somehow been buried and turned to stone. One major question remained: Were these the remnants of extinct plants and animals? From 1750 to 1800, a heated debate raged about the possibility of extinction. Until 1750, most naturalists believed that fossils represented species that were either known to be still living or would soon be found in some hitherto unexplored parts of the world. Isolated fossil bones of vertebrates, backboned animals, had been found in Ireland, Britain, France, and elsewhere, but none of these shook the faith of naturalists that extinction had not happened.

Everything changed in the 1750s when explorers in North America began to dig up the remains of elephants (mastodons and mammoths) and sent some of the bones to Paris and London. There, distinguished anatomists and naturalists, such as William Hunter, Georges Louis Leclerc, Comte de Buffon, and others debated the specimens. As more and more specimens were found, and as more of the Americas were explored, it became clear that they truly were remains of recently extinct forms. Baron Georges Cuvier of the Muséum d'Histoire Naturelle in Paris was instrumental in clarifying this question. He showed, in a series of books and papers from 1796 onward, that the fossil elephants and giant mammal bones from Russia, South America, and other parts of the world represented extinct species.

The Vastness of Geological Time

The mammoths, mastodons, and giant ground sloths of the late eighteenth century obviously were not very ancient fossils; the bones were still in good condition, and the skeletons were often quite complete. Geologists were happy to accept that some of these animals had perhaps died out only a few thousand years ago.

Other fossils seemed much more ancient, however. Their bone or shell material seemed to be filled with crystalline minerals, often calcite, quartz, or iron oxide, and these minerals must have taken some time to enter the pores and solidify. In many cases, the fossils seemed to come from plants and animals with no obvious living relatives. Geologists debated the true antiquity of the Earth during the late eighteenth century, but the outlines of what had happened in the past became clear only from 1800 onward.

The first breakthrough came in the writings of James Hutton (1726–1797), a Scottish agriculturalist and naturalist. In his *Theory of the Earth* (1795), he argued that the Earth was ancient, based on his observations of modern-day processes. Hutton observed the rates of accumulation of sediments and the rates of erosion by mountain streams in his native Scotland, and he believed that such processes must always have been equally slow. He then compared modern rates of processes with the vast piles of rocks he saw in cliffs around Scotland, and the depth and ruggedness of the mountains. At slow rates of deposition and erosion, these features must have taken vast amounts of time, presumably millions of years, to form. Unfortunately, Hutton's book is very dull; even readers in his day found it so. Luckily, his work was popularized by James Playfair, and the idea of an ancient Earth spread rapidly among naturalists and philosophers.

The antiquity of the Earth was a theme developed by later geologists, among them Charles Lyell (1797–1875), another Scot. In his highly influential *Principles of Geology* (1830), he spread his net wide and called on examples from all over Europe to show how landscapes had changed, coasts had moved vast distances, and fossils were in abundance everywhere. Lyell especially championed Hutton's application of modern processes to the interpretation of the past, the principle of uniformitarianism, or "the present is the key to the past." Between 1795 and 1830, practical field geologists had begun to produce geological maps, and to divide up the sedimentary rocks according to their fossils. This marked the beginning of geology as a science, and particularly the beginning of the modern international geological time scale.

Making Sense of the First Dinosaurs

The first dinosaur discoveries, from 1818 to 1840, began to paint a portrait of an astonishing fauna of giant reptiles in the Mesozoic. The dinosaurs, found first in England, and then elsewhere in Europe and, after 1850, on every other continent, showed early naturalists that whole assemblages of plants and animals, quite unlike anything now living, had existed in the past. How were paleontologists to interpret these extraordinary giant bones that seemed to be quite unlike those of any living animal?

Georges Cuvier provided the method in the new science of comparative anatomy, a discipline in which he established new standards of precision. He carried out a painstaking comparison of every bone of a fossil form, and noted similarities and differences between equivalent elements in the skeletons of a variety of extinct and living forms. He found that the shapes of bones indicated the purposes for which they were used and the relationships of the animals in question. By the 1820s, Cuvier had honed his skills in comparative anatomy to such a pitch of perfection that it was said he could identify any animal from a single bone, and that he could reconstruct any unknown fossil form from a single bone.

Bones of a large meat-eating dinosaur came to light north of Oxford about 1818, and they were taken to William Buckland (1784–1856), professor of geology at the University of Oxford, and dean of Christ Church. It was not uncommon for churchmen to combine their careers with science in those days. Buckland was shown some collections of bones and teeth of a large meat-eating reptile about 1818, but he could not identify the bones, and he showed them to Cuvier in Paris, and to other experts. In the end, Buckland classified the animal as a giant reptile, probably a lizard, and he estimated it had been 40 feet long in life. After six years of consideration, Buckland finally published a description of the bones in 1824, where he stated that they came from a giant reptile, which he named *Megalosaurus* ("big reptile"). This was the first dinosaur to be described.

At the same time, independently, Gideon Mantell (1790–1852), a country physician in Sussex, was amassing large collections of Mesozoic fossils. During a visit to a patient near Cuckfield, so the story goes, his wife Mary, who had come with him, picked up some large teeth from a pile of road-builders' rubble. Mantell realized the teeth came from some large plant-eating animal, and when he sent them to Cuvier, the great French anatomist assured him that the animal must have been a rhinoceros. Then, Mantell compared the teeth with those of other modern animals in the Hunterian Museum in London; a student there, Samuel Stutchbury, showed him that they were like the teeth of a modern plant-eating lizard, the iguana, though they were much bigger. So Mantell described the second dinosaur, named by him *Iguanodon* ("iguana tooth") in 1825, based on the teeth and some other bones he had found since.

Buckland and Mantell had great difficulty interpreting those early dinosaur bones since there were no modern animals like them. In the end, they decided that *Megalosaurus* and *Iguanodon* were giant lizards. The third dinosaur to be named, *Hylaeosaurus*, came to light in southern England in the same Wealden beds that had produced *Iguanodon*. It was named in 1833. Further dinosaurs were named in the 1830s, including two from the Triassic: *Thecodontosaurus* was named in 1836 by Henry Riley and William Stutchbury from Bristol, SW England, and *Plateosaurus* in 1837 by Hermann von Meyer from southern Germany. It was only later that paleontologists realized these were both prosauropods (basal sauropodomorphs) and, in fact, closely related. By 1840, a number of different dinosaurs had been named; these included a theropod (*Megalosaurus*), two basal sauropodomorphs (*Thecodontosaurus* and *Plateosaurus*), an ornithopod (*Iguanodon*), and an ankylosaur (*Hylaeosaurus*). In 1841, Richard Owen named the first sauropod, *Cetiosaurus*. But still no one knew what these extraordinary giant beasts were.

Richard Owen and the Mammalian Dinosaurs

Richard Owen (1804–1892), professor of anatomy at the Royal College of Surgeons in London, was set the task of reviewing everything that was then known about the giant reptiles of the land, sea, and air. He had some difficulty in understanding the land reptiles, as some were clearly crocodiles, others were small and lizardlike, but there were all those giant "lizards." What was he to do? At first, he tried to shoehorn them all into existing groups. Thus, some were overblown lizards, others were crocodiles, and still others were something in between. Eventually, he turned that idea on its head when, in 1842, he argued that some of them at least represented a new group, which he called the Dinosauria (meaning "fearfully great reptiles"), characterized by being terrestrial and huge and by having more than two vertebrae in the sacrum (most reptiles have two elements of the backbone, which attach to the hip girdle; dinosaurs have three to seven). Owen argued that the dinosaurs were bulky quadrupedal reptiles that had many advanced mammal-like features. But he included only *Megalosaurus*, *Iguanodon*, and *Hylaeosaurus* in the group; the others he left as giant lizards and crocodiles.

In 1851, a huge exhibition of technical, scientific, and trade goods was held in the center of London. The so-called Great Exhibition was the biggest show of its kind, and it displayed all the wonders of British industry—the steam engines, railways, massive new factory works and products, the fruits, minerals, and fibers from the British Empire. It was housed in a vast metal and glass structure, the biggest ever built. When the exhibition was over, the public wanted to preserve the vast glass house, so it was taken apart and set up at a new location in south London. There, christened the Crystal Palace, a new public park was created for the enjoyment and education of the citizens. Prince Albert, Queen Victoria's husband, was keenly interested in science, and he wanted the park to reflect all the latest knowledge. He turned to Sir Richard Owen, as he then was, and together they developed a scheme to illustrate the geological history of the Earth in the park. Great rock formations were to be brought in from around the land and set up in the correct geological sequence. Lakes were excavated, and massive hydraulic schemes were designed to operate the water levels, creating floods and tides.

The studio of Waterhouse Hawkins as it appeared in 1853, while the naturalist-artist prepared his sculptures for the grounds of the Crystal Palace (note the vermin scampering on the floor). The dominant figure represents the view of the dinosaur *Iguanodon* at that time, when only parts of its skeleton were known. The spike was applied errantly to the nose. These pieces are still on display. In the same year, Richard Owen, who had coined the term dinosaur, offered a toast to his learned colleagues inside the newly completed sculpture of *Iguanodon*.

Celebrating the near completion of the world's first dinosaur sculpture, Richard Owen toasted his peers at a dinner party recreated here by an artist of the time. Although the public had but a single, incorrectly reconstructed animal to view, imaginations were set afire and "dinomania" was born.

The high point of the geological garden was to be a set of life-sized models of the ancient creatures that Owen had helped to interpret. There were models of ancient fishes, amphibians, marine reptiles, flying reptiles, and, of course, dinosaurs. The sculptor Waterhouse Hawkins was engaged to design the structures. Each dinosaur was to be sculpted in concrete over a hollow framework built from steel and bricks. These were to be some of the first "artificial" sculptures in the world.

Everything was ready by the end of 1853, and a great dinner was held inside the model of the *Iguanodon*, before the top was fitted. At the beginning of 1854, when the public first saw the Crystal Palace dinosaurs, they were amazed. The first "dinosaur craze" was underway. Pictures of Owen's marvels were made and sold as posters. Copies appeared in the popular illustrated newspapers, and small-scale models, made from plaster or bronze could be purchased as souvenirs. Dinomania is nothing new!

But why did Owen make these first life models of dinosaurs so mammal-like, which today look very odd to us? Partly, it was because, at that point, he did not have any complete skeletons. The first five or six dinosaurs from Europe were based on isolated bones. More to the point, careful historical study by Adrian Desmond of London has shown that Owen had an ulterior motive: He had mixed views about evolution and certainly did not accept that life had evolved from simple to complex organisms—the principle of progressionism. He believed that the fossil record showed many cases of *degeneration*; that is, a kind of reversal of evolution, whereby plants or animals that were once highly complex and advanced, degenerated to a more primitive state. He believed that modern reptiles were degenerate, the mere inferior relics of a once much more glorious past. Consequently, we have large, possibly warm-blooded, mammal-like dinosaurs in the Mesozoic, and miserable, cold, slimy, creeping lizards and crocodiles today, sad remnants of a once-great dynasty.

Progressionism and evolution were hot topics in the 1850s, and dinosaurs played their part. The debate over progressionism is forgotten now; not so the one over evolution. Thus it is important to understand where and how evolution enters the picture in the history of dinosaur studies.

Evolution

The theory of evolution by natural selection was proposed by Charles Darwin in 1859, who had come to this theory after a long voyage of discovery in the early 1830s, both literally and figuratively. When he set sail as gentleman naturalist onboard the survey ship Beagle, he held the general views of his day, that life had been created, and that species did not change. He had, after all, been brought up as a traditional Christian, and he was training to become a minister of the church. He came back from his voyage of discovery as a convinced evolutionist. It is important to note that Darwin did not propose the idea of evolution; this view had been championed by distinguished French naturalists, such as Buffon and Lamarck, during the eighteenth century. Their idea was that species were not fixed, and that they could change in some way through time, although these early proponents of evolution were unclear about how this happened. That is all evolution means: literally "unrolling," or change.

Paleontologists throughout Europe had also been gathering evidence that pointed to some kind of change, or progression of life, through time. They found simple fossils in older rocks, then fishes, then reptiles, then mammals, which showed, they thought, some kind of sequence of development. Whether this progression meant that there had been a series of separate creations and extinctions or that ancient fossil forms had somehow changed was debated in a general way. This was the progressionism against which Owen railed.

During the voyage of the *Beagle*, Darwin saw and collected numerous examples of fossil mammals from South America, and these confirmed that the fossil forms resembled mammals still living in South America. If life had been created, how could there be apparent evidence of some relationship between extinct and living animals in one part of the world? When he visited the Galapagos islands, Darwin saw that the plants and ani-

The modern view of *Iguanodon*, as restored to life by paleontologist and paleo-artist Gregory Paul. It is a totally different animal from the original restoration.

mals there were like those on the South American mainland. How could that be, if life had been created? He saw also that the tortoises and finches on each of the dozen or so islands in the Galapagos group looked very similar, even as he observed that there were clear differences from island to island. When he showed his collection of bird skins to an ornithologist on his return, he was assured that the finches were all different species. Darwin had to admit the impossible: Species were not immutable. If species were not permanent, that meant they could evolve and split. This also meant that God had not created all life in one act. All life could have evolved from a single common ancestor, and the vastness of geological time was already available for this process to take place.

The final brick was fitted into Darwin's shocking new edifice when he read "An Essay on the Principle of Population," written by Thomas Malthus, an economist. It described how human populations always breed faster than the increase in available food, and that certain processes must come into play to maintain the correct level of population. Darwin saw that this idea applied to animals and plants, which all produce too many young to survive, and that, in general, only the strongest survive. The features that enable them to survive (bigger teeth, stronger legs, brighter feathers) must be inherited in some way, to be passed on to their offspring. In time, the makeup of the whole population may change, or evolve, in the direction of the features that most promote survival at the time. This is natural selection.

Looking at the diversity of life today, the countless millions of living species, while looking back over the history of life—admittedly, only patchily known in the 1830s—Darwin could see a single principle at work: Species could evolve and split. In time, those species themselves could change further and split, and over countless millions of years, a single population of simple organisms could have diversified into many species. A group such as the dinosaurs could have arisen from a single ancestor, which, if well adapted to the prevailing conditions, would survive and multiply. Over time, as new opportunities presented themselves, and with the vast potential for reproduction and variation, any kind of evolutionary change could be imagined: increase in body size, change from bipedalism to quadrupedalism, change in diet, or change in habitat.

Missing Links: Archaeopteryx

In his *On the Origin of Species,* Darwin devoted two chapters to paleontological questions. He realized that fossils could provide critical evidence that evolution had actually happened; that is, looking at geographic distributions of modern plants and animals, and talking to animal breeders, could show how evolution might have happened, but the time dimension was essential to show it really happened. But Darwin was also very concerned about the completeness of the fossil record (this is still an important question today); in other words, how much of the history of life is actually documented by the fossils we know: 90 percent, 50 percent, or less than 1 percent? It was impossible for Darwin to answer that question, and it is still a knotty problem. But what Darwin most hoped for was a spectacular "missing link," some fossil that was exactly midway between two living groups. He didn't have long to wait.

In 1860, quarrymen found an isolated feather in a limestone quarry near Solnhofen in Bavaria, southern Germany. The quarries produced so-called lithographic limestone, that is, a very high-quality limestone that broke naturally into perfectly flat sheets. It was used for making lithographs, printing plates for intricate engraved pictures in books. The rocks were known to be Jurassic in age, probably latest Jurassic (now dated at about 150 million years old). This was unmistakeable evidence for the oldest fossil bird by a long way, and it was named *Archaeopteryx lithographica* by Hermann von Meyer (1801–1869), the father of German vertebrate paleontology, who had named *Plateosaurus* in 1837. The name means "ancient wing from the lithographic limestone." The feather proved that birds had lived in the Mesozoic; it did not show, however, what kind of bird.

A year later, an even more dramatic discovery was made: a complete skeleton of *Archaeopteryx*. This was the perfect missing link, and Darwin could not have asked for a better one: It had the jaws and teeth, the long bony tail, and the strong hand with claws of a reptile, but it also had wings and feathers, so it must be a bird. Carl Häberlein, a physician in the nearby town of Pappenheim, obtained the specimen from the quarrymen. He was a keen fossil collector, and the poor workmen often paid for their medical services with fossils. But Häberlein recognized this was something better than the usual fishes and plant fronds that they brought him. He offered it for sale, and several major museums around Europe made strenuous efforts to obtain it. In the end, Richard Owen was able to buy it, and the rest of Häberlein's collection, for $1,000-plus. This first skeletal specimen of *Archaeopteryx*, now known as the London specimen, to distinguish it from others that were found later, shows the wings outstretched, the neck bent round, and much of the skull and braincase. In 1862, $1,000 was a great deal of money certainly, but now the specimen is essentially priceless.

A German paleontologist, Andreas Wagner, published a brief account of the skeleton in 1861, and Owen prepared a full account, which was published in 1863. Wagner bitterly opposed any kind of evolutionary interpretation of the specimen, and Owen followed suit, although perhaps less vehemently. He interpreted it as a bird, but did not draw the obvious evolutionary conclusions. His younger rival, Thomas Henry Huxley (1825–1895) had no such scruples. Huxley had already stood up as "Darwin's bulldog" to argue the case for evolution, and here was the perfect ammunition. Huxley saw that *Archaeopteryx* had the skeleton of a small theropod dinosaur, and he never wavered from that view, which is the current view, that birds evolved directly from small flesh-eating dinosaurs. The small theropod *Compsognathus* had also just been reported from the Solnhofen limestones, and it was the perfect comparative animal. Plucked of its feathers, Huxley declared, *Archaeopteryx* is a small dinosaur.

A second *Archaeopteryx* skeleton, complete with feathers, came to light in 1877, and was acquired by the natural history museum in Berlin. In the twentieth century, an additional five specimens were found, one in an old collection that had been described in 1855 as a pterosaur, along with four entirely new finds. An attempt in the 1980s to declare that *Archaeopteryx* was a fake, an amalgam of a genuine small theropod dinosaur skeleton with feathers printed around the sides, ignored the fact that there are numerous specimens; indeed, the finest and closest examination showed that they are genuine fossils.

Dinosaur discoveries up to 1850 had relied on chance finds, usually by quarrymen, in Europe. These discoveries continued, with important new finds in southern England, France, and Germany. Then, in the late 1850s, attention shifted to North America as the first reports were made of fantastic new discoveries across the Atlantic.

Joseph Leidy and the First American Dinosaurs

Dinosaur footprints had been discovered in New England earlier in the nineteenth century, although they were thought to have been made by giant birds. The first specimen was found, famously, by a schoolboy, Pliny Moody, on his parents' farm in Massachusetts about 1800. He found three-toed prints in red Triassic sandstone. (He is probably remembered partly for his discovery and partly for his splendid name.) Such footprints were common in the Triassic rocks of the eastern seaboard of North America, and collectors assembled many specimens. They were studied first by Edward Hitchcock, president of Amherst College, and state geologist for Massachusetts, who summarized his views in 1858 in his book *Ichnology of New England*. He called the footprints ornithichnites ("bird footprints"), and pictured their maker as a large flightless ostrichlike bird, perhaps 12–15 feet tall. His insight was the first inkling that Owen had been wrong in his vision of dinosaurs; but Hitchcock was right for the wrong reasons. The maker was indeed a large lightly built biped, but it was a dinosaur, not a modern-style bird.

The first dinosaur bones from North America were modest enough, a few teeth collected by an official geological survey team operating in Montana in 1855. They were described in 1856 by Joseph Leidy (1823–1891), professor of anatomy at the University of Pennsylvania. However, two years later, Leidy was able to report a much more significant find, a nearly complete skeleton of a large plant-eating dinosaur from Haddonfield, New Jersey, which he named *Hadrosaurus*. Leidy realized that *Hadrosaurus* was related to *Iguanodon*, though younger in age. Most significant was the fact that the skeleton was more complete than anything yet known from Europe, and it proved for the first time that this dinosaur at least stood on its hind legs. Owen's Crystal Palace models had already had their day: Leidy's announcement soon showed that many, if not most, dinosaurs had been bipeds. But he didn't get the pose quite right.

The skeleton of *Hadrosaurus* was mounted in the Philadelphia Academy of Natural Sciences by Joseph Leidy in the 1860s. He chose to show it in "kangaroo pose," that is, with its backbone nearly vertical, the tail bent hard along the ground, and the arms held out to the front. This was, however, a reasonable assumption for the time. Remember that by 1858, only 15 or so dinosaurs had been described around the world, and *Hadrosaurus* was the first relatively complete one. Among modern animals, Leidy looked at humans, ostriches, and kangaroos. Obviously, *Hadrosaurus* was not fully upright as we are; it had a long tail for some reason. The ostrich lacks this kind of tail, so he chose the kangaroo, which does have a powerful tail for balancing. We now know that dinosaurs could not easily have adopted the "kangaroo pose." Their tails, for one thing, did not have a natural bending point midway along the length, as in kangaroos. To mount the skeletons this way, the tail had to be broken and forced out of joint.

Leidy's *Hadrosaurus* inspired a strong wish among the citizens of New York that they should have their own Crystal Palace-style dinosaur museum, but bigger and better, of course, than Owen's establishment in London. Plans for a Palaeozoic Park in Central Park were prepared, and Waterhouse Hawkins, Owen's sculptor, came over from England to build the models. All the dinosaurs were to be shown in bipedal pose, a dramatic improvement on Owen's quadrupedal designs of only five years before. Hawkins set up his workshop and began to build the concrete dinosaurs. Drawings from the time show *Hadrosaurus* as a tall, slender biped. Other wonderful beasts were constructed. But a gang of thugs wrecked the workshop and smashed the models. This was a time of violence in local politics in New York City, and the ruffians were almost certainly members of Boss Tweed's gang, who were protecting their patch in the city. The wrecked frames and concrete debris were buried in Central Park, where they still reside, and so the first American paleontological theme park bit the dust.

Cope and Marsh and the Great American Bone Wars

The so-called North American bone wars began in earnest in the 1870s, with the rise of Edward Drinker Cope and Othniel Charles Marsh. Cope (1840–1897) was a Quaker, taught by Leidy in Philadelphia, and his interests spanned paleontology and herpetology, the study of modern amphibians and reptiles. Cope toured Europe in 1862 to improve his education and to visit the museums. He taught for a while at Haverford College in Pennsylvania, but did not enjoy that, and soon became a scientist of independent means, living off his family wealth. In his lifetime, Cope wrote thousands of technical papers on fossil and modern reptiles, and he named more than 1,000 new species, a record that will probably never be beaten (probably not a bad thing). Cope was brilliant and aggressive, but socially inept, and utterly obsessed. He was also impatient, and leapt from project to project. And though this gave him an encyclopedic knowledge of nearly everything in natural history, he often made mistakes. When Cope died, he donated his body to science, and his skull is sometimes reckoned to be the type specimen of the human species, *Homo sapiens*, his dearest wish.

Marsh (1831–1899) was also an enthusiastic paleontologist, but his education followed a more patrician course. He was educated at Yale University, and in the early 1860s spent three years in Europe. In 1866, his wealthy uncle George Peabody was persuaded to donate $150,000 to Yale University to establish a natural history museum. Thus was born the Peabody Museum. Marsh was appointed professor of paleontology, an unpaid position, and he relied on his uncle's fortune for financial support for the rest of his life. Marsh was much calmer than Cope, and he worked slowly and methodically, but he was somewhat aloof and overbearing. He employed many assistants, who in fact did much of his work for him. Whereas Cope wrote every word himself, many of Marsh's famous monographs were ghost written by poorly paid aspiring young naturalists he had taken under his wing.

Cope and Marsh met first during their respective European tours, and for a time they worked together in a friendly manner, both of them collecting and describing a range of fossil reptiles and mammals, first from the East Coast, then from the new territories of the Midwest of the United States. Their rivalry began about 1870. Cope had shown

Marsh some fossil beds in New Jersey but found later that he could not collect anything from the sites because Marsh was paying the workmen more for any finds they made. Then, during a visit to Philadelphia, Marsh noticed that Cope had mounted a skeleton of the marine plesiosaur *Elasmosaurus* with its head on the end of the long thin tail, instead of at the other end. Supposedly, when Marsh told him of his error, Cope's pride was so hurt that he swore lifelong enmity. The enmity was to be a constant and major force for the remaining years of their lives.

The Cope versus Marsh fight developed into the famous "bone wars" of the late nineteenth century. It might seem today that the North American continent would have been big enough for two dinosaur paleontologists. They certainly found enough new dinosaur sites and dinosaur specimens to keep both of them, and armies of assistants, busy for decades. But the ambitions of both men led them to compete with each other for the best finds. Each had enough money to buy fossils from local collectors and to employ teams of excavators who operated in the Midwest. The excavators at first were the tough workmen who lived in wild conditions while building the great railroads across America. These men were used to hard work, but they were not trained paleontologists. Nevertheless, Cope and Marsh found that they were resourceful, that they had a good eye for bones, and that they did not have to be paid much. When word came through that a survey team had found some fossil bones, Cope's and Marsh's agents would gather a team of fieldworkers and have them work day and night removing bones at speed. Some of these operations were in dangerous country, and the fieldmen were armed against attack. At times, they worked through winter, removing bones in appalling conditions of ice, snow, and blizzard. It's no wonder that a huge amount of damage was done to specimens by these speedy operations. Inferior bones were smashed in the field, and the good ones were hacked out at great speed, with none of the careful mapping, strengthening, and packaging that is commonplace today.

Marsh's crews can, however, be credited with inventing the "plaster jacket," the mainstay of modern dinosaur collecting. They found that bones fell to bits when they were lifted from the rock. Thinking of typical medical practice, they realized that a strengthening structure could be built rapidly by coating the bones in plaster of Paris. (This material comes as a powder and is simply mixed to a paste with water, then soon sets hard. By inserting strengthening materials into the plaster, bones can be permanently preserved. Sackcloth (burlap) is commonly used to wrap the casts, and, for larger bones, wooden struts are set into the plaster casts to assist dragging (like runners of a sledge) or as carrying handles.)

Dinosaur and mammal bones were loaded into boxcars and sent east by rail when possible, where Cope and Marsh fell on the packing cases, tearing them open, and describing the new dinosaurs and other wonderful beasts in haste. They rushed their manuscripts to their editors and published new dinosaur names as fast as the presses could roll. Each man had his own journal: Marsh, the *American Journal of Science*; Cope, the *Proceedings of the Natural History Society of Philadelphia*. A brief description of the new find would be written and delivered to the presses the same day. It would be typeset, checked, and published in two or three days. The competitors' papers often appeared at the backs of the monthly issues of those journals as supplements, tagged on in the day or two before they were mailed to subscribers. Even a day could make a difference, because a rule of biological nomenclature is that the first name to be given to a new plant, animal, or fossil is the name that everyone uses thereafter. So, if Cope could beat Marsh by a day, his name would stand, the other would fall. Working at such speed, of course, mistakes were made; and certainly neither man had the time to carry out a full-scale study of many of the wonderful new dinosaurs they were bringing to light. It took Cope or Marsh less than a week to unpack and describe a new dinosaur, whereas today, it typically takes three or four years to remove the bones from the rock, conserve them adequately, draw everything, carry out the intricate anatomical interpretations, and prepare drawings and photographs.

Thanks to Cope and Marsh, however, we have many of the famous North American dinosaurs: *Allosaurus*, *Apatosaurus* (*Brontosaurus*), *Camarasaurus*, *Camptosaurus*, *Ceratosaurus*, *Diplodocus*, *Stegosaurus*, and *Triceratops*. These men also opened up the famous dinosaur sites of Utah, Colorado, the Dakotas, and Montana. At the same time, their collectors investigated younger rock layers and turned up spectacular finds of fossil mammals and birds, which were also shipped back east, to be avidly described by Marsh and Cope.

The Mass Burial at Bernissart

Many new dinosaurs were found in Europe in the late nineteenth century, but one attracted much attention. This was the discovery of the mass grave of dozens of skeletons of *Iguanodon* found at Bernissart, Belgium. In 1877, coal miners working a deep shaft, more than 985 feet below the surface, came upon large bones in the roof of a cutting. The mining company contacted scientists in Brussels, who, by examining the isolated teeth, were able to identify the remains as belonging to *Iguanodon*. Normal mining operations were stopped, and paleontologists from the Royal Museum of Natural History moved in to supervise the careful excavation of 39 skeletons of *Iguanodon*, most of them essentially complete, as well as the skeleton of a meat-eating megalosaurid, plus fishes, turtles, crocodilians, insects, and plants.

Louis Dollo (1857–1931) was appointed museum assistant at the Royal Museum in Brussels in 1882 with the job of sorting, preparing, and describing the astonishing

Bernissart dinosaur collection. The job took him most of the rest of his life. He supervised the cleaning of the skeletons, not an easy task, since the bones were damp and cracked. The museum technicians had to use a terrifying cocktail of varnishes and glues to strengthen the bones. (This has led to endless conservation problems today. If the Victorian varnish is removed, the bones decay and crumble, but left alone they look unpleasantly shiny and damaged beneath the thick covering of glue.) For the first time in Europe, Dollo was able to reconstruct some complete dinosaur skeletons in natural pose, efforts that rivaled those of Leidy, Cope, and Marsh in North America.

Dollo announced his results to the scientific world through a series of dozens of papers about the Bernissart *Iguanodons*, and the associated faunas. He was one of the first paleontologists to consider the biology of the dinosaurs—how they lived—rather than merely giving them a name and moving on rapidly to the next specimen, as was more the habit of his contemporaries. The new *Iguanodon* skeletons were more complete than anything yet found in England, and Dollo was able to solve a long-standing problem concerning the correct location of a conical pointed bone. Mantell thought this was a nose horn, and indeed Owen reconstructed *Iguanodon* in the 1850s in his Crystal Palace models with the bone mounted on the snout. The new skeletons from Bernissart showed that the mystery bone was a specialized thumb claw, used, presumably, for defense, or in fighting over mates.

After Cope and Marsh: The Morrison Formation

Cope and Marsh both died in the 1890s, leaving behind a new generation of bone hunters. Their discoveries had also inspired the American people, and new museums were built in several cities there to house the huge skeletons. Millionaires stepped in to fund the work, and major expeditions went out all over North America, especially in the midwestern United States and in Alberta, Canada. The first focus of attention was the Late Jurassic Morrison Formation.

The early excavations by Cope's and Marsh's collectors had been motivated by a desire to excavate as many giant bones as possible in the shortest possible time, and the men who did the digging, and their methods, were often extraordinarily crude. New approaches were introduced in 1897 when the first American Museum expedition entered the area, and work on the Morrison Formation last century revealed a great deal about the life of the Late Jurassic.

The American Museum of Natural History continued operations in the Morrison Formation until 1905, and during that time sent many tons of bones back to the fledgling museum in New York. These provided a basis for one of the world's best dinosaur collections, and many of the Morrison dinosaurs collected around 1900 form the core of present exhibitions.

As soon as complete skeletons of sauropods were found, most dinosaur experts concluded that they walked on erect legs like elephants. But there were dissenters, including Oliver Hay who prior to WWI commissioned this sketch of *Diplodocus* crawling and basking on their bellies like herbivorous, snake-necked crocodiles.

Other museums were looking toward the Morrison Formation about 1900 as a quick source of spectacular dinosaur specimens. The Carnegie Museum had been set up in Pittsburg from donations by Andrew Carnegie, a Scotsman who had made his fortune making steel. Earl Douglass, on the staff of the Carnegie Museum, first devoted his efforts to collecting Tertiary mammals in Montana and Utah. In 1908, he was visited in the field by W. J. Holland, the director of the Carnegie Museum, and he suggested to Douglass that they should perhaps look at some Jurassic rocks in the vicinity. This they did, and very soon Douglass stumbled upon a perfect *Diplodocus* femur, lying isolated at the bottom of a ravine.

Douglass returned to the locality the following year, in arid canyonlands close to the western border of Colorado with Utah. His goal was to find the precise source of the *Diplodocus* femur, and to assess whether the site could be excavated. After many days of prospecting, marching up and down the dry canyons, Douglass hit paydirt. He spotted some vertebrae of a big dinosaur in articulation, proving that this was an undisturbed dinosaur bonebed. Douglass wrote to his director, Holland, and together they excavated the skeleton represented by the vertebrae, later to be named *Apatosaurus louisae* in honor of Carnegie's wife.

With this encouragement, Carnegie donated the funding necessary for large-scale excavations. Douglass decided to remain at the site, to work there full-time. He used the money to hire workmen locally and to buy heavy equipment for the task of removing rock and of moving the bones out. His wife, delightfully named Pearl Douglass, joined him in the remote corner of the Colorado-Utah border where the first bones were found, and the family remained there for many years. The nearest town, Vernal, Utah, was some 19 miles away, so Douglass set up a homestead and built a log cabin right on the bone quarry. His excavations lasted from 1909 to 1923. During this time, Douglass and his crews excavated hundreds of skeletons, which are now in the Carnegie Museum, and even after so much effort, the bonebed showed no sign of running out.

Douglass unearthed skeletons of the sauropod *Apatosaurus*, the first find from the site, as well as the longer sauropod *Diplodocus*, the plated *Stegosaurus*, the bipedal orbithopod *Camptosaurus*, and the predatory *Allosaurus*. These dinosaurs had been found by Cope and Marsh before, but no site in the Morrison Formation had produced multiple skeletons of each form in a single locality. Douglass collected 20 complete skeletons and isolated remains of a further 300 individual dinosaurs.

Douglass was at first puzzled by the immense richness of the site. How had such a diversity of dinosaur skeletons ended up in one place? He was sure that these animals could not all have come here and died by some catastrophe; there were too many of them, and the range of species represented was too great. Douglass suggested that he had chanced upon an ancient sandbar that lay in the middle of some great meandering river in the Late Jurassic, and that the skeletons were accumulated by normal river action. Carcasses of animals that had died upstream were washed along and eventually ran aground on the bar, where the flesh rotted, and where they were eventually buried under more sand brought down by the river.

One of the best finds from the quarry was a complete skeleton of *Diplodocus*, which Holland named *Diplodocus carnegiei* in honor of his patron. This greatly pleased Carnegie, who had always said he wanted his museum crews to find a dinosaur "as big as a barn." He was so pleased that he decided the world should see his new dinosaur, and he ordered that the whole thing should be cast in plaster, bone by bone. Duplicates of the skeleton were made and prepared for shipment to the leading museums of the world. Holland followed the specimens as they were sent out, and he and the dinosaurs were received with acclaim and obvious pleasure by the directors of the British Museum in London, the Musée d'Histoire Naturelle in Paris, the Senckenberg Museum in Frankfurt-am-Main, the Royal Museum in Vienna, the University Museum of Natural History in La Plata, Argentina, and the Museum in Mexico City. (The casts are still on show in most of these institutions; indeed, they are so large that they would be hard to dispose of.)

Further dramatic dinosaur discoveries in the Morrison Formation included a skeleton of the short-snouted sauropod *Camarasaurus*, collected by the Carnegie Museum team in 1922, near the end of their long series of excavations. This skeleton created a stir since it was small, a mere 16-plus feet long. Large sauropods were by 1922 quite familiar, but the new *Camarasaurus* was exciting, as it was a juvenile. During this time, the Carnegie Museum recognized the longer-term value of their quarry to the nation, and after much lobbying of the government by Holland, the site was named Dinosaur National Monument in 1915, and preserved for posterity. It is now a "living" museum, covered with a building, where a small part of Douglass' sandbar is still actively worked in front of visitors.

Dinosaurs of the Late Cretaceous: Canada and the United States

Following the leads set by Cope and Marsh, paleontologists began seriously exploring the rich dinosaur beds of Alberta, Canada, in the early twentieth century. The Red Deer River cuts deep canyons through the Late Cretaceous sediments of southern Alberta. Climatic conditions are desertlike for most of the year, and there is very little vegetation other than scrubby bushes. Every year, there are torrential rains, and huge rivers of water cut through the canyon walls, creating classic badland scenery (bad for farmers; good for fossil-hunters).

About 1910, two collecting groups set out along the Red Deer River to collect dinosaur skeletons. Although the fieldmen were working for money, meaning the more they found, the more they would earn, this collecting proceeded without the animosity that had existed between the parties working for Marsh and Cope. The two collecting teams were led by Barnum Brown, acting for the American Museum of Natural History, and Charles H. Sternberg, working for several institutions, but especially the Geological Survey, the National Museum of Canada, and the Royal Ontario Museum. All these institutions wanted to acquire dinosaur skeletons, and the Canadians in particular wished to build collections of their own dinosaurs.

Brown was there first, and he had invented a new collecting technique. He built a large wooden barge and floated downstream in 1910 and 1911, tying up here and there, and ventured up side canyons looking for shards of bone. When he found a good prospect, he and his team excavated the bones and loaded the plaster packages on to the barge. They were later offloaded and sent east to New York. The Canadian government, aware of Brown's discoveries, then hired Sternberg, a commercial collector, who operated at times with the help of some or all of his sons, George, Charles, and Levi. The Sternbergs set off for the Red Deer River in 1912, and they found rich dinosaur beds around Drumheller and Steveville. The Sternbergs and Brown returned in 1913, and trips continued until 1917. During this time, both teams found dozens of complete skeletons of dinosaurs at various levels in the Late Cretaceous, which are now assigned to the Santonian, Campanian, and Maastrichtian. This sequence through time is one of the most important aspects of the Alberta collecting, since it showed for the first time that it might be possible to track dinosaur evolution in one area through a long time span.

The Biggest Dinosaur Expedition

One of the biggest bone-hunting expeditions ever was mounted by German paleontologists in what is now Tanzania, then German East Africa. The fauna was large, consisting of three meat-eaters and six plant-eaters, and they created a sensation when their skeletons were first exhibited in Berlin about 1920. The first finds were made in 1907 at a locality called Tendaguru, four days' march inland from the important coastal port of Lindi.

The locality was discovered by W. B. Sattler, an engineer working for the Lindi Prospecting Company, as he searched for valuable mineral resources. He found pieces of gigantic fossil bones weathering out on the surface of the baking scrubland. Sattler reported back to the director of the company, who in turn told Professor Eberhard Fraas, who happened to be in the colony at the time. Fraas was a noted paleontologist, who had earned his reputation describing Triassic fossil reptiles from Germany; he had also worked on the early prosauropod *Plateosaurus*. Fraas visited the Tendaguru site, and he clearly recognized its huge potential. He collected some good specimens to take back to Stuttgart, where he showed them to various museum curators. Dr. W. Branca, director of the Berlin Museum was enthusiastic, and he set about raising funds for an expedition.

Branca and his colleagues succeeded in raising the then-huge sum of almost $1 million from the government, from various scientific organizations, and from wealthy local donors. This enabled him to mount a substantial expedition, which was planned to run from 1910 to 1911, under the leadership of Dr. Werner Janensch, curator of fossil reptiles at the Berlin Museum, and with the assistance of Dr. Edwin Hennig. In the end, the expedition ran into 1912, but under the leadership of Dr. Hans Reck.

The expedition was on a larger scale than anything mounted before. In the first season at Tendaguru, 170 native laborers were employed, and this number rose to 400 in the second season, and 500 in the third and fourth. These workers were accompanied by their families, so the German dinosaur expedition at Tendaguru involved an encampment of 700–900 people, causing major logistical problems in obtaining enough food and water, and in maintaining the operation.

The laborers dug numerous pits all over the site, which ran from Tendaguru hill to Tendaguru village, a site spanning about 2 miles. Most of the bones were very large, but they were fractured and had to be protected. The finds were mapped, measured, then

For a century, the enormous bulk of sauropods such as African *Dicraeosaurus* and towering *Brachiosaurus* were believed to limit them to a largely waterborne existence. This view has disappeared as it was realized that track-

ways show they were able to easily stride across Mesozoic landscapes, and that they were dinosaurian versions of elephants and giraffes built to browse on land.

encased in plaster for the long journey to the coast. Teams of workmen carried the bones on their heads or slung on poles for the four-day trek to Lindi, where they were shipped to Germany. In the first three years of the expedition, 4,300 loads of fossil bones were sent out, weighing a total of 200 tons. During the fourth year, a further 50 tons were shipped out.

In Berlin, the long process of cleaning the bones began, and this lasted for many decades. As the materials were cleaned up, the huge skeletons were mounted in the Humboldt Museum in Berlin, where they may still be seen; the new species were described up to the 1960s by Janensch, Hennig, and others. The Berlin specimen of *Brachiosaurus* is the largest complete dinosaur skeleton in the world.

China and Mongolia

Dinosaurs from Asia were found sporadically at first. Specimens from China were collected up to the 1920s by American, French, and Swedish priests and explorers. The first good Chinese dinosaur skeleton, *Mandschurosaurus*, was found in Manchuria, then under Russian control, and was sent back to St. Petersburg in 1917. Asiatic dinosaurs really hit the headlines in 1922, 1923, and 1925, when an American expedition, led by Roy Chapman Andrews, returned from Mongolia with spectacular dinosaur specimens: skeletons of the small ceratopsian *Protoceratops*, with nests containing eggs, and the extraordinary slender meat-eaters, *Saurornithoides*, *Velociraptor*, and *Oviraptor*.

The expedition had been sent out by Henry Fairfield Osborn, the director of the American Museum of Natural History, in the hope of finding early human fossils, but the team came back with evidence for one of the best dinosaur-hunting areas in the world. Mongolia was little known in the West, but Vladimir A. Obruchev, a distinguished Russian paleontologist, had found a rhinoceros tooth there in 1892, suggesting that rocks of the right age to hold fossil humans might be there. Andrews raised money and bought numerous cars and other equipment, and a caravan of vehicles and camels set off from China in early 1922, crossed the border into Mongolia, and headed north toward Ulanbaatar, the capital. The expedition was led by Andrews, and the chief paleontologist was Walter Granger. They made discoveries early on, first at a site called Iren Dabasu, on the road to Ulanbaatar, where they found Cretaceous mammals and dinosaurs. They visited Ulanbaatar briefly, then set off west into the Gobi Desert.

After some weeks of exploring the Gobi Desert and of making a few discoveries, the expedition turned east again, heading back to China. One day, the vehicles drew up on the edge of a large eroded basin formed in red sandstones, a site they named "Flaming Cliffs," and now officially called Bayan Zag. The collectors found abundant dinosaur bones and eggs, but they had to head home almost immediately. Fired by enthusiasm, the expedition returned in 1923, and they were able to spend adequate time at Iren

Dabasu and at Bayan Zag. They collected bones of several extraordinary new dinosaurs, including dozens of specimens of the small ceratopsian *Protoceratops*; and, most dramatic of all, these were associated with several nests containing elongate eggs arranged in neat circles. When Andrews and his crew returned to New York and announced the collections of *Protoceratops* with their nests, they created a sensation. Dinosaur eggs had been found as isolated remains in the Late Cretaceous of southern France in the nineteenth century, but these were the first complete nests.

The Americans returned to the Gobi Desert in 1925, and during three expeditions there they amassed large numbers of specimens. Apart from *Protoceratops*, they found small theropods such as *Saurornithoides* and *Oviraptor*. Later expeditions to Mongolia were mounted by the Russians in 1946, 1947, and 1949, and they found new dinosaurs, especially *Tarbosaurus*, a relative of the North American *Tyrannosaurus* and the hadrosaur *Saurolophus*, also a North American form. Polish expeditions operated in Mongolia in the 1960s, and the American Museum of Natural History began a second series of expeditions in the 1990s, all of which have been marked by astonishing new discoveries.

Waterhouse Hawkins never dreamed of dinosaurs like these. Among post-war finds in Mongolia were great *Tarbosaurus*, which might have retained display feathers on its reduced arms since it descended from smaller, presumably feathered theropods. Even more peculiar was *Therizinosaurus*, with its "Freddy Kreugar" hands and pot belly.

Later Finds: The 1940s, 1950s, and 1960s

Dinosaur paleontology went through a recession during the middle decades of the twentieth century, though major expeditions were run by Russian, Chinese, and Polish teams, and some important discoveries were made in more well-explored parts of the world. For example, John Ostrom found skeletons of *Deinonychus* in the mid-Cretaceous of Montana in 1964, and this marked the beginning of a revolution in understanding of theropod dinosaurs, and a revival of interest in Huxley's old idea that birds had evolved from theropod dinosaurs.

Further isolated discoveries were made up to 1950 in North Africa (Egypt, Tunisia, Morocco, Niger), Brazil, and Argentina. Initially, the work was stimulated by expeditions mounted mainly by European and North American institutions, keen to acquire exotic specimens and to expand their scientific reputations. Increasingly, during this time, countries in Asia, Africa, and South America established their own geological surveys, universities, and museums, and the focus shifted to locally based experts. By 1950, dinosaurs had been found on every continent, except Antarctica, and more than 500 dinosaur species had been named. This was a dramatic rate of discovery considering that dinosaur hunting on a large scale began only after 1850.

However, very little dramatically new work was published during these decades on dinosaur paleobiology. The view of dinosaurs had become rather stagnant. Most scientists, and the public, saw them as broken-down overblown animals that deserved to go extinct. Dinosaurs, metabolically, were giant lizards, cold-blooded, stupid, and slow. Paleontologists concentrated on economic areas of their subject, particularly the rapidly advancing field of biostratigraphy. Oil companies around the world were employing paleontologists to date the rocks they drilled through. Dinosaurs came to be regarded as kids' stuff, not serious, not interesting. A detailed survey of publications on one of the hottest dinosaur topics of all, their extinction at the end of the Cretaceous, revealed that only two or three papers on this topic were published each year up to the 1970s, when the annual total rose to ten to fifteen. In contrast, the total is now probably several hundred per year since 1980.

Dinosaur Renaissance

Bob Bakker, in his 1975 article in *Scientific American* entitled "Dinosaur Renaissance," drew attention to the fact that something amazing was happening in the field of dinosaurian paleobiology. Suddenly, dinosaurs were "sexy" again; bright young geologists and biologists wanted to work on dinosaurs. Publications about dinosaurs were burgeoning, as were articles in all the serious scientific journals. Museums were hiring paleontologists again; expeditions were setting out and finding amazing new species of dinosaurs. What had happened?

It was a three-step process: First came Bakker (1970s), then the impact extinction hypothesis (1980s), followed by computerized cladistics (1990s). All three shook up dinosaurian paleobiology. All three had been on the boil since the 1960s, and they burst on the scene neatly at the beginning of each of the last three decades of the twentieth century. All three were interdisciplinary. All three affected the professionals profoundly, yet were also played out largely in public.

Bakker was a student of John Ostrom's at Yale in the late 1960s. He was working on dinosaurian posture at the same time as Ostrom was completing his descriptions of the startling *Deinonychus* and beginning to formulate ideas about bird origins. Both Ostrom and Bakker saw that *Deinonychus* was no slouch. Here was a small dinosaur with powerful limbs, and obviously adapted for rapid movement. Its flick-knife toe claws were clearly not designed for scraping moss off tree trunks. This was a dinosaur that leapt and slashed its prey. It had stiffening ossified ligaments along its tail, which implied balancing: *Deinonychus* could stand on one leg, rotating, and positioning itself as it slashed with the other foot. This required fine coordination, excellent eyesight, and a powerful motor area in the brain. Later evidence suggested that *Deinonychus* hunted much larger prey, such as the ornithopod *Tenontosaurus*, presumably by pack hunting. Social behavior of this kind required communication and more brain power. Perhaps dinosaurs were not overblown lizards after all. Maybe Richard Owen had been right in 1842 to see dinosaurs as mammal-like in many ways (even if he held this view for the wrong reasons).

This was all very daring thinking. The message that something was afoot in professional dinosaurian circles came out to a wider public through Bakker's artwork. He had a fine eye and produced excellent vigorous pencil sketches. His first major piece was a frontispiece to John Ostrom's monograph on *Deinonychus*, published in 1969. Normally, a monograph such as this would be read with interest by professional colleagues, but would not impact a wider audience. However, *Deinonychus* was such a vigorous and vicious little beast, Ostrom's descriptions and illustrations of the skeletal anatomy were so thorough and stimulating, and Bakker's vision of *Deinonychus* in life was so startling, that the news spread. Pictures were published widely in popular science magazines and newspapers, generating a new excitement.

A key to the renaissance was the reposturing of dinosaurs. Bakker saw that *Deinonychus* had to be shown with a horizontal backbone, not vertically, as Leidy had thought. Dinosaurs should be modeled on seesaws, not kangaroos. The stiffening rods down the tail showed that the tail had to be held out vertically. This meant that there was no choice but to make the whole body horizontal. Another young student, Peter Galton, then recently arrived in Connecticut from England, showed that this new horizontal posture had to apply to the bipedal ornithopods, too. Some of them had stiffening ligaments over the hips and tail. A horizontal dinosaur meant business; it was a go-faster posture. The beast was heading somewhere, whereas Leidy's vertical kangaroo-dinosaur couldn't really budge. But it wasn't only posture that Bakker and other young turks changed. There was also the thermal physiology of dinosaurs.

Were the Dinosaurs Warm-blooded?

How can such a question be answered without a time machine and a giant rectal thermometer? The debate began virtually with the discovery of the group. In 1842, when Richard Owen coined the name Dinosauria, he speculated that these giant reptiles had been rather mammal-like in their physiology, and very different from living reptiles, such as lizards and crocodiles. Owen clearly thought that dinosaurs were able to control their body temperature to some extent, and to keep it high.

Modern animals divide into two main categories in terms of temperature control. The ectotherms, like fishes and reptiles, generally use only *external* means to control their body temperature. So lizards bask on rocks to raise their temperatures or hide in holes to cool down. The endotherms, like birds and mammals, can maintain warm body temperatures by internal means, by burning up food and by a complex feedback mechanism that heats and cools the body to maintain temperature at a precise level. In endotherms, a change of only a few degrees can be critical, as in humans, whereas ectotherm body temperatures may vary by 68°F or more each day.

The second distinction in modern animals is between poikilotherms and homeotherms. Poikilotherms have variable body temperatures, whereas homeotherms have constant body temperatures. Lizards are clearly poikilothermic ectotherms, and birds and mammals are generally endothermic homeotherms. But poikilotherm does not equal ectotherm, or homeotherm equal endotherm. The four terms are necessary because fishes are generally ectothermic homeotherms; their body temperature is constant, although controlled externally, since the temperature of the sea does not change much. Bats and hummingbirds are poikilothermic endotherms, since they can switch off their expensive heating system at night or in winter. But what were the dinosaurs?

Until 1970, dinosaurs were thought to be sluggish, cold-blooded reptiles, in other words, ectothermic and poikilothermic, despite Owen's classic ideas. Then, in 1970, Bakker, extending his revised view of dinosaurs, marshaled a range of evidence that dinosaurs had been endothermic homeotherms. He noted these points, among others:

- Dinosaurs have a complex bone structure with evidence of constant remodeling, a bone feature seen in modern mammals, but not in reptiles.

- Dinosaurs have an upright posture, as in modern mammals and birds.

- Dinosaurs evidently had active lifestyles, or at least the small theropods certainly did.

- Predator-prey ratios of dinosaurs show more in common with those of mammals than with those of modern reptiles.

- Dinosaurs are found in polar regions.

Bakker's collection of evidence shows some of the ingenuity paleontologists must employ in their efforts to understand the past, drawing on the physiology and anatomy of modern animals, as well as bone histology, paleoecology, and paleogeography.

There was a furious debate over Bakker's proposals in the 1970s and 1980s. Much of his evidence was equivocal and did not stand up to strong scrutiny. Further study of bone structures has shown that the dinosaur and mammal pattern is associated with large size and fast growth rather than simply with endothermy. Upright posture does not necessitate endothermy, nor does an active lifestyle (think of insects or small lizards). Predator-prey ratios for dinosaurs do suggest that the predators were endothermic, but serious problems emerged when trying to calculate such paleocological measures in a precise manner. Dinosaurs are found in regions that lay near the poles in the Jurassic and Cretaceous, but apparently there were no polar ice caps in the Mesozoic, and so conditions were not cold.

The debate isn't over. Most paleontologists accept that large dinosaurs were inertial (or mass) homeotherms, meaning they were certainly warm-blooded in a common-sense interpretation of the term, but they achieved their warm-bloodedness (homeothermy) simply by being huge; it takes a giant lardbag weeks to cool down and warm up. But the small theropods, like *Deinonychus*, might well have been mammalian endotherms, eating high-protein food in large quantities and generating high resting metabolic rates. New studies of isotopes in the 1990s have tended to confirm this: Different mass states of oxygen can give indications of ancient temperatures. In a study on *Tyrannosaurus*, ribs indicated temperatures some 39°F higher than in peripheral parts, like fingers and toes. Maybe this giant flesh-eater kept its core regions hotter than external temperatures. Or maybe not. Another set of studies in the 1990s revealed that dinosaurs lack nasal turbinates, these paper-thin bones that occur inside the nasal cavities of modern mammals and birds. They support complex infoldings of the nasal tissues, and they function as heat exchangers (radiators), warming up cold air as it is breathed in and cooling hot air before it is breathed out.

Impact Extinctions

The second boost to dinosaurian paleobiology came in 1980 with the publication of one of the most daring papers in earth sciences of the twentieth century. It was published by a Nobel laureate in physics, Luis Alvarez, and his colleagues from Berkeley, California, and proposed that the dinosaurs died out as a result of the impact of an asteroid, a giant meteorite, measuring 6¼ miles across. (The impact happened 65 million years ago, at the boundary between the Cretaceous and Tertiary periods, and it is known in short-hand as the "K/T event.")

Over the years, hundreds of theories for the disappearance of the dinosaurs had been proposed, yet none had gained general acceptance. The problem was that many of the ideas took no account of what actually happened; that is, it wasn't only the dinosaurs that disappeared, but also pterosaurs, marine reptiles, ammonoids, belemnites, rudists, and much of the marine plankton. Another problem was that the extinction of the dinosaurs was not really regarded as a serious scientific problem, but more a kid's problem or a kind of parlor game. Alvarez and his colleagues made people sit up and take notice, and within a few years, cosmologists, astronomers, physicists, geophysicists, volcanologists, mineralogists, climatologists, ecologists, mathematicians, and other "serious" scientists wanted to be in on the act. Dinosaur extinction had become a major interdisciplinary research problem, attracting thousands of dollars of research money.

This intensity of research on the K/T event reflected the breadth of Alvarez's vision. His idea was that the impact caused massive extinctions by throwing up a vast dust cloud that blocked out the sun and prevented photosynthesis; hence, plants died off, followed by herbivores, then carnivores. The dust cloud also prevented the sun's heat reaching the Earth, causing a short freezing episode.

There are three key pieces of evidence for the impact hypothesis: an iridium anomaly worldwide at the K/T boundary and associated shocked quartz and glassy spherules. Iridium is a platinum-group element that is rare on the Earth's crust and reaches the Earth from space in meteorites. At the K/T boundary, that rate increased dramatically, resulting in an iridium spike. Many localities have also yielded shocked quartz, grains of quartz bearing criss-crossing lines produced by the pressure of an impact, as well as glassy spherules produced by melting at the impact site. A catastrophic extinction is indicated by sudden plankton and other marine extinctions in certain sections, and by abrupt shifts in pollen ratios at some K/T boundaries. Alvarez had only the iridium spike at a couple of localities, one in Italy and one in Denmark, but that was enough for him to propose his dramatic idea. Work since has confirmed what he proposed in a dramatic way.

The impact model was strengthened by the discovery, in 1990, of the impact site, the Chicxulub Crater in Mexico. This is big enough (12.5 miles diameter) for a 6¼-mile asteroid, and there is strong evidence for impact fallout and tsunamis (tidal waves) all round the Proto-Caribbean. This is the smoking gun, but the debate isn't over yet. Paleontologists still have to explain how the impact actually caused the selective extinctions that occurred 65 millon years ago. There are also other bits of evidence that don't quite fit, or that may suggest a more complex set of events 65 million years ago. First, there was major volcanic activity at this time. The Deccan Traps in India represent a vast outpouring of lava that occurred over the 2–3 million years, spanning the K/T boundary. Vast eruptions like this can themselves cause extinction. In addition, it is clear that many of the groups that died out at the K/T boundary had actually begun to decline long before the boundary. Either they knew the impact was coming and turned up their toes in good time, or other processes were going on—major long-term volcanism can cause major climatic fluctuations, and climates were certainly becoming colder in North America.

Like the Bakker-endothermy phenomenon of the 1970s, the K/T impact hypothesis of the 1980s is still a wide open research topic, and far from being fully explained. The last of the triumvirate is the computerized cladistic revolution.

The 1990s: Computerized Cladistics

Cladistics might at first seem an arcane bookkeeping method, and hardly the stuff of revolutions. The methods were developed in the 1950s and 1960s by a German entomologist, Willi Hennig, and they were picked up only slowly in the late 1960s and 1970s by paleontologists. Initially, there were major arguments over the implications of the new methods, but they are now well established. Basically, Hennig presented biologists and paleontologists with a scientific method for reconstructing evolutionary trees. Until 1960, systematists (the scientists who try to understand the diversity of life) were regarded as half-artists, half-scientists. Only by years of careful study was it possible to get a feel for the true classification and pattern of evolution of a particular group. This kind of intensive training and careful work is still absolutely essential, but the method is now a science, not an art.

Hennig's great insight now seems so obvious that it sounds like nothing at all: Systematists should use only derived characters in classification. In sorting out a group of parrots, for example, it's not helpful to note that they all have wings and beaks—those are primitive characters found in all birds. Equally, it might *not* be helpful to classify them simply by color; after all, there are other birds that are red, green, or blue, but are entirely unrelated. Thus it's better to focus instead on specific features of the anatomy found in subgroups of a sample, perhaps six or seven species of parrots that all share a particular shape of the lower jaw or a specific structure on the thigh bone. Characteristics are coded in data matrices, with species on the x-axis and characteristics along the y-axis. In a typical study, the data matrix might contain 60–70 characters coded for 10–20 species. What to do next?

Hennig worked through his data matrices by hand, as did most systematists until about 1985. They looked for any pair of species that shared a lot of characteristics, then paired them off. Next they found a third species that was closer to that pair than to any others, and it joined the group (the clade). Then they added a fourth, the next most similar, and so on. In 1985, computer programs for doing this became available, though early versions had to be run on clunky old mainframe computers. After 1985, they were reconfigured for use on PCs (though the most popular, PAUP, worked best on Macs). Paleontologists quickly invested their $50 (the amount of research funding paleontologists can just about afford), and they set to. From one or two cladograms a year in the 1980s, typical dinosaur paleontology journals now contain hundreds.

The computerized cladistic revolution really took hold after 1990: Early dinosaur cladograms of the 1980s were reanalyzed; data matrices grew to gargantuan proportions; monthly, new cladograms of major dinosaurian groups appeared. Much attention focused on the theropods. Ostrom's intuition of the 1960s and 1970s that birds were dinosaurs was proved time and time again. *Archaeopteryx* shares dozens of intricately complex anatomical features of its skull and skeleton with dinosaurs. The debate moved on to which dinosaur was the most closely related to birds. The debate continues today, with radical cladistic revisions of theropods and sauropods published about 2000, and more in the pipeline.

Does this matter? Is it all a trend? Computerized cladistics matters because it is a *method*; it exposes the workings of the practitioners, and it is available to all. Debates about dinosaurian phylogeny up to 1980 were about prejudice. Evidence could be found for anything (indeed, some diehards who operate in the old mindset continue to look for obscure bits of evidence that birds are not dinosaurs). The computer cladist after 1990 had to lay herself or himself on the line and expose every detail of the reasoning. Anyone could pick over the evidence and tear it to shreds (and they did/do with glee). The computerized cladistic method encouraged a new kind of research study. Young graduate students embarked on ambitious systematic projects on dinosaurs, projects of a kind that would not have been undertaken 20 years ago. They tried to see every specimen of some particular group, often traveling very widely, and they scrutinized the specimens in more detail than had ever been done before in search of the elusive characteristics that might reveal the true phylogeny. The end products have an authoritativeness that is admirable.

The computerized cladistic revolution of the 1990s, added to the K/T impact revolution of the 1980s and the Bakkerian revolution of the 1970s, have all conspired to make dinosaur paleontology a burgeoning field. The newfound interest in these kinds of paleobiological, systematic, and theoretical studies has been matched by a startling renaissance in fieldwork.

Dinosaur Finds Since the 1970s

In North America, spectacular discoveries include some giant sauropods, whose precise size cannot be estimated, because the specimens are incomplete, but that have all been said to exceed the length of *Diplodocus* (98 feet): *Supersaurus* and *Ultrasaurus* were named from bones found in Dry Mesa Quarry, in southern Colorado in the 1970s. An even larger sauropod, *Seismosaurus*, was named in the 1980s from monstrous bones excavated in central New Mexico, at the southern limit of the Morrison Formation. This animal has not been fully described, but it is estimated to have measured over 130 feet in length.

Long-term collecting by John Horner and his colleagues in the Montana Badlands since the 1970s has turned up yet another series of dramatic discoveries. He located a dinosaur nesting ground where hadrosaurs, or duckbilled ornithopods, had apparently returned year after year to lay their eggs. Horner found numerous nests on an elevated patch of ground, and by digging through the sediments of the site, which he named Egg Mountain, he found that the same hadrosaur species had built nests here time after time. The nests were shallow hollows scraped in the ground, and the eggs were apparently tended by the parents and older offspring, since he found skeletons of all ages groups associated with the nests. He suggests that dinosaurs showed intelligent parental care, and he named the new hadrosaur *Maiasaura*, or "good mother reptile."

A third new area of research in North America has been the study of dinosaur footprints, and especially megatrackways. These are sites, often extending over tens or hundreds of miles, covered with dinosaur tracks. Martin Lockley of the University of Colorado has shown that there are Late Jurassic and Cretaceous horizons that can be traced along the western shore of the Mid-American Seaway, where great herds of dinosaurs (especially sauropods and ornithopods) hiked from north to south, perhaps migrating 600–2,000 miles each year as climates and food sources changed.

Dinosaurs had been excavated in Europe during the early decades of the twentieth century. Friedrich von Huene, for example, carried out excavations of mass burials of *Plateosaurus* in the Late Triassic of southern Germany during the 1920s. At the same time, Baron Franz Nopcsa, an unusual man by any score, studied the latest Cretaceous dinosaurs of Romania, while making brilliant interpretations of dinosaurian paleobiology. A dramatic discovery was the finding of *Baryonyx* in 1983. Against the odds, here was an extraordinary new dinosaur found within the commuter belt around London. *Baryonyx*, with its massive slashing claw and long, slender jaws with tiny teeth, was hard to interpret. It appears to have fed on fishes, and is most closely related to the North African spinosaurids.

The modern phase of dinosaur collecting in Asia began perhaps in the 1960s, when Polish paleontologists forged a link with their colleagues in Ulaanbaatar, and a series of Polish-Mongolian expeditions took place. These expeditions were spectacularly successful, turning up more specimens of the classic Mongolian Late Cretaceous dinosaurs *Protoceratops*, *Oviraptor*, *Velociraptor*, and *Tarbosaurus*, but also some new sauropod

This image of North American *Troodon* chasing *Parksosaurus* represents the current view of dinosaurs as sophisticated and often swift creatures. *Troodon* was a near-bird with a large brain, overlapping fields of vision, a killing sickle toe-claw that required great agility to use, and judging from fossils of relatives it was probably feathered. It is now known that a relative of *Parksosaurus* may have had a bird-like four-chambered heart! The fur-like covering of the herbivore is speculation based on its presumably high metabolic rate.

specimens, *Opisthocoelicaudia* and *Nemegtosaurus*, and an extraordinary specimen that has come to be known as the "fighting dinosaurs." This specimen preserves a skeleton of the lightweight flesh-eater *Velociraptor* with its claws seemingly locked into the bony headshield of a *Protoceratops*. Were the two fighting, *Protoceratops* perhaps defending its nest from a raid by *Velociraptor*, or is this merely a chance association of two skeletons? The Polish-Mongolian expeditions scored another coup in turning up dozens of beautiful skeletons of the tiny mammals that lived side by side with the dinosaurs. Previous expeditions had yielded a few mammal skeletons, but no one had previously collected carefully enough to be able to bring such beautiful, delicate fossils back to the lab.

Dinosaur collecting in Mongolia has continued since the 1970s by Mongolian and Russian paleontologists, and, in the 1990s, by a renewed collaboration between Mongolians and the American Museum of Natural History. The recent Mongolian-American expeditions have turned up *Mononykus*, billed as a large flightless bird, but possibly an aberrant theropod dinosaur. They have also shown that the theropod *Oviraptor*, far from being an "egg thief," was actually found near nests with eggs since it was incubating its own eggs. Discoveries in Inner Mongolia, which is part of northern China, have resolved a long-standing problem of saurischian systematics. Strange long sicklelike claws from Mongolia were named *Therizinosaurus* in the 1950s by Russian paleontologists. In the 1960s and 1970s, Polish and Mongolian paleontologists found a new group of plant-eating theropods, which they called segnosaurs. Discoveries by the Sino-Canadian expeditions to Inner Mongolia in the 1990s proved that these two groups were one, the therizinosaurs, and that these were some of the most bizarre theropods of all: not only plant-eating, but armed with long claws for scraping food together, and with a skeleton that made them look like a cross between a giant gorilla and a ground sloth—Godzilla truly lives in eastern Asia!

Dinosaur study began to expand seriously in China about 1965, with the establishment of several museums and the training of a new generation of paleontologists. Since then, dinosaurs have been found in Jurassic and Cretaceous rocks in China, and 100 or more species have been reported. The rate of collection of new dinosaurs in China shows no sign of slowing down. The latest efforts have focused on the spectacular Early Cretaceous sites of Liaoning, which have yielded spectacularly well-preserved plants, fishes, and reptiles. Birds with feathers have been found there; and, most astonishing, in the 1990s, several dinosaurs came to light that were equipped with feathers. These feathered dinosaurs included forms close to the German *Compsognathus*, therizinosaurs, and others.

Dinosaurs had been found in Argentina and Brazil before, but young paleontologists in both countries began to find more material in the 1960s, and the pace quickened especially in the 1990s, with dramatic new finds of the world's oldest dinosaurs in the Late Triassic of Argentina (*Eoraptor*, *Herrerasaurus*), and whole faunas of new forms from the Jurassic and Cretaceous. The Late Cretaceous dinosaurs of South America in particular are important, since some forms are unique to that continent. The latest finds from Argentina include *Giganotosaurus*, a theropod that was possibly larger than *Tyrannosaurus*.

In Africa, sporadic collecting has turned up dinosaurs from north to south and east to west. Recent American expeditions to Niger and other parts of the Sahara have shown that Cretaceous dinosaurs of Africa still had some affinities with North America, even though Africa was virtually an island by then. Unusual African forms include the spinosaurids, long-snouted theropods that might have been fish-eaters, distributed over North Africa, South America, and Europe. The Late Triassic and Early Jurassic dinosaurs of South Africa, first found about 1850, were further explored, and some new specimens came to light. Extensive collecting in the latest Cretaceous of Madagascar has turned up titanosaurid sauropods, abelisaurid theropods, and other forms that prove a close land link to South America, but putative isolation from the African mainland.

Polar dinosaurs have long exerted a fascination. Bakker used Arctic finds to bolster his view that warm-blooded dinosaurs lived contentedly in polar regions. While some pale-obologists pictured dinosaurs happily skipping through the snowdrifts, others preferred to consider that they migrated away in winter. New finds in Australia have reopened the discussion of polar conditions. Work by Pat and Tom Rich at Dinosaur Cove in South Australia has turned up a restricted fauna of mainly small dinosaurs. That part of Australia lay within the Antarctic Circle during the Cretaceous: Did they live through the icy winter or migrate north? This is part of a wider debate about Mesozoic paleoclimates: Were temperatures pretty uniform from Equator to poles, or were winters severe around the north and south poles?

The final piece of the jigsaw of dinosaur distribution was completed in the 1990s with the discovery of several dinosaurs in the Jurassic and Cretaceous of Antarctica. Dinosaurs are now known from every continent on Earth.

Chapter Two
Putting Dinosaurs Together:
Anatomy, Function, and Reconstruction

Introduction

One of the fun things about researching dinosaurs is figuring out how they functioned. How did they breathe, eat, and digest their food? How did they move their sometimes enormous bodies? What did they look like? How do we know when dinosaurs felt "out of sorts"?

This section is a quick course on the excavation, functional anatomy, and reconstruction of dinosaurs. Dan Chure, paleontologist at the famed Dinosaur National Monument in Utah, spins a tale about how one goes about finding and digging them up. Next, Guy Leahy peers inside the guts of dinosaurs to examine their respiratory and digestive tracts. Then, Per Christiansen updates us on what has been happening in the study of biomechanics since McNeil Alexander's essay appeared in *Scientific American*. I update my recent "Dinosaurs Past and Present" survey of the science and art of restoring dinosaurs to incorporate the flood of new data that has come along in the intervening years. Also covered in this chapter is the history of dinosaur art, with an emphasis on the master, Charles Knight. Finally, Rebecca Hanna reminds us that dinosaurs were once real, living creatures by surveying how they often suffered from disease and injuries, as recorded in their bones. Living was not easy back in the Mesozoic.

Digging Them Up

by Daniel J. Chure

Nick Hotton, the respected paleontologist at the United States National Museum, in 1964 wrote this about finding vertebrate fossils: "The first thing a collector does upon finding such an occurrence is to sit down and have a smoke." Although not stated in a manner acceptable in today's health-conscious environment, Nick did succinctly capture the emotional aspect of discovering a dinosaur. Sometimes, the initial discovery is just a few bone fragments scattered on the surface. At other times, one may stumble across most, or all, of a skeleton, or a bonebed that may extend for miles! In a way, every dinosaur quarry presents its own set of problems and difficulties, with some worse than others. However, the same steps are followed, regardless of those idiosyncracies—discovery, excavation and documentation, transport, preparation, study, and publication.

But how did we even learn about the places to go to look at these dinosaurian treasures sticking out of the ground at our feet? Maybe a sharp-eyed citizen saw the fossils and reported them, or a construction crew uncovered them and stopped work. It can, and does, happen. More often, however, finding dinosaurs is the result of planning, hard work, and a good dose of luck. We take out our geological map for the area where we will be doing fieldwork and focus our efforts by looking in rocks of the right age, from the Mesozoic era (approximately 230 to 65 million years ago). There's no point in looking for dinosaurs in Pleistocene sediments (unless one is looking for birds). Next, we narrow the search by looking in rocks formed in an environment likely to have preserved dinosaur remains, such as river and floodplain deposits. (Dinosaur carcasses did float out to sea and sink to the bottom, but they are only rarely found in marine sediments, so we cross those formations off our list.) Finally, we go to those areas where rocks of the right age formed in the right environments and are well exposed. This is generally a badlands area with little vegetation. It's nearly impossible to find a dinosaur fossil in a deciduous forest or in rocks buried underneath a shopping mall parking lot. In North America, the features we have identified as giving us the best potential for finding our dinosaur would direct us to the arid regions of the western United States and Canada. Sure, dinosaurs are known from the East Coast, but discoveries there are few and far between.

Once in the field, we start looking for the clues. We walk the rock outcrops we have identified as holding promise. We move along the base of hills, looking for fragments of bone or the glint of a piece of tooth enamel. Once spotted, the trick is to find the source. We look for other fragments higher up on the hill slope and follow them up. Maybe our prize is on the top. Maybe the bone chips stop in the middle of the slope, indicating that the specimen is weathering out in the middle of the hill.

Okay, we've made our discovery and have finished our smoke. Our first need is to assess the significance and extent of the discovery. That's not as easy as it may sound. To which group of dinosaurs does the discovery belong? How complete is it? Is there a skull? These questions may be hard to answer in the field. Some complicating factors include the location of the specimen (is it on the surface or sticking out far down the side of a cliff?) and the nature of the enclosing rock (are the sediments soft and relatively easy to dig into, or is it in a depressingly hard sandstone?).

Let's assume that we can tell that the specimen is significant and that we want to collect it for our institution. How do we get it out of the ground? First we need to delineate the specimen by removing enough rock to see as much of the specimen as we can in the field, but not so much that the specimen will be jeopardized during collection and transport. This may be possible using only hand tools or small pneumatic chisels. Overburden can be taken off using pneumatic tools, such as jackhammers, powered by portable generators; in extreme cases, bulldozers and even explosives can be used (trained and certified individuals only please!).

The bones will inevitably be cracked, fractured, or broken, and they will need to be stabilized. Various types of chemical consolidants can be applied to seal the bones and penetrate cracks to help hold pieces together. After that, trenches are dug to begin outlining the block or blocks to be removed. In the case of a small fossil, one block may suffice. In the case of disarticulated elements or large skeletons, the specimen may need to come out in several, or even many, blocks. One simply can't haul out a 100-foot-long sauropod in a single block. Also, how the specimen will be moved can affect the size and weight of the block. For example, if all blocks must be carried out on human backs, they will necessarily be small.

Once the blocks are formed by trenching, it is time to start jacketing the fossil. Although fossilized, the bones are brittle. Vibrations from excavation, lifting, and transporting the fossil can result in it breaking apart and being destroyed (or becoming an immensely aggravating 3-D jigsaw puzzle). Here the approach is the same as putting a cast on a broken arm or leg: enclose the fossil in something tight-fitting and rigid to protect and strengthen it. The standard fare is to use strips of burlap dipped in plaster. However, putting that compound directly onto the bone will make it very difficult to remove back in the lab without endangering the specimen. The solution is to put a separator on the fossil before applying the burlap. The best separator is toilet tissue. It need not be Mr. Whipple's extra-soft Charmin; virtually any paper will do. Here's the process: Put sev-

Digging up dinosaurs is hard, hot, grungy work. Here Gregory Paul maps some bones of a Wyoming sauropod. In the foreground, a massive femur is partly encased in plaster, while Robert Bakker and an assistant work on other remains of the Jurassic in the summer of 1978. The technologies used here date back to the 1800s. At 4:00 PM, masses of mosquitoes regularly drove the workers from the site, just one among the many practical problems involved in dinosaur digs.

eral layers of strips on part of the block and slightly dampen them with a moist brush. This will allow the tissue to conform to the nooks and crannies of the block. Then mix up the plaster, dip in the burlap, and begin applying them with hands and fingers to make it fit as tightly and closely to the fossil as possible. By repeating this process it is possible to eventually get the top and sides of the block nicely jacketed.

But what about the bottom, which is still part of the rock outcrop? Just trench underneath the jacket until it's loose enough to roll over. And don't forget to support the jacket with rocks while you are under it cutting it away from the outcrop! Once loose, it's time to turn it over and finish jacketing. If the jacket up to this point is tight and well fitting, and if the rock is fairly well consolidated and solid, the flip should go off without a hitch, other than the occasional hernia. However, in cases where the jacket is poor or the sediment is weak, the contents might spill out of the bottom when it is rolled. A Kodak moment. . . .

Once the block is all jacketed and nothing has spilled, the worst is over. Now we just need to get our dinosaur back to lab to begin the months or years of preparation in order to study and exhibit the specimen. But it's two miles to the nearest road, and the smallest jacket weighs 300 pounds. If we have a four-wheel drive truck, we can drive to the site and use blocks and tackles or winches to get the blocks onto the truck bed. Maybe the jackets can be strapped to a sled and hauled out with ATVs. But what if we can't get a vehicle to the site? Our dinosaur might be at the headwall of a narrow ravine. Or maybe we're working in a National Park and can't get permission to drive off established roads. In many such circumstances, a helicopter may be the only way to get the specimen to the vehicle that will transport it. Luckily, our expedition is well funded and we can afford the helicopter time, which is not cheap.

Of course, during all of this work, which can extend over many years, it is essential that we keep good records. We must give all bones and jackets field numbers and must accurately map and photograph their location. We take careful notes of the orientation and (sometimes tentative) field identifications. A quarry map will show the relative position of all bones found and collected. We need a complete photographic record. Also, we must outline the jackets and list the bones in each jacket. All this documentation is critical both for scientific study and for planning our preparation priorities back in the comfort of the lab. Also, future researchers can and will use this data in studying specimens. Quarry maps produced over a century ago are still valuable today; hopefully, the documentation at our quarry will be of a quality to be valuable to researchers in the next century.

All that is left is to drive our treasure back to our museum—unless, of course, we have been working overseas, in which case we work with various government bureaucracies until the paperwork is finished, then crate the jackets and send them back on ships or planes. Nothing to it.

Although our modern expedition now uses trucks instead of horses and helicopters instead of buckboard wagons, it is still basically the same drill as that used by the great vertebrate fossil collecting expeditions of the nineteenth century. Over the last decade or so, however, some high technology has crept into the field, often because of difficult or seemingly insurmountable problems keeping the paleontologist from his or her dinosaur. Dave Gillette, currently at the Museum of Northern Arizona, was the among the first to try new technologies to solve his dinosaur problem—and what a problem it was. Gillette had discovered the remains of a new sauropod dinosaur. Sauropods are the biggest of all dinosaurs, but Gillette had found the biggest of the big, the first and only known specimen of *Seismosaurus*, the earthshaking dinosaur. Here were the remains of a dinosaur over 100 feet in length! But they were buried under 10 feet of extremely hard (one might even say *damn hard*) sandstone. To make matters worse, the fossil was within a wilderness study area in New Mexico. The latter fact made the use of vehicles and power tools difficult, as there were some who felt those tools violated the wilderness concept. Gillette did get to use the power tools, but he had to minimize the impact, and wanted to finish the excavation as quickly as possible, even though that would not be too quick regardless.

A key issue was to figure out how the skeleton was buried and in which direction it ran under the sandstone. Gillette used his contacts at the Los Alamos National Laboratory to try and come up with some way to look into the rock, see the bones, and plan the excavation accordingly. We've all seen the seismic imaging done by Sam Neill's character's crew in the beginning of the movie *Jurassic Park*. And no, nyet, nein, the images we can get are not that detailed. However, any information would be a help.

Gillette and his Los Alamos colleagues tried a wide range of techniques. Ground-penetrating radar sends radio waves into the ground; the reflected waves are picked up by a recorder, which gives a picture of wavy lines, something like a seismograph. Photon free-procession magnetometry measures the magnetic field of the Earth and detects differences as small as one-millionth the intensity of the magnetic field of our planet. Clustered high readings indicate a concentration of the magnetic field, possibly due to fossil bones. Gamma emission scintillation measures the gamma ray emissions from uranium that is concentrated in many fossil bones during the fossilization process. However, rock absorbs these emissions, and if the overburden is too thick, as it was at the *Seismosaurus* site, there are no high readings to identify the presence of bones. Acoustic diffraction tomography uses the fact that sound waves travel at different rates through different materials. Low-density sandstone propagates sound waves at a slower rate than the high-velocity waves through high-density fossil bone. The acoustic waves are generated by a 8-gauge shotgun, which shoots a lead slug into the ground. The waves are picked up by sensors that are placed in water-filled holes drilled at least as deep as (and preferably twice as deep as) the fossil-bearing level. And therein lies a problem. One must drill the holes without knowing where the fossil is and thus possibly drilling into and damaging the very bones one is looking for. Finally, even bone-witching (the paleontological equivalent of water-witching or dowsing) was tried.

Unfortunately, the success of such high-tech wizardry was mixed, and none was particularly successful. However, Ray Jones of the University of Utah has come up with a remote sensing technique that has proved successful in locating subsurface bone. It is a variation of gamma emission scintillation. By using a shield on the gamma-emission detector, Jones is able to identify a point source for high emissions, rather than a more general reading of an area. By gridding off a site and taking readings from each square, Jones can localize sources of high emission. He has used his technique with great success in rocks ranging from the Late Jurassic to the Pleistocene. In one case, he found an exquisitely preserved *Allosaurus* skull in an abandoned excavation site at Dinosaur National Monument. At Hagerman Fossil Beds National Monument, he located many

Cope and Marsh would have loved to have one of these! A recent advance in dinosaur location and excavation is ground-penetrating radar. Here it is successfully used to locate the complete skull of an *Allosaurus* that had drifted away from the complete skeleton it belonged to.

fossil horse remains in a bonebed. In the latter case, his predictions were tested by excavating all of the grid and finding that his high reading closely matched the distribution of bones. But Jones' technique has limitations. Not all fossil bone produces gamma emissions, and not all sources of highs are fossil bones; the source could be fossil wood or just a concentration of uranium in the rock. A critical factor is the depth of burial; if the fossil is underneath too much overburden, the emissions will be too weak to detect. So while Jones' "dino dowser" is not the answer to all dinosaur hunters' prayers, it is the most successful remote sensing machinery devised to date.

At the beginning of the movie *Jurassic Park*, the audience is treated to the sight of a field crew excavating a complete and articulated "raptor" skeleton using only some 3-inch-wide paintbrushes. If only! In real life, most specimens are incomplete and scattered, and they present formidable problems of excavation and collection. Still, it is worth all the effort, for only through this process can we continue to explore dinosaurs and the world they lived in. That world is one of excitement, discovery, and surprise, but it gives up its secrets only grudgingly.

Noses, Lungs, and Guts

by Guy Leahy

One disadvantage of studying fossil animals is that 99 percent of the time we have only the skeleton (frequently incomplete) to work with. Therefore, attempting to infer how extinct organisms operated as living animals is heavy on speculation and light on direct evidence. Spectacular exceptions, such as the juvenile theropod dinosaur *Scipionyx* are extremely rare and treasured finds. Even where such unusual preservation is lacking, however, the bones themselves can tell us much about the internal anatomy of dinosaurs. Though the internal organs themselves have not preserved, they leave traces of their former existence in the form of ridges, scars, and excavations. These traces enable paleontologists to reconstruct features such as the nasal cavity, lungs, and digestive organs with surprising confidence.

Nasal Cavities

The first function we think of in regard to the nose is the sense of smell, or olfaction. The olfactory chambers are located away from the main pathway for inhaled and exhaled air, since olfactory tissues are very delicate. In many reptiles, birds, and most mammals, the surface area within the snout to detect odors is considerably increased by the development of lateral projections of cartilage, or bone called conchae. Though some variation exists, in general, those species with a good sense of smell possess more elaborate conchae than species whose sense of smell is poor. Humans, for example, fall into the later category, while dogs have extensive development of their olfactory conchae.

Though olfaction is a critically important function, it is not the only task the nose performs. The nose traps and filters dust, pollen, and bacteria, preventing these particles from entering the lungs. The nose also humidifies and warms air upon inhalation, and upon exhalation reclaims this added heat and water before it is lost to the external environment. In vertebrates with low resting metabolic rates, such as crocodiles, lizards, and snakes, heat and water recovery is of little importance. These animals lose very little heat and water to the environment, due to low metabolic heat production and low breathing frequency. In birds and mammals, however, this loss of heat and water may pose significant problems. Birds and mammals have much higher metabolic rates and breathing frequencies than reptiles, so loss of heat and water is potentially much more extensive.

Birds and mammals solve this problem by having a separate set of conchae designed to recover heat and water. These conchae are located directly in the respiratory airstream, and come in various shapes. Respiratory conchae are very effective in recovery of heat and water from exhaled air. In various mammals and birds, up to 70–80 percent of the heat and water that would otherwise be lost from exhaled air is recycled.

In birds, the respiratory conchae usually take the form of a coil, where in mammals they are more varied, ranging from simple struts to the spongelike conchae found in seals. Respiratory conchae are well developed in most birds and mammals, but they are weakly developed or absent in a few. Among birds, respiratory conchae are absent or vestigial in pelicans, gannets, cormorants, and frigate birds. In mammals, respiratory conchae are weakly developed or missing in whales, manatees, elephants, and many primates (including humans).

Because respiratory conchae are found only in endotherms, it's reasonable to conclude that any extinct species that possessed respiratory conchae would also have elevated metabolic rates. Conchae are very delicate structures, though, and seldom fossilize. In mammals, conchae are paper-thin sheets of bone. In birds, the conchae are usually composed of cartilage, which almost never leaves any fossil trace.

Though conchae rarely preserve, they may leave behind distinctive ridges or knobs that indicate their presence in life. Because conchae also take up space, their presence may be inferred by the diameter of the main nasal passage. Ridges that appear to have supported respiratory conchae have been found in various therapsids (mammal ancestors), suggesting the appearance of endothermy began over 200 million years ago.

Using high-tech equipment such as computed tomography (CT) scans, scientists at Oregon State University looked for evidence of respiratory conchae in dinosaurs. They examined two predatory dinosaurs (*Nanotyrannus* and *Ornithomimus*), plus the duck-billed dinosaur *Hypacrosaurus*. Unlike therapsids, these scientists concluded there was no trace of respiratory conchae in these dinosaurs, and the nasal cavity was too small to house them. These researchers concluded that dinosaurs possessed rates of respiratory water loss more similar to crocodiles than mammals and birds, and therefore had not evolved high metabolic rates like those of endotherms.

Other scientists, however, examined the nasal cavities of other dinosaurs (sauropods, ceratopsians, hadrosaurs, ankylosaurs, and other herbivorous dinosaurs), and concluded these taxa did have large enough noses to possess respiratory conchae. In addition, structures that appeared to support respiratory conchae in life were found in at least two (ceratopsids and ankylosaurs). Thus, this research was consistent with the hypothesis that some dinosaurs did have the elevated breathing rates characteristic of endotherms.

So where does this leave us? First, let's return to the purported function of respiratory conchae, recovery of heat and water. Do respiratory conchae recover both heat and water, just one, or neither? The problem here is that recovery of heat upon exhalation results in an unavoidable condensation of water, due to the temperature drop. It's possible, therefore, that the recovery of water is coincidental and not under active control. There is considerable evidence that endotherms actively control heat loss through their nasal cavities. For example, the temperature of expired air is much cooler in rabbits whose fur has been shaved compared to unshaved rabbits. Birds exhibit the same type of thermoregulatory control. The question remains, however: Is water conservation actively controlled?

Recent experiments with seals suggest that water is not reclaimed independently of heat; these results are in line with the idea that water conservation via respiratory conchae is a consequence of body temperature regulation, and is coincidental. If body temperature regulation is the prime function of respiratory conchae, what does that tell us about the apparent distribution of such structures in dinosaurs? Regarding theropods, there appears to be a paradox between what the nose says and other structures. Several small theropods including *Sinosauropteryx*, *Beipaiosaurus*, *Sinornithosaurus*, *Protarchaeopteryx*, and *Caudipteryx* seem to have been covered in life with a coat of downlike fibers. In *Beipaiosaurus*, these fibers exceeded 2 inches in length on some portions of the body. In addition, *Protarchaeopteryx* and *Caudipteryx* possessed true feathers, which attached to the tail and forearms. Besides providing strong evidence that birds are descendants of theropods, these feathers and downy fibers suggest that small theropods needed an insulating coat to preserve body heat. External coats of fur, down, or feathers are exclusive to endotherms; no living ectotherm has (or, for that matter, needs) such insulation. Evidence that small theropods were insulated stands strongly in contrast to the conclusions that such theropods did not have respiratory conchae and, therefore, did not have high metabolic rates.

What about the lack of respiratory conchae, then? As noted earlier, respiratory conchae are exceedingly delicate structures, so the odds they will fossilize is very small. It may simply be they were present but have not preserved. Do the reported small nasal passageways render respiratory conchae impossible? In a word, no. Some birds and mammals (echidnas, anteaters, kiwis, petrels, and fulmars, as examples) have very narrow, tube-shaped nasal chambers, which, despite their small diameters, house respiratory conchae. Other research suggests that the original measurements of nasal chamber size were made in a region of the snout too far forward to represent the actual area where respiratory conchae would be expected. Other indirect evidence suggests that some

Predatory dinosaurs such as these *Allosaurus* regularly dissected their prey, albeit in a crude and mindless way. We cannot examine the interiors of dinosaurs so directly, but we can combine fossils and intelligent analysis to restore their organs.

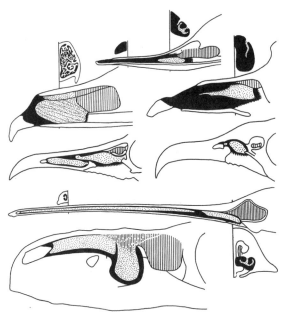

A look at nasal cavities and the conchae that they contain in side view and cross section. Vertical lines indicate the olfactory scrolls that lie behind the main airflow; the respiratory conchae that modify the main airflow are stippled. Most mammals have large, complex respiratory conchae like the dog on the upper left. Mammals with smaller, simpler respiratory conchae include the spiny anteater at top, and primates such as the baboon at upper right. A similar pattern is seen in birds; most like the gull at center left have a pair of well developed respiratory scrolls, but these conchae are reduced in some birds of prey and the long-snouted kiwi, in which the nasal passage is especially narrow. At the bottom is the nasal passage of a juvenile *Tyrannosaur*, showing that the nasal passage was long, and may have been large enough to contain bird-like respiratory scrolls.

theropods possessed respiratory conchae. During embryonic growth, the last respiratory conchae to develop attaches to a bone in the front of the snout called the premaxilla.

One researcher who has studied the development of respiratory conchae in birds suggests that the evolution of this conchae is related to the enlargement of the nostril and premaxilla. In most theropods, the premaxilla and nostril are small, whereas, in oviraptorosaurs and segnosaurs, both the nostril opening and premaxilla are very large. Because oviraptorosaurs and segnosaurs possessed feathers and other skeletal features of birds, it's possible the enlargement of the anterior nasal cavity was linked to the development of respiratory conchae, as in birds. More research is needed, as it's not possible at this point to state conclusively that theropods lacked respiratory conchae.

As noted earlier, some herbivorous dinosaurs did have enlarged nasal cavities big enough to house respiratory conchae, and at least two (ceratopsians and ankylosaurs) appear to possess ridges that are located in the right place to be attachment sites for respiratory conchae. It seems likely that some dinosaurs did possess respiratory conchae. Small herbivorous dinosaurs, however, have not yet been found with these structures, and some researchers have speculated that respiratory conchae in large herbivores is more strongly linked not to heat or water regulation, but to keeping the brain cool. Brain tissue of many vertebrate species is very sensitive to temperature, and high brain temperatures have been known to cause brain damage. It would make sense to protect the brain from damage due to excessive heat. Many large herbivorous mammals have a dense vascular net near their brains. This net uses cooled blood from the nasal cavity and face to keep brain temperature below body temperature. Many birds possess a similar vascular bundle that appears to act in the same way.

However, many mammals and birds do not seem to tightly regulate their brain temperatures. Some mammals and birds, for example, seem to function quite well even when brain temperatures exceed 108.5°F. In addition, some mammals do not possess a vascular net, even though they have well-developed respiratory conchae (rabbits, horses, rhinos, marsupials, and monotremes, as examples). Many mammals and birds that keep their brains cool do so without using their respiratory conchae. Rabbits, for example, use cooled blood from their ears, while pigeons and other birds use venous blood from their large eyes to keep the brain cool. If respiratory conchae were important for brain cooling, one might expect they would be very large in those mammals with the biggest brains—whales and primates. However, such conchae are very reduced or absent in these groups.

In addition, the relative brain size of ceratopsians, ankylosaurs, sauropods, and other herbivorous dinosaurs was much smaller than the brain sizes seen in small theropods, many of which possessed brains comparable in size to those of birds. If respiratory conchae evolved for brain cooling, it doesn't make sense they would occur only in those dinosaurs with the smallest brains. Additional evidence against the idea that conchae evolved as brain coolers is that within the therapsid-mammal lineage, they first appear in primitive therapsids whose brains are no larger than present-day reptiles.

As with birds and mammals, respiratory conchae in herbivorous dinosaurs seems best correlated to the need to conserve metabolic heat. The development of conchae may have been more important in sauropods, ceratopsians, hadrosaurs, and ankylosaurs, since fossilized skin impressions of these groups show that unlike small theropods, their skin was scaly and uninsulated. The evolution of conchae may have represented part of an alternative strategy to control metabolic heat loss. It may seem excessive for multi-ton animals to evolve structures to regulate heat loss, but these multi-ton animals started out as hatchlings, weighing only a few pounds. One of the mysteries of dinosaurs is how they were able to grow as rapidly as birds and mammals if uninsulated, due to the negative effect of low body temperature on growth. The evolution of conchae may have been one strategy by which baby dinosaurs could more tightly control metabolic heat loss.

The Lungs

As the primary site for gas exchange, the lung is a critically important organ in regard to meeting metabolic demands, and lung structure varies quite a bit among terrestrial vertebrates. Snakes, sphenodontids, and most lizards possess a fairly simple lung design. Extensions from the lung wall produce a series of simple partitions. Varanid lizards, turtles, and crocodiles exhibit a somewhat more complex lung, where a central tube (bronchus) connects with a number of separated chambers. In lizards, the lungs are ven-

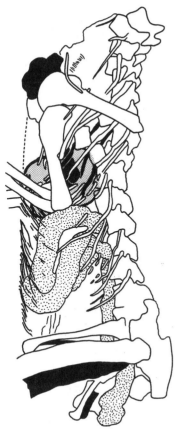

The baby theropod *Scipionyx* demonstrates spectacular preservation of soft tissues. In this sketch bones from the far side of the specimen are solid black. The heart, lungs and stomach are absent, but the traces of the trachea in the throat are present. So is a poorly preserved liver, indicated by regular stippling. Best preserved are the intestines, which progress to their terminus behind the pelvis.

tilated by movements of the ribs. Turtles, whose ribs are immobile, breathe via movements of the limbs. Crocodilians use both rib action and a special set of muscles that attach to the pelvis and liver. These muscles pull back on the liver to help draw air into the lungs.

Because mammals and birds have much greater resting and exercise metabolic rates, their lungs are much more complex. The mammalian lung also has a central bronchus, from which branches off a sequence of ever smaller segments (like the branches of a tree). At the ends of these "branches" are clusters of grape-shaped sacs called alveoli. It is in the alveoli where gas exchange occurs. Bird lungs are very different. In reptiles and mammals, respiratory airflow is bidirectional. Air flows in and out the same direction. In birds, airflow is primarily one-directional. Inhaled air passes through the trachea, down the bronchus, and into a highly intertwined net of air- and blood-filled capillaries called parabronchi. Some inhaled air also passes into large balloonlike extensions of the lung called air sacs. Upon exhalation, some of these air sacs compress air into the parabronchi; exhaled air then passes through the parabronchi back though the bronchus. Bird lungs are stiff and capable of little expansion. It is the air sacs, in conjunction with rib movements, that provide the means to ventilate the lung.

The next time you eat turkey, take a look at the number of holes and excavations in the vertebrae. Break open the humerus or femur, and you'll notice they are hollow. This is because the air sacs actually invade bone, and this invasion creates hollow spaces. Birds have five different air sacs, and the bones each one invades are very consistent from species to species. The cervical air sac invades the cervical vertebrae, the thoracic vertebrae, and ribs. The clavicular air sac invades the pectoral girdle, sternum, sternal ribs, and wings. The next two air sacs (anterior and posterior thoracic) do not invade any bones. The abdominal air sac invades the pelvis, sacral vertebrae, and leg bones.

The energetic requirements of mammals and birds during exercise are extremely high. Most running mammals and birds exhibit exercise metabolic rates 6–12 times higher than resting. Flying birds and bats may have exercise metabolic rates over 20 times resting values, and extremely athletic mammals and birds like rheas, greyhounds, and pronghorns, may have exercise requirements more than 30 times resting rates. Because the oxygen demands of mammals and birds are so much greater than reptiles, it might be possible to shed light on dinosaurian metabolic rates by discovering evidence of lung design.

One group of researchers who studied the small theropods *Sinosauropteryx* and *Scipionyx* concluded these carnivores possessed crocodilian-like lungs, powered by a liver-piston system. This conclusion was based on interpretations of several pieces of evidence. First, examination of the *Scipionyx* specimen under ultraviolet light and photographs of *Sinosauropteryx* suggested the body cavity was divided into distinct abdominal and thoracic compartments, similar to crocodiles. The pubis (one of the pelvic bones) was forwardly placed and expanded at the tip. This expansion was inferred to provide an attachment site for the muscles that powered a crocodile-like liver piston. In *Scipionyx*, traces of what were thought to represent these muscles were described. The positions of the trachea, liver, and colon in *Scipionyx*, were also inferred to be consistent with a crocodile-like lung.

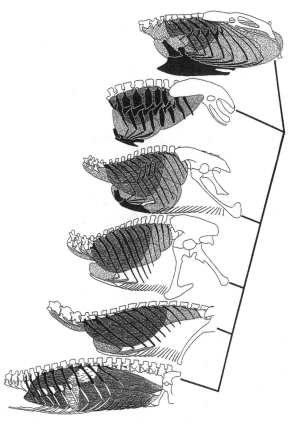

Reconstructed evolution of a bird-like respiratory system in theropods. Lungs are indicated by wavy lines, air sacs by stippling. In early theropods such as *Coelophysis* the chest ribs were still long and the posterior ribs fairly short, suggesting that abdominal air sacs were at best weakly developed. More advanced theropods such as *Allosaurus* had the shorter chest ribs and longer posterior ribs seen in birds, implying a similar reduction of the size of the lung in favor of expanded air sacs. Bird-like theropods such as dromaeosaurs even had sternal plates larger than those of some flightless birds, such as the kiwi. In flying birds the extremely enlarged sternum helps ventilate grossly inflated air sacs. Quite different from theropods and birds are crocodilians, in which muscles anchored upon the pubis help pull on the internal organs in order to ventilate the lungs. (Bottom to top: crocodilian, coelophysid, allosaur, dromaeosaur, kiwi, duck)

Other researchers who examined the same specimens, however, came to different conclusions. The "division" of the body cavity into distinct segments was due to misinterpretation of repair work on *Sinosauropteryx*, a specimen that was never directly examined by the researchers. The position of the trachea, liver, and colon in *Scipionyx* was consistent with either a crocodilian or birdlike lung; the identification of traces of crocodilian-like liver piston muscles was incorrect; and the pelvis of *Scipionyx, Sinosauropteryx,* and other theropods was not at all similar to those of crocodilians.

Let's step back and look at the theropod skeleton to see what clues to lung function are present. Because there is strong evidence that birds are direct descendants of small theropods, it's reasonable that any aspects of theropod respiratory anatomy that resemble birds are the result of the same structural adaptations for high metabolic rates. First, it's clear the neck and anterior trunk vertebrae are highly excavated. These excavations are identical in appearance and position to those in birds that are hollowed by air sacs. This suggests that most theropods possessed cervical air sacs like those of birds. The forelimb skeleton is not excavated, and because the anterior and posterior thoracic air sacs do not invade bone, there is no evidence for any of these air sacs. In some dinosaurs, like *Oviraptor,* pelvic and tail bones are excavated, suggesting the occurrence of abdominal air sacs in a few theropods.

Theropod rib design is also consistent with the presence of a birdlike lung. In crocodilians, the trunk ribs are long, and the trunk is flexible. The ribs bear a single head that attaches to the vertebrae. These single-headed ribs allow the wide range of movement needed to produce volume changes in the lung. Lumbar vertebrae bear no ribs, which provides a space for volume changes in the abdomen produced by the liver-piston breathing mechanism. The pubis is mobile, and this mobility assists in inflating the lung upon inhalation, as the muscles that attach to the pubis and liver shorten, increasing thoracic volume.

Bird lungs are ventilated very differently. The lung itself is very stiff, and volume changes during inhalation and exhalation are minimal. The anterior trunk ribs are short and serve only to anchor the lung. Ribs from the mid-trunk to the pelvis are long and bear double, not single heads. Such double-headed ribs prevent damage to the lung during ventilation and serve to produce volume changes in the air sacs, rather than the lung. The sternum is attached to the trunk ribs by a series of short sternal ribs. These ribs form a hinged articulation with the sternum and help to move air in and out of the air sacs.

Theropod ribcages share many features with those of birds. In early theropods such as *Coelophysis,* all the trunk ribs are long, so the lung was probably more flexible than in birds. However, excavated neck and trunk vertebrae indicate the presence of cervical air sacs. In more derived theropods, such as *Allosaurus,* the front chest ribs have shortened, and those ribs in the lumbar region have lengthened, suggesting a less compliant lung and more extensive development of air sacs. In highly derived, birdlike theropods, such as *Sinornithosaurus* and *Oviraptor,* the anterior trunk ribs are very short, whereas the mid-trunk and lumbar ribs are very long. Large ossified sternal plates are present, and the sternum of *Sinornithosaurus* bears articulating facets for five hinged sternal ribs. All ribs in these theropods are double-headed.

In both rib design and the pattern of bone excavation, theropods are similar to birds, not crocodiles. Because the avian lung is designed for very high aerobic performance, it's reasonable to conclude that most theropods possessed birdlike metabolic rates.

What about other dinosaurs? Interestingly, many of the same aspects of rib design and vertebral excavation seen in theropods are seen in sauropods such as *Apatosaurus* and *Brachiosaurus*: highly excavated vertebrae and ribs, short trunks, short anterior trunk ribs, and long posterior ribs, all with double heads. The striking similarity of sauropod trunk design to that of theropods strongly suggests they, too, had developed a birdlike lung-air sac system. However, because this system evolved independently of theropods, there is no guarantee sauropods had the same number or arrangement of air sacs as theropods or birds.

Ornithischian dinosaurs are harder to figure out. Dinosaurs such as *Triceratops, Anatotitan*, and *Stegosaurus* exhibit no evidence of birdlike lungs. Their vertebrae and ribs are not excavated; they bear single-headed ribs on nearly all vertebrae; and the trunk ribs are long. Ornithischians appeared to have possessed a lung that was much more flexible than theropod lungs. Exactly what type of lung ornithischians had remains open to speculation. One interesting structure to discuss is the prepubis, a large anterior extension from the pubis. Some ornithischian fossils show scarring of the last long rib of the trunk closest to the prepubis. It has been suggested this may indicate a stout ligament or muscle attached from this rib to the prepubis, which may have helped to support the abdomen. Alternatively, the prepubis may have provided an attachment site for a mammalian- or crocodilian-like diaphragm muscle. Future discoveries will hopefully shed light on this aspect of ornithischian anatomy.

The Guts

With spectacular exceptions such as *Scipionyx,* internal organs such as the liver and intestines rarely preserve. However, information gleaned from the skeleton and other fossil evidence enables us to make some reasonable speculations about the digestive anatomy of dinosaurs. Many of us have probably seen chickens or other birds swallow grit or small rocks. What function does this serve? Birds lack teeth, so they cannot process food in their mouths the way mammals can. Instead, the ingested grit is passed to a specialized organ called a gizzard. When food enters the gizzard, the stones assist in pulverizing the food item, in concert with contractions of the gizzard's highly muscular walls. Thus, the gizzard stones, or gastroliths, substitute for teeth. Crocodilians also

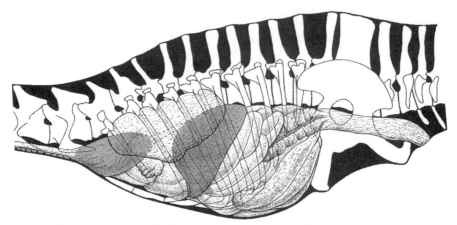

A hypothetical reconstruction of the internal organs of a sauropod. The shaded areas represent the crop, followed by the gizzard, which was probably the site of stone gastroliths. The rest of the digestive tract must have been massive. The lungs are shown tucked up high in the ribcage, as they are in birds.

deliberately swallow stones, but they possess no gizzard, and it is unclear whether the stones aid digestion or act as ballast.

Interestingly, several dinosaurs have been preserved with clusters of stones inside the abdominal cavity, which are likely to represent gastroliths. To date, probable gastroliths have been found in prosauropods, sauropods, nodosaurs, ornithomimids, *Psittacosaurus*, and *Caudipteryx*. With the partial exception of *Psittacosaurus*, all of these dinosaurs possess either simple or no teeth. In prosauropods, sauropods, ornithomimids, and *Caudipteryx* the head is small and was likely used only to procure the food that was likely swallowed either whole or with minimal processing. Nodosaurs possessed large skulls, but their teeth were small and not well designed for chewing. The presence of gastroliths in dinosaurs whose capacities for oral processing were limited is consistent with the idea that dinosaurs, like chickens, used swallowed stones to help process food. Particularly informative examples of gastroliths are preserved in the body cavity of the sauropod *Seismosaurus*. In *Seismosaurus*, the gastroliths are preserved in two clusters. The first cluster was situated near the neck base, while the second cluster was located several feet behind the first. Both gastrolith clusters consisted of highly rounded stones, the smooth texture being the probable result of repeated grinding against each other.

By analogy with birds, the location of these gastroliths is consistent with the presence of both a birdlike gizzard and a separate organ called a crop. In most birds, the crop is used primarily to store food for later digestion. However, one living bird, the Hoatzin, possesses a very muscular crop, which it uses, like the gizzard, to process food. The *Seismosaurus* gastroliths are consistent with the presence of a similar digestive arrangement.

Compared to reptiles of similar size, mammals digest food 10 times faster and digest much more food in the same time. Mammals accomplish this by having intestines that are much longer than those of comparable size, and the intestinal surface area is much greater. This increased gut size is needed to provide the high food requirements of endotherms; however, such capacious guts are also expensive to maintain. If dinosaurs possessed an elevated metabolism, one might expect they would also have a large gut to process food rapidly.

Sauropods, for example, exhibit long ribs and deep pelvic bones. This implies an expansive gut capacity. In addition, the front end of one pelvic bone (the ilium) flares outward, further increasing gut space and providing a support for large abdominal muscles. Some theropods, such as dromaeosaurs and segnosaurs, have one bone of the pelvis (the pubis) rotated backward. This would also increase available room for a large gut. Segnosaurs, like sauropods, also have a flared ilium. A backward-pointing pubis is also found in almost all ornithischian dinosaurs. The most bizarre and extreme gut adaptations are found in ankylosaurs and pachycephalosaurs. These dinosaurs have extremely wide ribcages, to the point of making hippos appear svelte by comparison. Pachycephalosaurs went one step further by separating the halves of the ischial bone so the intestines could expand behind the pelvis. Such prodigious guts may appear comical, but they make sense if high food intakes are required.

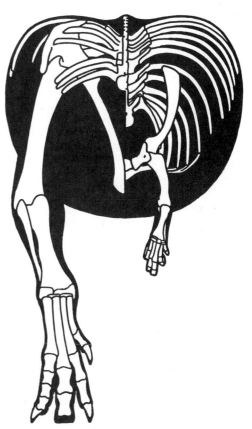

As a whole, skeletal clues to the respiratory and digestive anatomy of dinosaurs suggest these creatures possessed metabolic and aerobic capacities more similar to those of mammals and birds, rather than reptiles. This view is consistent with the discovery of insulating feathers on theropods such as *Sinornithosaurus* and *Caudipteryx*. It's also consistent with the microstructure of dinosaur bone, which indicates dinosaurs grew very rapidly, like birds and mammals. Science will probably never be able to resurrect dinosaurs as fantasized in *Jurassic Park*. However, utilization of new research techniques, such as CAT scans and oxygen isotopes, promise to provide us an even more accurate picture of dinosaurs as living animals.

The plumpness of pachycephalosaur bodies indicates that they contained large digestive tracts. Even the transverse processes at the base of the tail were elongated, apparently to support a major expansion of the guts behind the pelvis.

Dinosaur Biomechanics

by Per Christiansen

The past 20 years have seen the emergence of a host of paleobiological techniques for probing the secrets of dinosaurs, including the subjects of mass prediction and locomotion. For example, the sizes of some dinosaurs, mainly sauropods, have been subject to wild exaggerations, which in some cases could have been avoided by the application of proper methods. Advances in our understanding of animal locomotion have allowed suggestions of running performance in dinosaurs to rest on a much firmer foundation than was possible prior to the pioneering work of R. Alexander and others. Despite the fact that all phylogenetic analyses over the past 20 years have shown birds to be living theropods, extant mammals have repeatedly been shown to be better analogs for many dinosaurs, mainly owing to convergent similarities in limb function and the fact that large animals tend to be built similarly. Many herbivorous dinosaurs, in particular, bore quite substantial resemblance to various groups of larger mammals, probably due to the strictly physical limits imposed on all living things by the laws of physics, primarily gravity.

Prediction of Mass

The body mass of an animal is of great value to biologists. For instance, it enables them to compute how much force the skeleton must resist during fast locomotion. Additionally, body mass is related to many biological parameters, such as metabolic rate, growth rate, life span, number of young produced, migration distance, territorial size, prey size (in the case of carnivores), and more. No wonder that much work has been devoted to refining the methods for calculating the body mass of extinct mammals (whole books have been dedicated to that), as many analyses of paleobiology are highly dependent on a reliable mass estimate. With dinosaurs, however, the situation is a bit different.

Traditionally, a commercially available plastic scale model of a dinosaur is used, which is then submerged in water. This method is based on the well-known Principle of Archimedes, that an object immersed into a fluid will be buoyed upward by the same amount of fluid it displaces. However, the commercially available models (from the Museum of Natural History in London) are usually too inaccurate to use for this kind of work.

A more complicated but also more reliable way of obtaining a mass estimate is to rely on illustrations. This method is new and has been applied to dinosaurs only a few times. The best skeletal restorations are produced by skilled artists such as Gregory Paul. An illustration can be broken down into separate three-dimensional squares and cones, as appropriate, by drawing vertical lines across it. Measuring the height and diameter in several places of such a square or cone facilitates computation of the volume of this particular section. Each section is then multiplied by the scale to get the final volume for this section in the real animal. This is done separately for head, neck, body, legs, and tail. The total volume is found simply by adding the sections.

To converge the computed value into body mass, one needs a measure of the density of the dinosaur's body. Many extant animals have an overall density about equal to that of water (1,000 kgm-1). However, a number of dinosaurs—for example, most sauropods—had vertebrae that were extensively modified for weight-saving, with great external excavations and internal cavities, which probably contained air sacs connected to the lungs, as in extant birds. Since bone is by far the densest tissue in a body (around 2,000 kgm-1), this implies that an overall density close to that of water is probably slightly too high. An overall density of around half that would seem appropriate. It is evident that the commercially available models produce quite different results from the other two methods. Bearing in mind how important body mass estimates are for many analyses, it should be clear that the method of using illustrations is to be preferred. For example, when doing analyses on the strength of the bones of *Brachiosaurus*, estimating the amount of energy it would have needed per day, or how fast it might have reached adult size, it matters greatly if the animal weighed 36 tons or 59 tons.

Limb Morphology

Some dinosaurs, especially the large theropods, appear different from extant large mammals. Nevertheless, mammals and dinosaurs share a number of advanced features with respect to their limbs and, thus, probably locomotion as well. All dinosaurs walked with their limbs directly underneath their bodies, and their limb joints clearly indicate that the limbs worked predominately to exclusively in the fore-and-aft, or parasagittal, plane. Sometimes considered characteristic of all mammals, this is really typical only of most larger forms (above 20 or so pounds). Small mammals most often live in burrows or under logs or stones, and they forage in a much more three-dimensional environment. They are probably in need of more flexible limbs.

Sauropods are often compared to elephants simply because of their massive size. Ankylosaurs, with their large, heavy stomach and short, stumpy legs, appear somewhat hippolike, albeit with massive dermal armor. As with hippos, their legs were flexed; their muscle scars were larger than in sauropods or elephants; they walked on their toes (dig-

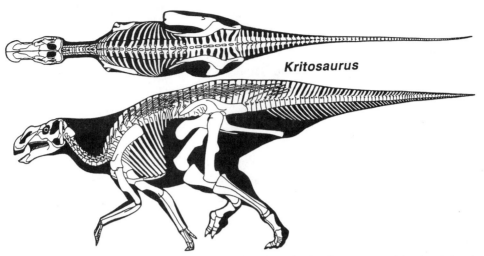

Kritosaurus

Mass predication by means of an illustration is performed by drawing lines across a lateral and dorsal skeletal restorations, and computing the volume of each section. In this case an outline figure is used, based on an illustration of the hadrosaur *Kritosaurus* done by Gregory Paul. Since legs are usually not available in multiple views, they are treated as cyclinders. As this method relies on carefully drawn illustrations and not on plastic models, it is likely to be the most accurately used method to date. The *Kritosaurus* has a volumes of 2885 liters (= 2.74 tons), a specific gravity of 0.95, an overall legnth of 24 feet, and a surface area of just under 200 square feet.

Brachiosaurus

The limbs of huge *Brachiosaurus* show the nearly columnar nature of sauropod legs, designed for bearing great weight at slow speeds. Note the extremely abbreviated finger bones and the immobility of the short foot. The shortness of both the olecranon process behind the elbow and the cnemial crest in front of the knee provided little leverage to the rather slender limb muscles.

itigrade); and their limb proportions are rather like those of hippos, as noted below. This indicates that they could have been capable of a moderate trot, as are extant hippos. Ceratopsids, however, are more powerfully built, and their limbs and bodies bear closer resemblance to those of rhinos. Smaller theropods looked similar to extant ground birds in a number of respects, and their limb proportions are often quite gracile, with long shin (tibiae) and foot (metatarsal) bones, similar to the limbs of extant fast-running mammals. Although less extreme in large theropods, their limb proportions, too, are similar to those of extant fast-moving mammals.

Anatomical adaptations for speed have been extensively studied in mammals, but there is still no unifying explanation of exactly what determines top running speed. Many small mammals are capable of fast running, relative to their size, without notable adaptations for it. It appears, however, that larger animals are in need of quite extensive anatomical redesigning if they are to be fast-moving. If one wants to run fast, this can be done either by moving the legs rapidly or by taking longer strides. The former is an unwise strategy for an animal that regularly runs fast, as it leads to rapid energy depletion. Extant fast-moving mammals and large ground birds tend to look alike in a number of respects, having large limb muscles, long legs, and especially long lower legs and metapodials (middle feet); and they walk on their toes. The advantage of walking on the toes is that the metapodials can be incorporated into the leg, and the advantage of long legs is that one may take long strides. The advantage of a long lower leg and metapodials compared to the upper leg may seem less obvious, but probably has to do with the energy involved in moving a structure (such as a bone) around a pivotal point (such as a limb joint).

When running fast, it is an advantage to minimize energy consumption. In addition to long lower legs, fast-moving animals usually have proximally placed limb muscles, which operate the lower limb by means of long tendons. This arrangement means that the mass that has to be accelerated and moved about during locomotion is less. A quick way to learn this is to go for a run with weights strapped to the thighs and then to the ankles. Although uncomfortable, the weights strapped to the thighs will not seriously affect one's performance, while the weights strapped to the ankles will result in rapid fatigue.

Thus, to run fast, the limbs should be long to facilitate long strides and have long distal sections to increase outspeed, but they also require substantial force to be able to support the mass while running. The limb muscles should be proximally placed but must also be fairly large, thus taking up more space. The trade-off is that longer bones are less resistant to breaking. Overall, it is clear that theropods, and to a lesser extent smaller ornithopods, fit the above criteria well. Their limbs are long, and the distal parts especially are elongated. As in extant mammals, the adaptations for speed are most pro-

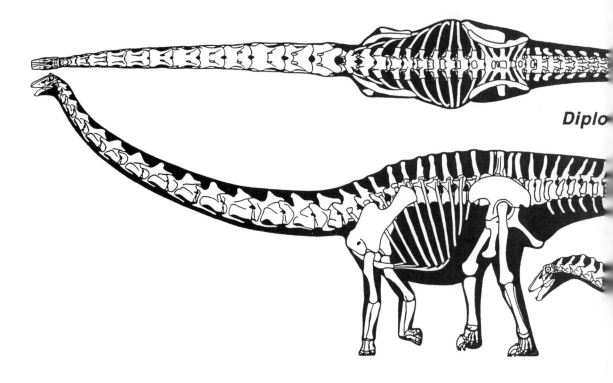

Diplo

The forces that are exerted on a bone during locomotion can be divided into an axial component (F_{ax}) which compresses the bone and a transverse component (F_{trans}), acting at right angles; to compute the bone strength indicator value one needs to measure a cross section of the bone where it is thinnest (usually around the middles) and the section modulus is calculated from the measurements. Furthermore, the distance (x) from this cross section to the end of the bone must be measured, as the forces act on the ends of the bone, but set up stresses in

nounced among the smaller to medium-sized forms, while larger forms, such as quadrupedal ornithischians, often are so heavy that it would require huge muscles and heavy bones to run fast. Large muscles use up a lot of energy, and large bones are energy-consuming to move about. Thus, many large dinosaurs, as most large mammals, had some adaptations for running fast, but were probably much less fleet than smaller forms.

Bone Strength

If one wishes to run fast, the bones must be strong enough to withstand it. The ratio of peak bone stress encountered during fast locomotion and/or jumping compared to the stress that would damage the bones (yield stress) is similar among extant mammals as different as mice and elephants. The ratio between yield and peak stress is called the safety factor, and in mammals this value is around 2–4, implying that the maximal stress

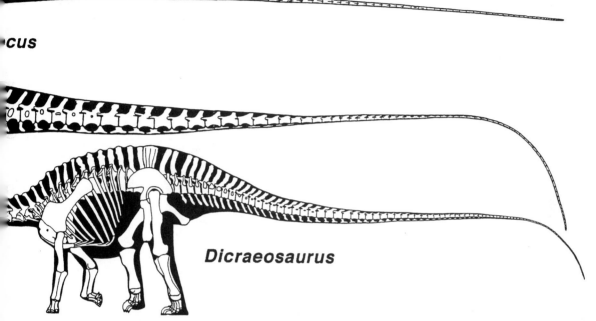

cus

Dicraeosaurus

the bone shaft. Next one needs the mass of the animal and the percentage of mass supported by the limb pair from which the bone comes. In this case, the bone is a thighbone (femur) from a *Dicraeosaurus*, a small long-necked sauropod resembling *Diplodocus*, with a mass of 6.25 tons. *Diplodocus* supported most of its body mass on the hindlimbs (75%). The bone strength of *Dicarosaurus*' femur is 12.8 Gpa $^{-1}$ comparable to that of an elephant's femur.

set up in the bones is 25–50 percent of the stress that would break them. But why are the safety factors so similar? The reason is probably that limb bones are centrally important for survival, so it is probably wise to have a fairly large safety factor. On the other hand, too-thick limb bones are costly to move about, indicating that the bones should not be disproportionately thick.

Fast locomotion places a tremendous stress on the limb bones, as the exchange of energy with the ground must equal body weight (mass times the gravitational constant). When an animal is standing motionless, all four (or two) feet contribute to carrying the body mass; but when the animal starts to move, some feet are off the ground some of the time, and the peak forces exerted must increase correspondingly. The faster an animal moves, the shorter time the feet will touch the ground and, conversely, the greater the forces affecting the bones. Really fast-moving animals may exert peak forces four times their body mass! Additionally, when running, the limb bones are held at an angle to vertical,

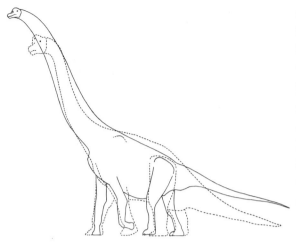

The dotted outline represents the profile of a commercial model of *Brachiosaurus* often available in museum gift shops. It is compared at the same scale to the solid outline of an accurate, technical restoration. Because the commercial model is too bulky, it produces an excessive mass estimate.

and are thus subjected to bending and torsional forces, which act to distort them about their long axes.

The forces act on the ends of the limb bones (at the joints) and set up stresses in the bone shafts. These forces are complex and change during a stride. However, peak forces are set up in the middle of the stride, where the foot supports the greatest part of body mass. These forces can be divided into an axial component, acting along the long axis of the bone and setting up a compressive stress, and a transverse (bending and torsional) component, acting at right angles to the axial component, and distorting the bone about its long axis. As the transverse component distorts the bone about its axis, instead of just compressing it, it is much more important in setting up stresses in the bone. If one wants to break a stick, it is easier to try to bend it by placing the hands near the ends of the stick than to try to split it down the middle.

Naturally bending and torsional stresses are set up all along the bone, but they are greatest around the middle of the bone because this section is farthest away from the joints; and, thus, the lever arm is longest around midshaft. Additionally, bones are frequently thicker around the ends. The proportions of the bone shaft also matter. If two bones have the exact same circumference but differ in shape, the bone with the largest amount of tissue in the plane of bending will be strongest. Bone proportions usually reflect the nature of normal bone stress. In elephants, the torsional moments are reduced, as the limbs are columnar and mainly loaded in compression and mediolateral bending. Thus, the mediolateral diameters of the shafts (especially the thighbone) exceed the anteroposterior (fore-and-aft) diameters. This is also the case in sauropod bones. This makes bone geometry an additional parameter for evaluating how dinosaur limbs were held and moved.

From the above, we may calculate how strong the bones are. This implies that a lower value of the "stress indicator" will mean less bone stress. For quadrupedal animals, one must also take into account how great a percentage of body mass is supported by the limbs in question. Thus, in an animal carrying 40 percent of the mass on the forelimbs and 60 percent on the hindlimbs, body mass is multiplied by 0.4 and 0.6 when computing the bone strength of its fore- and hindlimbs, respectively. The fraction of mass supported by the fore- and hind-limbs of extinct animals may be computed from suspension of scale models or, alternatively, from the values calculated when doing mass prediction from illustrations.

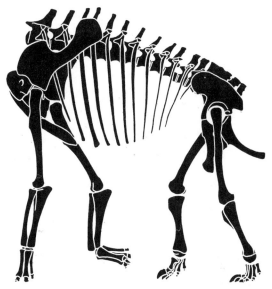

Gregory Paul's skeletal reconstruction of *Brachiosaurus* shows the thick, sturdy leg bones of the immense Sauropod, which explains how the great weight was supported as well as why it was a slow-moving animal.

Conversely, the reciprocal of the stress indicator can be used as a measure of the strength of the bones. This value is usually called the strength indicator. A greater strength indicator value implies that the animal could place greater stress on the bones without risking breakage. Since the bone safety factors are so narrow among extant mammals, it is likely that dinosaurs did not deviate significantly from this pattern. Thus, bone strength is yet another parameter for evaluating how athletic dinosaurs could have been, but it must be combined with analyses of morphology as well. Needless to say, a reliable body mass estimate is of crucial importance here.

In sauropods, the bone strength is rather elephantine, as we might expect, albeit with some differences. Long-limbed sauropods, supporting a greater percentage of mass on their forelimbs, such as *Brachiosaurus*, had weaker forelimbs, and the diplodocids had stronger forelimbs, owing to their short arms and low percentage of mass supported by the forelimbs. Overall, the conclusions about the locomotion of sauropods drawn from morphological comparisons with elephants are supported by bone strength analyses. Ceratopsid bone strengths are comparable to those of extant rhinos and large bovids, again supporting the conclusions from comparisons of morphology.

Sauropods: Elephant Analogs

In many ways, dinosaurs, particularly the quadrupedal herbivores, resemble large mammals to a great extent. Sauropods are often compared to elephants, which seems appropriate, as both groups included the largest land-living vertebrates. Extant elephants do not approach the sizes of many sauropods, but the African elephant will on occasion reach 7–8 tons in mass. However, larger forms once existed; the largest mammoths may have reached 10–15 tons in mass. Likewise, a species of hornless rhinoceros called *Indricotherium* lived in Asia some 30 million years ago, and recent finds of this colossal animal indicate that it stood around 18 feet tall at the shoulder and could have weighed well in excess of 15 tons.

When going a bit more into detail with sauropod anatomy, the elephant analogy becomes even stronger. Sauropod limb bones were about as strong as those of elephants, indicating that they were built to withstand similar levels of physical activity. Sauropod limb joints indicate that their limbs were kept rather straight, which is similar to the limbs of elephants. Having straight limbs implies that the limb bones are mainly loaded in compression and that bending and torsional moments are dramatically reduced, a great way of supporting large body mass. Limb muscle scars in sauropods are also not as large as in animals capable of running, which again is similar to the limbs of elephants. However, as elephants move their limbs—much like an inverted pendulum— they are in need of strong muscles at the shoulder and hip, and in this respect sauropods appear to have been similar as well.

The hand and foot bones of sauropods are rather short, as are the forearm and shin bones, implying that sauropods were not anatomically adapted for fast locomotion; and the foot bore a large posterior heel pad. Again, this bears resemblance to the limbs of elephants, and overall the similarity is so substantial that one may conclude that they probably moved in a similar fashion to elephants. There are, however, a number of differences to indicate that sauropod locomotion diverged from that of elephants in some interesting details.

The olecranon process at the elbow of sauropods is extremely small, and only titanosaurids later evolved a somewhat larger process. The process is also reduced in elephants, but not nearly as much. This indicates that sauropods had weaker elbows than do elephants. The hand of sauropods was not elephantine; it was perfectly upright and formed a semicircle. The fingers of most sauropods were extremely reduced, consisting of small, knoblike bone to the left, but not crossing each other. Such a forelimb anatomy indicates that the forelimb was probably used mainly for support of mass and that forward propulsion was markedly reduced. This is quite different from the condition in elephants, in which the forelimbs contribute considerably to forward propulsion.

Sauropod hindlimbs, however, appear more powerful than elephant hindlimbs. The hips of sauropods are quite powerfully constructed, and the large muscle insertion crest on the tibia (shinbone) was substantially more well developed than in elephants. The foot was elephantine in that it was platelike in shape and bore a posterior heel pad. The large claws would have assured firm anchorage, even on more slippery surfaces, and overall the hindlimb of sauropods appears to have provided most of the force for forward propulsion. This would have been further enhanced because of the large percentage of body mass supported by the hindlimbs of the tail-heavy sauropods. Thus, in sauropods the power needed for forward motion was mainly supplied by the hindlimbs, an unusual condition compared to small-tailed mammals.

Sauropods have often been claimed capable of rearing up on their hindlimbs. This is indicated by a number of things. Some sauropods, particularly diplodocids such as *Diplodocus* and *Apatosaurus*, have long neural spines around the hip area. This would have provided the hypaxial muscles, which lie alongside the neural spines, and would have provided the lift, with a lot of leverage. Additionally, since their center of gravity was just anterior to the hips, they could easily have reared up by placing a hindfoot a little forward of the center of gravity and using their large tails as counterweights. The bone strength of the femur of a bipedal diplodocid is not much lower than if the animal walked on all fours, with the exception of the enormously long-necked *Barosaurus*, which supports more mass on its forelimbs (41 percent) than the other species.

Such a posture, however, would have raised the head far above ground and created a long distance between the head and the heart. Recent analyses suggest, however, that diplodocids could not have raised their necks vertically (see the following essay by Gregory Paul, "Restoring the Life Appearances of Dinosaurs"). At a glance, the blood pressure in the tripodal position appear staggering for the very long-necked *Barosaurus*. It is twice as much as has been verified in the giraffe, which among animals today has by far the highest blood pressure. The giraffe can move its long neck about rapidly, as its jugular vein is voluminous, thus acting as a blood reservoir when the animal lowers its head. Additionally, when the animal lowers its neck, its heart rate drops immediately. Many of its arteries have very thick walls, as does the large heart, and the skin around the legs is quite tight. When contracting the leg muscles, the tight skin forces the fluid that has been forced out of the thin capillaries back into the venules, thus preventing fluid buildup in the legs.

How did sauropods cope with blood pressures twice as high? The answer is that we simply do not know. At present, we cannot explain how an animal could have had such a massive blood pressure without suffering fatal fluid buildup, especially in the legs, or how it avoided rupturing all the arteries in its head when lowering the neck. The easiest answer would be that they were incapable of raising their long necks and never used a tripodal stance to feed off treetops. Attractive as this sounds, it is a red herring. *Brachiosaurus* was built like a huge giraffe, and recent analyses suggest that this animal did indeed have a giraffelike posture, carrying its very long neck nearly vertically in the normal walking posture. At about 42 feet in height, the head-heart distance would also imply a blood pressure of twice that of a giraffe. If *Brachiosaurus* could do it, why not the diplodocids? It is, however, clear that in this respect sauropods differed dramatically from elephants, which have nowhere near the heart rate nor the blood pressure of

The limbs of elephant-sized *Triceratops* were not particularly elephant-like. Instead they were more similar to those of rhinos, being flexed at the shoulder, elbow, knee and the flexible ankle. The large olecranon process of the elbow and the cnemial crest of the knee provided high leverage to powerful leg muscles, and the massiveness of the leg bones indicates that they were stressed for movement at high speeds.

giraffes, let alone a *Brachiosaurus*. In many respects, the elephantine sauropods did not resemble elephants at all.

Ceratopsids: Rhinoceros Analogs

The large ceratopsids have traditionally been considered dinosaurian rhinos, mainly owing to their cranial ornamentation. The entire skeleton is also rhinolike, with a heavy, rather barrel-shaped body, and powerful legs. The ceratopsid upper arm bone (humerus) is sturdy, and its bone strength is high. This is necessary, as the forelimb clearly had substantial shoulder and elbow flexure, thus positioning the humerus at an incline.

The shoulder and hip regions of ceratopsids are massively constructed, and the limb muscle scars are well developed. This all indicates a powerful musculature, as in rhinos, necessary for running with a body mass of several tons. When looking at the limb proportions, ceratopsids were not indicative of fast locomotion. Large animals need fairly short and massive legs to support their mass, unless the legs are columnar, as in sauropods and elephants. Such legs, however, are impossible to run with. Ceratopsids also walked on their toes, as rhinos, and overall the rhino analogy seems substantial.

But is this really the case? Were ceratopsids particularly rhinolike in their limb anatomy? Probably not. The reason for the similarities between ceratopsids and rhinos most likely is that they are roughly similar in size. If one wants to run with a body mass of several tons, there are limited ways to anatomically adapt to doing so. Rhinos and ceratopsids are probably similar because both have faced the same problems of maintaining the ability to run at a large body size. (Large extinct mammals such as brontotheres or arsiniotheres also appear rhino- or ceratopsid-like, but were unrelated to both.) Rhinos and ceratopsids evolved from smaller ancestors that did not look like this, but they ended up looking alike in many respects because they all grew big while maintaining the ability to run. This beautifully illustrates the principles of evolution.

Galloping tricertops. Analyses of limb morphology and bone strengths indicate that *Triceratops*, like the other ceratopsids, was probably capable of a slow gallop, as extant white rhinos. Due to its larger size *Triceratops*

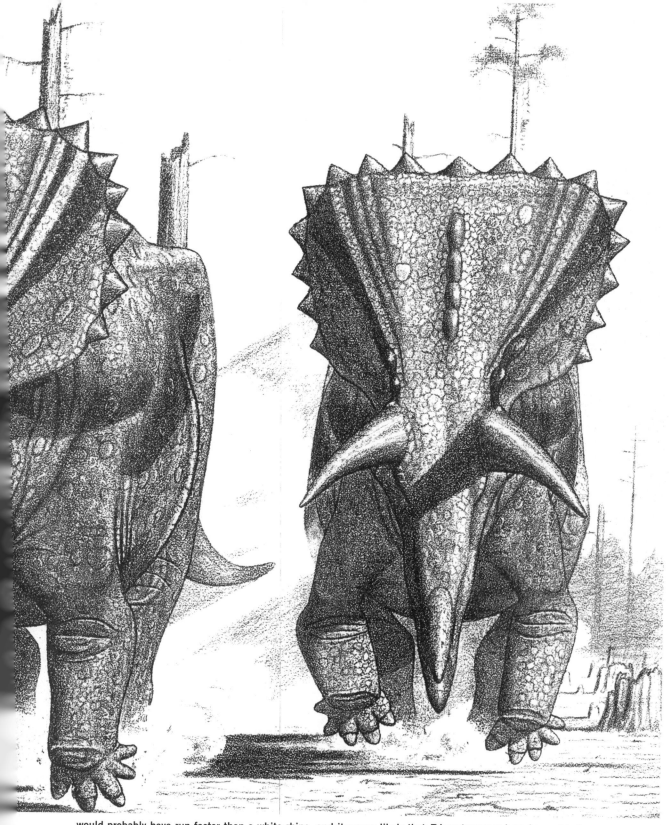

would probably have run faster than a white rhino, and it seems likely that *Triceratops* may have been able to run around 35 kmh[1].

Restoring the Life Appearances of Dinosaurs

by Gregory S. Paul

Long ago people imagined that dragons inhabited distant places, and artists made drawings and sculptures that showed the public what it was thought they looked like. Though science has since proved that dragons are not real, it has more than made up for the loss by showing that dragonlike dinosaurs did once inhabit the Earth. Science also supplies artists the information they need to restore the life appearance of the extinct beasts. In some cases, dinosaurs can be restored with remarkably high fidelities, almost as accurately as some more recently extinct animals. This is especially true of those dinosaurs preserved as "mummies," that is, with most of the skin preserved as impressions, and of the feathered dinosaurs whose plumage is preserved.

How to Start a Dinosaur Restoration

A high-quality life restoration begins with a good skeletal restoration. Restorations based on previously published skeletal restorations or outline skeletal sketches usually prove to be seriously flawed. Also important are multiple-view restorations of at least one representative of each major group. These make obvious anatomical errors not always apparent in side view, and they detail the subject's three-dimensional structure.

It is not possible to restore the muscles of extinct animals in great detail, but much can still be done. Some of the more important muscles follow consistent patterns of origin and insertion upon the bones in most vertebrate groups, and some of the important muscle scars on bones can be positively identified. Muscle function also defines the limits as to where muscles can be placed. It is helpful to profile the muscles in black around the skeletal restorations.

When restoring dinosaurs, think birds, then think big mammals. But don't go too far with this. Skin impressions show that over their avian-mammalian form was, at least in the big species, a reptilian veneer of scales, hornlets, and frills. Internally, mammals are not the best models for dinosaur muscles because dinosaurs had simpler crocodilian and birdlike muscles, and they had many features not found in other animals.

This *Tyrannosaurus* series shows the steps involved in restoring the appearance of a dinosaur. The beginning is a skeletal restoration, preferably in multiple views. This example is a composite of two specimens. Next is a muscle restoration that fleshes out the contours of the animal. The level of detail in such muscle restorations is partly hypothetical, in the manner that many details in the finished life restoration are speculative. The skin is largely inferred from that of other large theropods. The arm-borne display feathers are speculative, but plausible in view of their presence in smaller relatives.

Heads

Dinosaur skulls were like those of reptiles and birds in that they lacked facial muscles, and the skin was directly appressed to the skull. This is quite unlike mammals that have extensive facial musculature, and it is this absence that makes dinosaurs easier to restore. Bony eye (sclerotic) rings often show the actual size of the eye. Most dinosaurs had large eyes, but relative eye size decreases as animals get bigger, so it is wise to avoid the tendency to make them overly large in big species. The eyes in juveniles and small species are proportionately much bigger, but even here far too many restorations overdo it. Consider that the eye of the ostrich, the biggest among living terrestrial animals, does

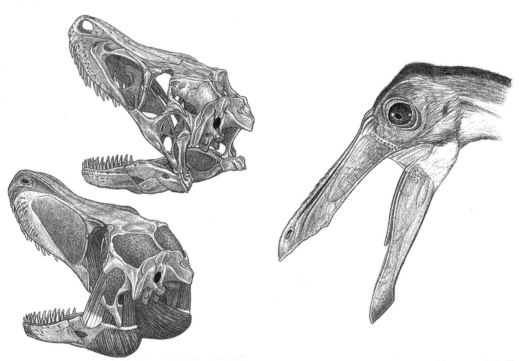

A partial restoration of a dromaeosaur skull shows features common to many dinosaurs. Facial muscles were absent. The most prominent skull muscle was the powerful posterior pterygoideus that wrapped around and deepened the back third of the lower jaw. Jaw muscles bulged gently out of the skull openings behind the eye sockets. The outer ear was a deep, small depression between the bones that supported the lower jaw and the jaw-closing muscles at the back of the head. In saurischians such as this one, sinuses probably filled the large opening in front of the orbit. In many theropods, small horny bosses and ridges should have adorned key features on the top of the head.

Gallimimus had very large eyes set in a small skull. But the eyeballs were partly covered by eyelids, so they probably did not appear disproportionately large.

not appear that large. Whether the pupils of dinosaur eyes were circular or slits is not known. The latter are most common in nocturnal animals, and either may have been present in different species.

Vertebral Column

Most dinosaur necks naturally articulated in a birdlike S-curve, but some examples were straighter. Powerful side neck muscles often formed a contour over the weaker underside muscles. The size of the muscles under the neck and throat are among the most difficult to estimate.

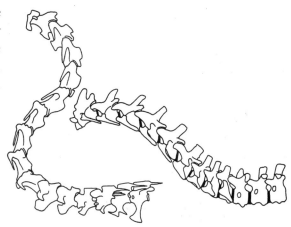

The necks of living birds, like that on the left, articulate in a strong S-curve. The necks of many dinosaurs, such as the theropod neck on the right, articulated in a gentler S-curve.

In many reptiles, mammals, and modern birds, the trunk vertebrae articulate in a nearly straight line. Dinosaurs and early birds differed in having more strongly beveled trunk vertebrae, which always formed a dorsally convex arch, sometimes very strongly so. Some recent restorations of ceratopsids with swaybacks are incorrect in this regard. Small zygapophyses and, in many cases, ossified interspinal ligaments indicate that dinosaurs had stiff backs. The hip vertebrae generally continued the line of the dorsal column.

It cannot be overemphasized that of the many thousands of dinosaur trackways only a handful show tail drag marks. This establishes that all dinosaurs normally carried their tails clear of the ground, even the whip-tailed sauropods and club-tailed ankylosaurs. However, this does not mean that the base of the tail always emerged straight out from the hips as many dinosaur artists now chronically show them. In some dinosaurs, the base of the tail was directed downward. Pretensed ligaments probably carried the tail erect with little effort. In this regard, dinosaur tail muscles and fat may not have bulged far beyond the limits of the bones, so they would not overload the tail. This is confirmed by hadrosaur tail-skin impressions. Most dinosaurs had fairly flexible—but not serpentine—tails.

Ribcages, Bellies, and Hips

As they are in crocodilians and birds, the front ribs are always swept back in articulated dinosaur skeletons of all types. But because this fact is little noticed, many dinosaur skeletons continue to be incorrectly mounted with vertical front ribs. This is a serious error because the chest was slab-sided; and because the rib heads are offset, swinging them forward overbroadens the chest and misarticulates the shoulder girdle. The belly ribs tend to be more vertical, but this condition is variable. Dinosaur trunk vertebrae and ribs formed a short, fairly rigid body with the shoulder and hip girdles close together, so the trunk musculature was rather light, like that of birds.

A few dinosaurs had exceedingly tall vertebral spines. It is not clear whether these were thin, sail-like finbacks, or helped support thick fat deposits. The latter is not likely in predators, which tend to keep their forms as light as possible.

All herbivores must have large digestive tracts that constantly contain the large bulk of plant material needed to sustain the healthy gut flora critical to plant digestion. Spacious posterior ribcages and hadrosaur mummies confirm that herbivorous dinosaurs had big bellies, so restorations that show them with hollow bellies are in serious error. In sharp contrast are big predators that gorge at a carcass, then fast until hungry enough to hunt again, so in restorations, you must give theropods hollow bellies when you show them hungry and on the hunt. The pubis and ischium, the two downward-projecting pelvic bones, were probably not submerged in flesh as often shown, but neither should the ischium or its tip project strongly out, as some restorations show. Such restorations may have been inspired by the externally prominent ischia of kangaroos, but their ancestors had the usual reduced mammalian tail and are not a good model for dinosaurs.

Body Posture

Traditionally, bipedal dinosaurs were restored with a strongly erect-bodied, tail-dragging posture reminiscent of partly bipedal primates. This view has been abandoned. Today, the debate is whether they walked and ran with their bodies nearly horizontal like birds, or tilted up about 20°. Dinosaurs that did the latter included tall-shouldered sauropods such as camarasaurs, the long-necked Chinese forms, and especially brachiosaurs and therizinosaurs. In these dinosaurs, the pelvis was rotated backward relative to the trunk vertebrae. This allowed the hip and tail to remain horizontal while the trunk vertebrae emerged at a sharp upward angle and carried the shoulders above hip level. This is an important characteristic because it kept the relationship between the pelvis- and tail-based muscles and the legs the same as in level-backed dinosaurs.

Most theropods and bipedal ornithischians did not have such retroverted hips. If the hips were tilted up along with the body, the ischia would have been rotated forward until they were between the legs. Some of the leg-retracting muscles would have been ineffective in this position. This is true even of small birdlike theropods that had backward-directed pubes. Most bipedal dinosaurs, therefore, walked and ran with, at most, a slight upward tilt to the body.

Quadrupedal dinosaurs walking out trackways. On the bottom, African *Brachiosaurus* moving in one of the footprint series made by latter Texas sauropods, either brachiosaurs or titanosaurs. Note the vertical, elephant-like posture of the legs. In the center, *Triceratops* walks in a ceratopsid trackway found in the same time and place as the horned dinosaur. The trackway shows that ceratopsids placed the hands directly under the shoulder joint, not sprawling to the sides. At top, *Sauropelta* is shown in a nodosaurid trackway found in the same time zone.

Arms and Legs

In dinosaurs, the shoulder joint was immediately in front of the ribs, as in reptiles and birds, not astride them as in mammals. The best-articulated specimens, especially hadrosaur "mummies," show that the majority of the shoulder blade overlapped the ribs to maximize their muscular support. Because the anterior ribs were backswept, the front end of the shoulder girdle was under the neck-trunk juncture, not well in front as too often shown.

In side view, the scapula blade of most dinosaurs was fairly vertical, as in most tetrapods, not strongly horizontal as is frequently restored. Articulated skeletons show that shoulder girdles were variable in position and mobile in quadrupedal dinosaurs. Dinosaurs lacked a clavicle brace that fixed the shoulder bones in place. This can be explained as a means of freeing the shoulder girdle to glide fore and aft in a groove on the side of the sternal complex. Scapular rotation is important to dinosaur artists because the shoulder joints swing fore and aft relative to one another during the limb cycle, a visible mammal-like feature.

The shoulder joints of most bipedal dinosaurs were open to the side, and the lower arm bones, the radius and ulna, could rotate around each other. These flexible articulations allowed the arm to reach out to the side and forward, and to grasp objects.

The long-running debate over whether some quadrupedal dinosaurs had sprawling arms is ending because their trackways consistently preserve a narrow- rather than

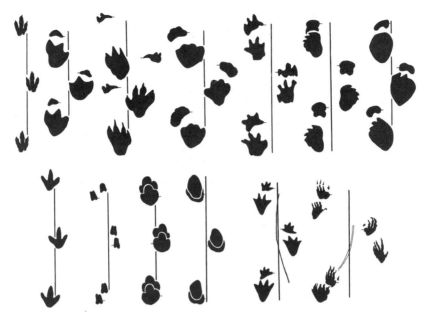

Dinosaur trackways show the same pattern, a narrow fore- and hindlimb gauge with moon-shaped hands (marked by a short line) that never faced inward. This is more similar to the pattern observed in mammals and birds than to the wider gauge trackways of sprawling reptiles. The relatively narrow crocodilian trackway was made with a semi-erect "high walking" gait that included stronger side-to-side undulations of the body than stiff-backed dinosaurs. The exact placement of the hands and feet varies in dinosaurs. Hands tended to be further apart in prosauropods and ceratopsids than in other quadrupedal dinosaurs. In some sauropod trackways both the fore and hindfeet fall nearly on the midline; in track-ways attributed to brachiosaurs and titanosaurs the feet were separated by as much as a foot's width. Note that the reptiles left tail drag marks that are absent in dinosaurs, mammals and birds. Also note that hindfeet often came down on at least part of the hand print. (Top row: large theropod, iguanodont, prosauropod, ankylosaur, sauropod, sauropod. Bottom row: moa, wildebeest, rhino, and elephant, crocodilian, Komodo monitors.)

wide-gauge forelimb posture. The shoulder joints of quadrupedal dinosaurs faced down and backward; indeed, the upper surface actually faced a little inward. These facts show that the feet were placed under the shoulder joint, and that the elbows did not project strongly out to the side. The hindfeet were also placed under the body, and the hindlegs were erect. However, it is a mistake to draw dinosaurs with perfectly vertical limbs because the elbow and knee were bowed outward a little, especially as the arm was swung forward during the recovery stroke. The shoulder joints of some partly or fully quadrupedal dinosaurs, such as ornithopods, ankylosaurs, and ceratopsians, were open enough to allow the arm to swing out to the side more than in stegosaurs and sauropods. All dinosaurs had cylindrical hindleg joints that limited their motion to fairly simple fore- and aft-arcs. This was like birds, and it means that dinosaur limbs, especially the hindlegs, were not as supple and flexible as those of mammals.

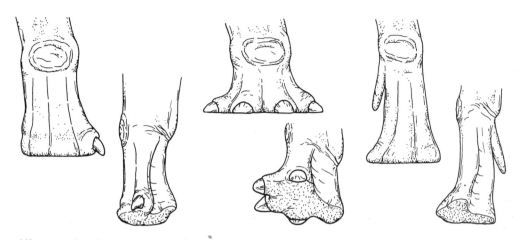

Life restorations by Gregory Paul for the front feet of a variety of four-legged plant-eating dinosaurs, from sauropods (left), to ceratopsians (center) to hadrosaurs.

Fore-and-aft arm posture varied among the dinosaurs. Most quadrupedal dinosaurs had backward-directed shoulder joints and humerus heads that wrapped far onto the back of the humerus. This flexed the shoulder and elbow, so the humerus sloped down and backward. The humerus could swing forward as far as vertical, but no further. In sauropods, stegosaurs, and nodosaurs, the shoulder glenoid faced more downward, and the humerus head did not wrap as far backward. The humerus was vertical and the elbow was straight in these elephant-limbed dinosaurs.

The upper end of the humerus bulged out a little; and in many, but not all dinosaurs, a large crest of the humerus formed a prominent contour along the upper and front edge of the arm. Hand-flexing and -extending muscles bunched around the elbow and operated the hand via tendons; the wrist formed a bulge. The wrist was held stiff and straight during the propulsive stroke. The wrist flexed to clear the ground during the recovery stroke; in doing so, the bones probably partly disarticulated as they do in some mammals.

Dinosaur handprints faced a little outward, or, less often, straight ahead, but never inward, as occasionally occurs in reptiles. All dinosaur hands were either digitigrade or unguligrade; they walked on the fingers or tips of the fingers. None had flat-footed plantigrade forefeet. An interesting fact is that no dinosaur had hands that looked like their feet. Due partly to different bone structures, their forefeet always lacked a heavy central pad and are hollow behind, giving them a half-moon shape. Sauropods, stegosaurs, iguanodonts, and hadrosaurs united the fingers into a hooflike hand by encasing them in a single pad, with most or all of the single hooves lost. A distinctive character of theropods, prosauropods, some sauropods, and some ornithischians was the big-clawed, inward-divergent-when-extended thumb weapon. It was often held clear of the ground when walking.

This sequence shows how the limbs of a 6-ton *Triceratops* should have worked during a hypothetical gallop. The elbows and knees were flexed, and helped add power to the gait. In the arm, the humerus never extended forward past vertical, in the hindleg the femur never retracted backward past vertical. Note that the gallop is rhino-like, not bounding in the manner of a smaller creature.

The fore-and-aft posture of the hindleg varied among dinosaurs. In most dinosaurs, leg posture was birdlike. The knee could not be straightened because doing so would have disarticulated the joint. This is true even of such giants as *Tyrannosaurus, Triceratops,* and *Shantungosaurus.* The orientation of the hip and knee joints show that in most dinosaurs the knee always stayed flexed, and the femur never retracted much past vertical. Overall limb action must have been very like that of fast birds and mammals, with the femur providing the main propulsive stroke. It is common for artists to restore the femur retracted well past vertical in running dinosaurs for dramatic effect, but this is a mistake. Indeed, there has been a recent tendency to take the new view of active dinosaurs too far, such as showing big ceratopsids galloping like small gazelles. If they did gallop, they did so with the more ponderous motion of rhinos.

In sauropods and stegosaurs, the knee was constructed so that it remained fully articulated when it is was straightened out. The femur could be retracted past vertical, and leg action must have been very like the vertical legs of elephants.

The tail-based caudofemoralis muscle, which helped retract the femur, formed a prominent contour; its profile is seen under the more superficial muscles in the muscle restorations. The ilium, the upper pelvic plate, was visible in living dinosaurs the same way that the pelvic bones of a cow can be seen under the skin. In most dinosaurs, the ilium was very long and supported thigh muscles that were flattened from side to side, but were very broad from front to back. One of artist Charles Knight's greatest errors was to show dinosaurs with narrow, reptilelike thighs. These occurred only of the most primitive dinosaurs, which had shorter hips than later, more advanced dinosaurs. In many dinosaurs, the knee's large crest and the birdlike feet show that a powerful "drumstick" of toe-extending and -flexing muscles operated the feet via long tendons. As in birds, the bulging ankle joint must have been very prominent, as was the Achilles tendon running behind the ankle to the foot. Most dinosaurs' toes probably drooped during the recovery stroke, like those of big ground birds. Exceptions to the normal pattern are again the sauropods and stegosaurs. There is a tendency to restore the limbs of these slow giants as heavily muscled pillars like the fat-limbed brachiosaurs in the movie *Jurassic Park*. This is a mistake. Slow-moving sauropods and stegosaurs should have had long, lightly muscled legs, like those of elephants. The ankle joint was buried in the great foot pad. The sauropod's and stegosaur's short, broad metatarsal bundles and toes backed by a large pad indicate a fixed ankle, and their very short toes were probably immobile. Like the hands, no dinosaur's foot was plantigrade; all were digitigrade.

Most theropods, all prosauropods, and some ornithopods had large, curved hindclaws. Those with robust tails may have balanced upon them and kicked out like dinosaurian kangaroos. Trackways show that bipedal dinosaurs often walked with their toes pointed in a little, as pigeons do. Thousands of trackways show that bipedal dinosaurs always strode when walking and running; they never hopped like kangaroos.

External Appearance

Although traces of dinosaur skin are rare, much more is known about their external appearance than is often realized. Impressions of skin are known from a number of large dinosaurs, including theropods, sauropods, armored dinosaurs, ceratopsids, and ornithopods, most especially the complete hadrosaur mummies. Even some dinosaur footprints show the impressions of scales. The fossil impressions show that most dinosaur scales were nonbony, flat, mosaic-patterned and never overlapping, and semi-hexagonal in shape. In general, large scales are surrounded by a hexagonal ring of smaller scales, forming rosettes that are themselves set in a sea of small scales. The size of the scales ranges from a substantial fraction of an inch to an inch or two across. In some dinosaurs, subconical scales an inch or two across projected from the skin. These were often arranged in irregular rows. Prominent frills and skinfolds, like those of lizards, are sometimes preserved and may have been fairly common in various dinosaurs. A pelican-like throat pouch has been found under the jaws of the ornithomimid *Pelecanimimus*, and soft crests, combs, dewlaps, wattles, and other soft

display organs may have been more wide-spread than we realize (but the flaring neck frill on the dilophosaur in *Jurassic Park* was fictional). Little is known about the skin on dinosaur heads. Large scales or horny coverings may have adorned ridges, hornlets, and other prominent locations. Beaks were lengthened by horny coverings. Flat armor plates were also covered with hard keratin. Long horns, spikes, erect armor plates, and claws were probably lengthened by a third to up to twice original length by horn-sheath coverings. Toe claws were probably worn at the tips due to contact with the ground.

It is now known that at least some small predatory dinosaurs were insulated with a feather-like plumage.

Color

This aspect of restoration is the most asked about, least knowable, and least important. Color patterns have never been preserved on any dinosaur, so we know little about this part of their appearance. Often I do not decide the dinosaur's color until I have completed the background and see what looks best. Most movie dinosaurs have been dull-colored, but some or all dinosaurs may have been more colorful. A light tone might appear advantageous

As an increasing number of baby dinosaurs are found they are becoming a common subject of the artist. This restoration of the South American prosauropod *Mussaurus* is based on a nearly complete hatchling skeleton.

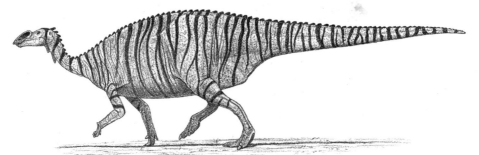

The bold striped pattern arbitrarily placed on this *Muttaburrasaurus* is speculative, but plausible.

for large animals in hot habitats, but tropical naked-skinned animals such as elephants, rhinos, humans, monitors, and crocodilians are dark-pigmented as a screen against ultraviolet radiation. Dinosaurs probably had full-color vision like reptiles and birds, and it has been suggested that dinosaurs should have borne dramatically colored camouflage patterns like those often seen on World War II aircraft. This may be correct, and the scales of big dinosaurs were better suited to carry bold patterns like those of giraffe and tigers than is the dull gray, nonscaly skin of big mammals. On the other hand, large reptiles and birds tend to be earth-tinged despite their color vision. Small dinosaurs are the best candidates for bright-color patterns, like those of many small lizards and birds. Archosaurs of all sizes may have used specific color displays for intraspecific communication or for startling predators. Crests, frills, skinfolds, and taller neural spines would be natural bases for vivid, even iridescent, display colors, especially in the breeding season. Dinosaur eyes were like those of birds or reptiles, not mammals, so they lacked white surrounding the iris. Dinosaur eyes may have been solid black or brightly colored as in many reptiles and birds. But except for the improbability of gaudy colors such as bright pink or purple on the big species, any color pattern is both speculative and possible.

The Dinosaur Groups

Saurischians

Saurischian skulls had a large opening set inside a shallow depression in front of the eye sockets; sinuses probably bulged gently out of this depression. The sharp rims of saurischian jaws were more similar to those of lizards than the rounded rims of crocodilian jaws with their exposed teeth. It is, therefore, more probable that lizardlike, nonmuscular lips covered most or all the teeth when the mouth was closed, rather than the teeth remaining visible.

It has been argued that the overlapping neck ribs of sauropod dinosaurs locked the neck into an inflexible unit. This conclusion was based on the example of crocodilians, which have very short necks. Almost all saurischians had such rib overlap, so if this were true, then all theropods, and even *Archaeopteryx*, would have had rigid necks. Such inability to flex long necks is improbable. It is more probable that the saurischians' slender neck ribs were flexible and could slide past one another to enhance neck mobility.

Early Predatory Dinosaurs

Unlike latter theropods, the ilium remained short in *Eoraptor, Staurikosaurus,* and *Herrerasaurus,* so the thigh muscles were rather narrow. Also, the inner toe was still complete and parallel to the other toes. Their short, slender, fingered arms mean that full bipedality had developed.

Tridactyl-Footed Theropods and Protobirds

Theropods were, in the main, a uniform group, quite birdlike, and fully bipedal with short trunks, long deep narrow hips, and long, narrow, three-toed (tridactyl) feet in which the inner toe was disconnected from the ankle.

Therizinosaurs may have had cheek tissues covering their tooth rows. Theropod cheeks were broader than the narrow snout. This was taken to an extreme in tyrannosaurs, especially *Tyrannosaurus*, troodonts, and dromaeosaurs, which had very broad cheeks and partly forward-facing eyes. However, the upper jaws of dromaeosaurs were probably not quite as extremely narrow as some crushed skulls imply. Deinonychus had a long snout, but its nasals were not depressed, as in *Velociraptor*. The suggestion that the snout of *Dilophosaurus* was too weak to kill prey led to the fictionally small-spitting dilophosaurs in *Jurassic Park*. Actually, the snout was braced internally by skull bones, and the teeth were very large. The mandibles of oviraptorosaurs were so narrow at the jaw joint that the gullet probably passed below this level, resulting in a deep throat pouch.

Theropods did not have raptorlike orbital bars shading their eyes. However, their skulls did bear a varying array of crests and horn bosses. It is not certain whether *Ornitholestes* or *Proceratosaurus* had a short nasal horn, like *Ceratosaurus*, or a long nasal ridge, like recently described *Monolophosaurus*. Some theropods had rugose nasal surfaces that probably supported a low horn ridge, including ornithomimids and tyrannosaurs. Others had sharply rimmed and prominent outer edges of the nasals that were probably enlarged by horn ridges; this is seen in *Allosaurus* and was taken to an extreme in crested *Dilophosaurus*. Virtually all theropods and protobirds had a small hornlet, or boss, just above and before the orbit, and sometimes another just above and behind the orbit. In *Carnotaurus*, these combined into a hyper-enlarged brow horn.

A few theropods, such as *Ceratosaurus*, had nearly straight necks, but in most examples, the neck was so strongly S-curved that they probably could not completely straighten out. Among most big theropods, tall occipital crests atop the back of large skulls imply powerful, bulldog upper neck muscles. The small-headed, slender-necked ornithomimosaurs and therizinosaurs had light neck muscles. In most theropods, the base of the tail tilted up a little, and the tail was fairly supple. Exceptions were the tails of dromaeosaurs, troodonts, and *Archaeopteryx*, which could be flexed 90 degrees upward at its base, but were partly stiffened along the rest of their length. Distally rigid were the extremely abbreviated tails of an oviraptorid, *Caudipteryx* and *Protarchaeopteryx*. Carnotaurs had a uniquely flat-topped back because the winglike side processes of the neck, trunk, and tail vertebrae were so long and tilted upward that they rose above the level of the central neural spines.

Specially formed elbows and pulleylike wrist bones allowed dromaeosaurs, troodonts, caudipterygians, oviraptorosaurs, therizinosaurs, and *Archaeopteryx* to fold up their long arms, although not as tightly as modern birds. More advanced theropods had only three long fingers, of which the thumb was the strongest. *Sinosauropteryx* shows that compsognathids had three fingers, not the two some have supposed, and its thumb and claw were exceptionally stout. The two short fingers of tyrannosaurs remained flexible. *Carnotaurus* had a strange, shrunken hand with immobile fingers.

Theropod trackways show that they walked with a very narrow gait. Narrow-hipped theropods waddled only a small amount, like narrow-hipped ground birds. The exception was the therizinosaurs, whose extremely broad hips helped support an enormous belly. The knees and ankles were flexed, even in the largest examples. The three central toes, which are long in even the biggest species, were underlain by birdlike rows of small pads. The feet of therizinosaurs had reevolved four complete toes, which had very large claws. The toe claws were relatively smaller and blunter in allosaurs, tyrannosaurs, and ornithomimids than in other theropods. These probably delivered quick, ostrichlike kicks, instead of bouncing on their slender tails. Although most theropods were designed for travel on the ground, the long-clawed fingers and toes probably enabled small adults and juveniles to climb.

A couple of small patches of small mosaic scales have been found on the tails of tyrannosaurs. The best theropod skin has been found on *Carnotaurus*; it included irregular rows of conical scales set among the usual flat mosaic scales. *Ceratosaurus* had a row of small, irregular, bone-cored scutes running atop its vertebral spines.

Prosauropods

Prosauropods were beginning to develop beaks at the front of their mouths, and perhaps cheeks at the back of their mouths. Slender birdlike necks carrying small heads suggest light neck muscles. The long-armed prosauropods could walk either on two or four legs, and some trackways show them walking with the hands placed somewhat further apart than the feet. Arms and legs were flexed. The long-backed prosauropods may have run with a bounding gallop, but the heavily built, shorter-footed melanorosaurs probably galloped less and trotted more. The ilia of prosauropods are shorter than in most other dinosaurs, showing that they still had narrow thigh muscles. Prosauropods had five free fingers, of which only the inner three bore claws. The slightly outward-facing hindfeet had four clawed toes. When feeding, the herbivorous prosauropods probably reared up on two legs to browse high in trees.

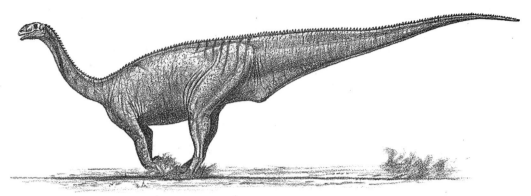

Plateosaurus may have run with a bounding galloping gait similar to that practiced by some juvenile crocodilians. The short pelvis was able to support only a narrow set of thigh muscles.

Sauropods

It is possible, but speculative, that the large nostril openings of most sauropods supported inflatable organs for visual and sound displays like those of elephant seals. The large bump usually restored on the lower jaw of dicraeosaur sauropods was not really there. In restorations, *Apatosaurus* used to be given a short, tall camarasaur-type head rather than the narrow, tapering diplodocid-type it actually had. Conversely, the diplodocid-like head initially applied to *Mamenchisaurus* was wrong; it really had a shorter, taller head like other Chinese sauropods. The heads of sauropods are sometimes drawn too narrow, especially those of brachiosaurs, which were fairly broad. The greath depth common to the front tips of most sauropod mandibles suggests that they helped support deep throat pouches, especially in diplodocids.

Sauropods had gently S-curved necks, the least so in diplodocids. Diplodocids lacked an upward bend at the neck base, and probably had horizontal necks. Dicraeosaur necks could not be elevated much above shoulder level. Because the skull articulated at a sharp angle with the neck, dicraeosaurs seem to have walked with a perpetual hang-dog look. Recent conclusions regarding other sauropods' necks differ radically. On one hand, calculations of neck-bone stresses indicate that brachiosaurs held their necks in a vertical, swanlike posture. On the other hand, computer-based restorations have concluded that all sauropods held their necks level, or even sloping downward. The latter posture would put the herbivores at risk of not spotting attackers and so appears illogical. A problem with estimating neck posture is that it is highly sensitive to the thickness of the cartilage separating the vertebrae, especially the discs. The computer-generated studies have assumed that the discs separating the vertebrae were thin; but so closely spacing the neck vertebrae jams the aft rim of one vertebra's centrum into the base of the rib of the following vertebra in some sauropods. It is therefore probable that at least some sauropods had thick intervertebral discs. The thicker the discs were, the more upwardly flexed the neck was. The neck bases of some primitive, short-necked sauropods were bent only a little upward. In most sauropods, the base of the neck was flexed sharply upward, so the head would have been held high above the body. With up to 19 vertebrae, sauropod necks were at least fairly flexible; how much so is controversial. Certainly, they were not as supple as bird necks; giraffes may be a better living model. The absence of rib overlap may have further enhanced neck mobility in diplodocids. It may have been difficult for tall-shouldered sauropods to reach down to drink.

Since sauropod skulls were small, and their birdlike, air-filled neck vertebrae quite light, their necks should have been relatively lightly muscled. Each neck vertebra was probably visible in life, as in giraffes. Some sauropods had normal, single neural spines on their vertebrae, but some sauropods are famous for their V-shaped neck and trunk spines. That a cable ligament lay between these spines is unlikely because it would have had minimal leverage in such a low position. The space between the V-spines may have been partly hollow; the doubled spines probably served to double the attachment area for intervertebral muscles and ligaments. *Brachiosaurus* was unusual in having withers—tall neural spines over the shoulders—that may have supported a set of low camel-like nuchal neck ligaments. The long-necked Chinese sauropods also had low shoulder withers. Most restorations incorrectly show sauropod necks as simple cylinders in cross-section. They were actually subtriangular along most of their length, first because of the great width of the low-set neck ribs, and second, because as the neck became wider, progressing toward the body, the trachea and esophagus were probably increasingly tucked up between the neck ribs.

Diplodocid sauropods had deep, fairly slab-sided abdomens. Most sauropods had broader bellies, and camarasaurs, brachiosaurs, and especially titanosaurs had very wide hips and abdomens. It has recently been realized that sauropods had flexible gastralia on the underside of the belly, as did other saurischians. The base of all sauropod tail vertebrae articulated in a gentle upward arch. Because they lacked any leverage, the whiplash vertebrae of diplodocids and titanosaurs must have drooped from the fleshy section of the tail like a whip. The front vertebrae of

A muscle study of giraffe-like *Brachiosaurus*. It is important not to overdo the leg muscles of sauropods.

titanosaur tails had ball-and-socket joints that indicate they were flexible to the point that they may have curled their tails forward over their heads like whips. Computer studies suggest that sauropod whip tails could produce a supersonic cracking sound! *Shunosaurus* had a small tail club to whack enemies with; whether some of its Chinese relatives also did is less clear.

Sauropod legs were straight-jointed and columnar, like those of elephants, and the top speed was similarly limited to a fast amble. The divergent, big-clawed thumb weapons of most sauropods were separate from the main unit of united fingers, except in brachiosaurs, whose thumbs were not separate and were much smaller. The three inner toes bore increasingly massive banana-shaped claws that swept down and outward. Theropods approaching the hips of sauropods were in danger of a nasty kick.

Because they are so reminiscent of elephants, it is tempting to restore sauropods with thick, wrinkled skin. But skin impressions show that they had typical reptilian-dinosaurian scaly skin. Some titanosaurs were lightly armored. It has recently been shown that diplodocids had a row of tall, iguana-like scale spines running down their backs and tails, even onto the whiplash. The spectacularly long double spines of the neck and anterior trunk vertebrae of *Amargasaurus* have been restored as supporting dorsal fins, but parallel skin sails would have interfered with flexion of the neck, and the spines are not flattened from side to side. The circular cross-sectioned, sharp-tipped spines were probably spikes lengthened by horn coverings. These could have been used for display, to protect the neck, and for combat by curling the neck ventrally and pointing the front spikes forward. Amargasaurs may have even generated a sound display by clattering the spikes against one another! Sauropod-theropod combat must have been impressive affairs in which tail whips and clubs, foot claws, and biting mouths were used to fend off attackers. The tail of large sauropods weighed many tons, as much as the theropods they were used against! I find the elephantine sauropods, *Brachiosaurus* most of all, among the most elegant and majestic of all creatures.

Ornithischians

The plant-eating ornithischian dinosaurs were characterized by rounded beaks, the edges of which, especially the lower one, were often serrated. There is currently a debate over whether the tooth batteries were covered by mammal-like cheeks covering the sides of the mouth. Some argue that the shallow depression that ran along the tooth rows instead supported horny coverings. However, it is difficult to see what functions such coverings would serve in most ornithischians. If cheeks were not present, much or most of the plant material would fall out of the mouth as they were chewed, especially in hadrosaurs whose chewing tooth surfaces actually sloped out and downward. The shallow jaw rim depressions are similar to the cheek bases seen in some mammals, such as giraffes. The cheeks may have been elastic tissue that served to press food back into the tooth rows.

The posterior ribs were always in front of the prepubic bones, projecting forward from the pelvis, never outside them. All ornithischians had backward-projecting pubic bones under the pelvis, a feature that swung the guts back between the legs.

Early Armored Ornithischians

Scelidosaurus and *Scutellosaurus* were primitive, quadrupedal ornithischians that lacked the skeletal specializations of stegosaurs and ankylosaurs, but had extensive armor coverings. *Scelidosaurus* is known to have had normal dinosaur skin.

Stegosaurs

Most stegosaurs were like other ornithischians in being fairly long and low in overall form. Famous *Stegosaurus* differed because it was taller and not as long. The shoulder girdle was set fairly far back, so stegosaurs had fairly long, slender necks, instead of the short, stout ones seen in most restorations. The neck had a gentle U-shaped curve, and the head was not held high. The trunk vertebrae had a strong downward curve. The tail base was upwardly arched. Stegosaurs probably reared up to feed, using their surprisingly diplodocid-like tails as a prop. The tail was neither supple nor rigid and appears to have been powerfully muscled. The spiny tails were probably powerful weapons.

Primitive *Huayangosaurus* retained moderately flexed limb joints suitable for running; the other members of the group had elephantine arms and legs capable of only a fast walk. Stegosaur hands and feet were exceptionally short and small. The inner two fingers had flat hooves; the outer three were buried. Their three toes had blunt, vertical, rather elephant-like hooves and were backed by a large elephantine hind pad.

New, complete *Stegosaurus* skeletons have answered many questions about the strange armor of this dinosaur. They show that there was a pouch of small, hexagonal-shaped armor nodules set in rows beneath the jaws and throat. Widely spaced small armor pieces were set on the flanks and hips; their exact position is unknown. Other stegosaurs probably had similar armor. It has been argued that the big armor plates of *Stegosaurus* were arranged in paired rows, in a single row, or in alternating rows. The last view was the traditional one, and the new specimens prove that the plates really did alternate in two rows. However, there were only four large plates on the tail, not the eight plates shown in older restorations. The plates and spikes were probably enlarged in life by horn coverings. Other kinds of stegosaurs may have had paired armor plates, but this is not certain. Most stegosaurs, but not *Stegosaurus*, had a large spine on each shoulder. It used to be thought that this spine was on the hip, but complete specimens have shown

A dinosaurian classic, *Stegosaurus* remains one of the strangest examples. The throat ossicles are based on articulated examples, and the arrangement of the dorsal plates and spines follows the latest data, including a complete specimen excavated in part by the artist.

that this is not true. There is no evidence that *Stegosaurus* had more than two pairs of tail spines. Most restorations show its tail spikes simply sticking straight up on the end of the tail, though a new alternative shows them all horizontal, but the spikes probably diverged from one another like a pin cushion, making an effective weapon.

Ankylosaurs

Ankylosaurs, which include the tail-clubbed ankylosaurids and clubless nodosaurids, have been among the most difficult subjects for the paleoartist. They are rather rare, and complete skeletons with armor in place are especially so. Their species identity and relationships are confusing, hindering attempts to combine partial skeletons to make a whole animal. Finally, the structure of ankylosaur skeletons is most peculiar, making it harder than usual to figure out how they go together. Many past restorations have been rather formless caricatures with inaccurate armor.

Ankylosaurs were unusual among dinosaurs in having relatively small eyes. In ankylosaurids, the upper eyelid is often armored, forming a protective shutter over the eyes. The head was directed downward from the short, horizontal neck, especially in nodosaurids.

The hips of ankylosaurs are extremely broad, to such a degree that is hard to believe. Some recent restorations have narrowed the hips, but the long ribs at the front of the pelvis had to be arbitrarily shortened to do so. The aft ribs of ankylosaurs were fused to the vertebrae, and they arced so far out to the sides that the middle of the ribcage was almost as broad as the hips. The mounted skeleton of *Talarurus* seems to have a narrower, more hippolike body, but this is an illusion due to breakage of the ribs from the vertebrae. The ribcage narrowed so strongly progressing forward to the chest that the shoulder blade was twisted along its length in order to conform to the curve of the ribcage. It is, therefore, possible that ankylosaur shoulder blades were less mobile than in other quadrupedal dinosaurs. Because ankylosaur bodies were long, and the legs short, they were low-lying forms. The result was a tabletop body that one could almost have lunch on. In front view, the appearance can only be called ludicrous; there is nothing similar alive today.

The tail sloped gently downward from the hips and curved until it was carried horizontally—there are no tail drag marks. Nodosaurid tails were fairly supple along their entire length. The last half of ankylosaurid tails were rigidly braced and inflexible. This helped carry the tail club, which was porous bone, not nearly as heavy as the mineralized fossil is.

Trackways show that despite their enormous girth, ankylosaurs had surprisingly narrow-gauge limbs. Legs were moderately flexed at the elbow and knee—less so in nodosaurids than in ankylosaurids—and a slow-trotting run is probable. There were five free fingers, of which the inner three or four appear to have had small hooves. Trackways show that the slightly outward-facing hindfeet were backed by a fairly large, rhinolike pad. The three or four toes were separate and bore well-developed hooves.

Although the armor of ankylosaurs was well developed, it was not as extensive as that of tortoises or the recently extinct glyptodont mammals of the Americas. In particular, the armor protecting the flanks of ankylosaurs was not a series of big plates or spikes, as shown in some past restorations. The combination of well-developed legs, limited armor, and tail weaponry suggests that ankylosaur defense was often more active than in slow glyptodonts and turtles. Nodosaurids with long shoulder spikes may have tried to spin around to keep theropods in front of them. Ankylosaurids may have ran away from attackers, keeping them behind while they swung their clubs back and forth to ward off assaults, hoping to land a bone- or joint-smashing blow to the theropod's delicate legs.

A few ankylosaurs, such as Australian *Minmi*, were small, but most were big. When the body proportions are correctly drawn, the articulated armor remains, and skin impressions make ankylosaurs among the most restorable of dinosaurs.

Heterodontosaurs

The long, strong forelimbs of these early, graceful little ornithischians suggest that they were able both to run bipedally and gallop quadrupedally with their long arms. The best-articulated skeleton clearly shows a strong downward arch of the trunk vertebrae. The thumb had a big claw and diverged inward a little; the outer two fingers of the five-fingered hand lacked claws.

Pachycephalosaurs

The dome-headed pachycephalosaurs were among the most bizarre of dinosaurs, and the only ones more interesting in top view than in side view. This is because of their incredibly broad ribcages and unbelievably long tail-base side processes. No other dinosaur has anything like the latter. The absence of chevrons under the front of the tail is another peculiarity. The unusually spacious tail base must have supported an unprecedented migration of part of the digestive tract to behind the hips. The tail base is down-narched a little, and its end is stiffened by a criss-cross of ossified rods. The four-toed hindlimbs are like those of ornithopods.

Primitive pachycephalosaurs such as *Homalocephale* probably slowly butted heads and tried to push one another back with their flat-topped, rather weakly reinforced heads. The advanced pachycephalosaurs with thick, rounded domeheads have often been drawn running at each other and smashing their heads together like bighorn sheep. But

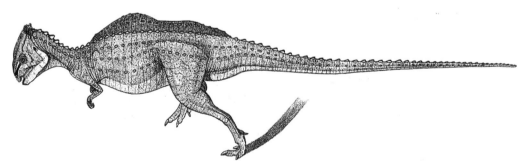

Pachycephalosaurs such as *Homalocephale* are notable not only for their domeheads, but for their exceptionally broad bellies and tail bases, which probably contained massive guts. This oblique view highlights these features.

bighorn sheep have flat-topped heads that are stable when they collide. The rounded heads of domeheads would have bounced off one another at unpredictable angles, so it is more probable that domeheads rammed the side of the bodies of their enemies. However, some researchers argue that the bone of the domes was not strengthened for collisions, and that they were used for visual display only.

Psittacosaurs, Protoceratopsids, and Ceratopsids

A feature common to all these dinosaurs was a large head with a narrow, deep, parrotlike beak, and very broad, triangular cheek bones that have prominent bosses ending in a small hornlet. These parrot-beaked dinosaurs gave wickedly powerful bites and probably snapped at predators.

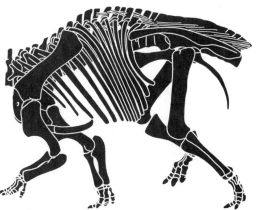

The complete, articulated trunk of *Anchiceratops* illustrates better than any other the peculiar configuration of the ceratopsid ribcage. As in other dinosaurs, the chest ribs were swept backwards. The posterior ribs were bunched so tightly together that they formed a rigid unit that was attached directly to the pelvis, perhaps to form a kind of armor for protection against the horns of other ceratopsids. The backbone was also massive and braced by ossified tendons. All this combined to form a thorax of exceptional strength for rapid locomotion and intense combat.

Psittacosaurs were rather like other small, largely bipedal ornithopods in overall design, except that they had broader bellies. Fairly long forelimbs suggest they were at least partly quadrupedal. The hand had four small fingers, of which the outer was very reduced; the inner three had small blunt hooflets. The four toes had blunt claws.

The heads of protoceratopsids and ceratopsids were unusually large. No other dinosaurs had heads as large relative to their small bodies as did the protoceratopsids; and the heads of ceratopsids were up to nearly 10 feet long! In more advanced protoceratopsids, the head frill was large and supported powerful jaw muscles. The large frills of ceratopsids probably did not support jaws muscles but were covered by tough skin. In ceratopsids, an array of hornlets and bosses decorated the rim and the midline of the frill, as well as before the nasal horn in some species. During restoration, care should be taken to apply the proper number of these little adornments in each species.

Ceratopsid necks were not as short as is usually shown, and it was longest in *Anchiceratops*. The neck muscles were attached to the back of the braincase in the normal manner, not to the top of the frill, as has occasionally been shown. Chasmosaurs were unusual in having moderately developed withers that probably anchored nuchal ligaments that helped support the head. Hip and tail orientation was unusually variable, although the tail was eventually directed downward in all cases. The posterior dorsal

This muscle study of *Chasmosaurus* shows the enormous size of the thigh muscles, which imply a powerful loco-motary performance. Also notable are the head supporting nuchal ligaments anchored upon tall shoulder withers, most unusual for a dinosaur.

ribs were strongly curved fore and aft (specimens seeming to show otherwise are partly disarticulated), and one of these ribs connected in life to the tip of the prepubis.

Ceratopsid trackways indicate that the hands were separated by two hand widths from one another; the hindfeet nearly touched the midline. Limb joints were flexed in all ceratopsians, somewhat less so as size increased. Protoceratopsids walked on all fours and may have both run bipedally and galloped quadrupedally. Ceratopsids probably could trot; whether they also galloped is less certain but possible. Despite the ceratopsid's massive heads, the hindlimbs remained more robust than the arms and bore most of the weight because their bellies were slung between their hindlegs. Their trunk was also short, so ceratopsids may have reared up like bears, possibly to reach the occasional choice food item, or—like an enraged bear—to present a most intimidating image to rivals and to tyrannosaurs. The five fingers and four toes of protoceratopsids and ceratopsids were all free. The fingers and toes bear fairly large, flat hooves, except as usual the hoofless outer two fingers. The hindfeet were backed by a fairly large, rhinolike pad in the bigger ceratopsids.

Although rhinolike in body form, large skin patches show that ceratopsid skin was a reptilian mosaic of large scales surrounding even larger scales. Combined with all the horns, hornlets, and bosses, these large scales present a very interesting body surface texture for the artist.

With its skull adorned by an enormous frill rimmed with hornlets, *Pentaceratops* well illustrates the complex topography of the horned dinosaurs.

Psittacosaurs and protoceratopsids were probably nasty little snappers that dashed about in small groups, aggressive even on the defense, and eager to bite whatever they could sink their parrotlike beaks into. In this view, one would not want to come across one or more of them any more than one would approach a wild boar without a rifle. It's hard not to imagine protoceratopsids squealing like pigs—then again, it's just as likely that they chirped like birds.

Notable features of big ceratopsids are the great strength of their skeletons and the size and power of their leg muscles. Traditionally, ceratopsids were portrayed as fending off attacks with wide, splayed forelimbs and lowered heads. Whether such a slow, static defense would have warded off the enormous, fast-moving tyrannosaurs is a dubious proposition. Perhaps ceratopsids started their defense by rearing up like a bear and tipping their long frills forward in the hope of over-awing the threat. If that failed, the enormous leg muscles were used to propel fast, immensely powerful charges with horns and beaks stronger than those of rhinos. The charges could be repeated until the theropod gave up. The tyrannosaur's goal was to deliver a crippling bite to the rear quarters, either by ambush or by intimidating the horned dinosaurs into a chase the ceratopsids could not hope to win.

Ornithopods

Both the largest and most uniform of the ornithischian groups, it is an important one because some species can be restored with exceptionally high accuracy.

Most ornithopods, except hadrosaurs, had orbital bars above the eyes that gave these most unfierce of herbivores threatening "eagle eyes." Many iguanodonts and hadrosaurs had large skull crests that were probably for both auditory and visual display, and, therefore, may have been colorful. There has been argument over the strange crest of *Tsintaosaurus*, but two specimens show it is a real structure.

The necks of ornithopods were flexible, especially those of iguanodonts and hadrosaurs. Hadrosaurs, *Ouranosaurus*, and *Tenontosaurus* had very strongly downarched anterior dorsal columns. Most specimens are preserved this way, and straightening the back violates the anatomy. It has been suggested that ornithopods with strongly down-curved backs supported the head with deep withers, similar to the neck and shoulder muscles and tendons of horses and other ungulates. In most ornithopods, the tail drooped immediately behind the hips; but in iguanodonts and hadrosaurs, the tail base was arched up just a little. In most small ornithopods, the end of the tail was stiffened by ossified rods; in iguanodonts and hadrosaurs, almost the entire vertebral column behind the neck was greatly stiffened by ossified rods and restrictive articulations.

In most small ornithopods and camptosaurs, the forelimbs were too short to use at any but the slowest speeds. In *Tenontosaurus*, *Muttaburrasaurus*, iguanodonts, and hadrosaurs the forelimbs were long, running organs; and trackways prove that the forelimbs were used at least occasionally. The common hindprint-only trackways do not prove that they were walking bipedally, because the hindfeet may have wiped out the foreprints. Trackways also show that the feet fell close to the midline. The long-forelimbed species' fastest gait was probably a trot, as the forelimbs are too slender to support a gallop. In most pre-iguanodonts, the five fingers are free and bear hooves on the inner three. The four long toes are also free and are underlain by small individual pads. Trackways and a new hand mummy show that the iquanodont's and hadrosaur's three central fingers and hooves were incased in a single hooflike sheath. The thumb was a great spike weapon in camptosaurs, muttaburrasaurs, and iguanodonts. The outer digit is free and unhooved. As for the hindfeet, trackways and mummies show each of the three short toes underlain by a single diamond-shaped pad, backed a fairly large central pad.

The newly described iguanodont *Lurdusaurus* is a massively constructed form that resembled giant ground sloths, but not enough of the skeleton has been figured to allow a restoration. Small mosaic scales are known on tenontosaurs, iguanodonts, and hadrosaurs. The hadrosaur mummies are both extraordinary and underappreciated for

Here it is all put together, dinosaur and surroundings. The duckbill *Parasaurolophus* is reconstructed using the skin of mummies of close relatives; whether a frill connected with the crest is not known. The dinosaurs are shown feeding on herbaceous ground cover amidst a stand of dawn redwoods. The trees (which now grow wild only in China) were based on a dense stand growing in the National Arboretum in Washington DC, where they have dropped their intermediate branches. The season is the dry part of the year, when the deciduous conifers drop their needles.

the nearly complete information they provide on surface topography. They show a serrated frill made of individual hornlets running along most, if not all, the vertebral column. Small, nonbony hornlets are found on the flanks or belly of some hadrosaurs. Vertical wrinkles mark some of the ribbon frills and other areas of the body. Most prominent of these are the large vertical skinfolds enwrapping the neck base, shoulder, and upper arm. These wrinkles were not caused by drying of the carcass because they are always present, always vertical, and not found elsewhere on the body. Virtually complete mummies make *Edmontosaurus* and *Corythrosaurus* the two most accurately restorable dinosaurs. These peaceful herbivores defended themselves by living in large herds, running away when attacked, and biting and kicking when caught.

Mesozoic Landscapes

Many times, dinosaurs are placed in landscapes meant to represent the long-ago world they lived in. Dinosaurs lived in many different kinds of places. Some of the smaller dinosaurs, such as many of those from central Asia, lived in deserts covered by dunes and with plants confined to oases and streams. Many of the big dinosaurs lived in seasonally wet-dry areas with open forests and denser river forests. The giant sauropods especially liked such semi-arid forests. Other dinosaurs lived in rainier places with denser forests and swamps. Some dinosaurs even lived near the poles, where the trees were short, and the warm summer days lasted for months, to be followed by long dark winters with snow and ice.

Restoring ancient landscapes is more difficult than drawing the dinosaurs themselves. This is because the extinct plant communities were very complex and diverse, and there are few complete plants preserved from the Mesozoic. Often, all that we have to study are pollen grains, loose leaves and stems, loose flowers and fruits, and broken logs and stumps stripped of their branches and leaves. We do not know what many Mesozoic plants looked like, so it is hard to illustrate them in a picture. It does not help that no paleobotanist has made the effort to produce a well-illustrated guide to Mesozoic plants (especially vexing since they often complain about the plants in dinosaur art).

If a time traveler were plopped down in the average Mesozoic landscape, they might well be surprised at how nonexotic it looked. Many of the plants living then were not dramatically different from those living today. And, as Triassic times gave way first to the Jurassic and especially the Cretaceous, the plants became increasingly modern. At all times when dinosaurs lived, there were no dry land grasses like those so common now. Equisite rushes were common in wet areas. In the Triassic, Jurassic, and Early Cretaceous, the most common plants were ferns, cycadlike plants, gingkoes, and various conifers.

Some of the ferns were tall trees like those found in the tropics nowadays, but most were little plants. Ferns covered the ground that today would be covered by grasses and herbs. In drier places unable to support trees, but not so dry as to be deserts, it is believed that there were fern plains that went on for hundreds or thousands of miles, like the grass steppes of today. The cycadlike plants looked similar to plants that still live in the modern tropics. They have long fern-style leaves sprouting from the top of a short trunk. The gingko trees survive as one Chinese kind, which have almost become extinct in the wild, but is now a common city tree.

The most common tall trees in the Mesozoic were conifers, which included yews and archaic pines. Many of the extinct conifers had fairly broad, soft leaves; some had the long stiff needles found on most living conifers. Podocarp conifers, which prefer wetter places, are still common in New Zealand. Araucarian conifers are often shown in dinosaur scenes, especially the monkey puzzle tree with its bare telephone-like trunk and umbrella-shaped crown of branches. However, it is not certain how common this form was in those days. Another common Mesozoic conifer was the dawn redwood. Once rare—it was discovered only this century in China—it has now been planted in many places around the world. This tree likes wet places, and it drops its leaves in the winter. Dawn redwoods were especially common in higher latitudes in the Mesozoic, where there was little winter sun. True redwoods were also common in the Mesozoic. They lived along streams and in wetter climates. The fossil trunks of Mesozoic redwoods and some other conifers are sometimes very large, and suggest trees 30–60 yards or more in height!

In the Middle Cretaceous, flowering plants first appeared. These were little shrubby plants that lived along streams. Social bees pollinated the simple, probably white, flowers. During Late Cretaceous, flowering plants became more common, spread into drier habitats and into the water, and became more diverse. In the fresh waters and along shorelines were lilies and cattails. On land, herbs covered the ground where once were ferns, and there were dogwoods, magnolias, and oaks. However, most flowering plants remained shrubby in the Cretaceous; tall, flowering trees were not so common.

There do not seem to have been many closed canopy forests or rainforests in the Mesozoic, perhaps because it was too dry; or it may have been because the giant herbivorous dinosaurs were wrecking the big trees and keeping the forests open so they could better move around and eat the other trees. When illustrating dinosaur landscapes, it is good to show the broken stumps and trunks and the piles of branches brought down by the big plant-eating dinosaurs.

Most dinosaurs that we know of were fossilized because their bones were covered by the sediments carried by large streams and rivers. So the dinosaurs found in museums lived in flat floodplains on which large rocks were absent, and from which mountains and other high places were not visible. There are exceptions. Some of the smaller dinosaur beds were formed in small highland basins and rift valleys surrounded by cliffs, hills, or mountains. Sometimes, a volcano could be seen smoking on the horizon, but volcanos were not dramatically more common then than they are today, so one should check the geographic geology of any particular formation before including them. In some places, thousands and even millions of dinosaur trackways have been found on shorelines that extended for dozens and hundreds of miles. These super-trackway sites were usually on the coasts of the oceans.

One way to do a complete Mesozoic scene and avoid the problems of extinct plants is to make distant plants out of focus, as in a telephoto image. A more common alternative is to include as much sky as possible, and to make the plants as few and as distant as possible. The sky should be appropriate for the climate; for example no cold winter sunset for a tropical Mesozoic climate. Some paleoclimatologists think that most of the world was somewhat cloudier when the dinosaurs lived, but this is not certain.

Instead of doing an entire landscape, it is sometimes possible to use a photographic print for part of the scene, and paint in the rest, including the dinosaurs. In this case, it is important to make sure that the kind of ground, water, and plants in the photograph are appropriate for the scene being restored (no dry grasses!). As computers become more powerful, it will be increasingly possible to do complete dinosaur scenes in digital media.

A Quick History of Dinosaur Art

by Gregory S. Paul

As towering a figure in dinosaur art as Charles Knight was, his is not the only story. The first significant artist to apply his skills to the field was Benjamin Waterhouse Hawkins. Burdened by the lack of adequate skeletons, he made quadrupedal, lizard-like, life-sized dinosaur sculptures for the British Crystal Palace grounds in the 1850s that were the most spectacular examples of early dinosaur art. The sculptures still survive. The discovery of much of a hadrosaur skeleton in the eastern United States a few years later allowed Hawkins to make a more correct restoration in the 1860s of the dinosaur as a kangaroo-like biped. Alas, plans for a major display of prehistoric life in Central Park were thwarted by the notorious Boss Tweed organization (the statues

In 1869, Waterhouse Hawkins was well on the way to completing his dinosaur sculptures for a prehistoric display in New York's Central Park. In his studio the kangaroo-like skeleton and life study of *Hadrosaurus* are shown completed; they would soon be destroyed in an act of political vandalism.

A horned *Triceratops* and towering *Tyrannosaurus,* met face to face in Charles R. Knight's most influential mural—not pictured—painted in the late 1920s. Such paintings still set high standards for today's paleo-artists. Relying on his vast knowledge of anatomy and his vivid imagination, Knight rendered many detailed images of prehistoric animals. For example, although the model *Stegosaurus* he created in 1899 bears too many plates by current standards, it carries them in the alternating pattern now accepted.

built for that display remain buried at the south end of the park, awaiting future excavation).

The opening of the dinosaur quarries in Belgium and the western United States in the 1870s provided artists with the first complete big dinosaur skeletons: iguanodonts, stegosaurs, and huge sauropods. Although the skeletons further boosted public fascination with the beasts, no artist came forward to dominate the scene at this time. Few museums had significant dinosaur exhibits, and none seemed to have found it worthwhile to invest in major dinosaur art, so no great works date from this period. This situation changed only with the appearance of Charles Knight at the turn of the twentieth century, but even he did not receive large-scale commissions during the first two decades of the 1900s. When the projects did arrive, it may have been because of the success of dinosaurs as the stars of the then new form of media.

A favorite subject of the old school of dinosaur artists was to portray a frustrated theropod gnashing its teeth as it stood impotent on a shoreline, watching its sauropod prey escape destruction by swimming into deep water. In reality, big predators are usually more than happy to catch and dispatch their prey in water, and theropods were probably no different. This illustration of a pack of *Allosaurus* assaulting an *Apatosaurus* was the first to capture this logical shift in thinking.

Dinosaurs hit the silver screen in the 1910s with the animated short *Gertie the Dinosaur* and famed D. W. Griffith's *Brute Force* in which a *Ceratosaurus* menaced cavemen. A decade later, the first feature film centered around dinosaurs, *The Lost World,* based on the Arthur Conan Doyle novel, was shown to the public. The stop-motion dinosaurs were crude, but they offered crucial experience to a young Willis O'Brien. He was responsible for the animation in the 1933 classic *King Kong*. RKO's big production of the year, the *Star Wars* of its day, *King Kong* amazed and horrified audiences not yet used to sound film. Whether any prehistoric film has rivaled the romance and visual artistry of this work—which featured a wrestling match between the 18-foot-tall super-ape and a *Tyrannosaurus*—is open to question. *Kong* was produced and scored by the same men who would later make *Gone With the Wind*, so it was appropriate that the gated wall from the former movie's set was consumed as part of the "burning of Atlanta" scene in the latter.

The Great Depression put a stop to most major conventional art projects. (The economic disaster combined with World War II to suppress dinosaur science so completely that it would not recover for decades.) A partial exception was Sinclair Refining Company's dinosaur-based advertising campaign, which led to the commissioning of some dinosaur artwork for promotional purposes. The company's "dino logo" still adorns its service stations west of the Mississippi. Walt Disney never produced an animated feature film centered on dinosaurs, but the beasts did make a famous cameo appearance in *Fantasia*. Knight's career was also winding down in the 1940s, although he illustrated one of the then-rare articles on dinosaurs and other prehistoric life in a 1942 issue of *National Geographic*. Nevertheless, dinosaur art began to pick up in the same decade. As America entered the war, Rudolph Zallinger began his five-year project to paint the great mural at the Yale Peabody Museum. A brilliant piece of art, it became an iconic representation of the modernist consensus of dinosaur biology. The dinosaurs were posed even more formally and statically than in Knight's work. The notion that dinosaurs were dynamic, social, birdlike forms was barely conceivable. Zallinger would continue to present this view of dinosaurs in a series of postwar books. He was fond of placing cinder cone volcanoes in his scenes, not always where they should have been. He particularly liked what I call "Zallinger twilights": thin, high-altitude clouds subtly lit by the rising or setting sun.

The primary challenge to Zallinger came from across the ocean. In Czechoslovakia, Zdeneck Burian began an extensive series of illustrations of prehistoric life. His dinosaurs were firmly set within the scope of modernist consensus. Following the school of Knight and Zallinger, Burian set all of a dinosaur's feet almost always on the ground; the flicked-back wrist of his *Monoclonius* was a notable exception. *Brachiosaurus* was portrayed immersed deep in the waters of a dramatic cliff-sided gorge of the type seen in dam-flooded canyons and fjords, rather than the flat floodplain landscape the animal actually lived in. Burian also followed other artists of his time in not executing multiview skeletal restorations, so his dinosaurs' forms were often more impressionistic than accurate. Burian had a particular and rare affinity for creatureless prehistoric landscapes, which are among his best evocations of Mesozoic times.

The 1960s saw a return of major motion pictures featuring dinosaurs (not to be confused with the nondinosaurian monsters of the Godzilla persuasion). The climactic scene of *Valley of the Gwangi* was a battle between *Allosaurus* and a circus elephant (see the movie to find out which wins). In *One Million Years B.C.*, a brontosaur strides across the barren landscape. (I remember how paleontologists of the time carefully tried to explain to the public that sauropods were amphibious forms that spent little time on dry ground.) In 1964–1965, the Sinclair Oil company funded the first major exhibition of (rather medicore) full-scale dinosaur sculptures since Crystal Palace days at the New York World's Fair.

The march of time brings us to one of the best, and least known, artists of the old school of dinosaur restoration. In the 1960s, the new facility at Dinosaur National Monument needed illustrations of the dinosaurs being exposed on the rock wall of the now-covered quarry. Bill Perry approached the drawing of dinosaurs with the same sensibility that he

Three-dimensional dinosaur art also progressed to a new level in the closing decades of the 1900s, as exemplified by this fine sculpture of the new ceratopsid *Einiosaurus* by Gregory Wenzel.

applied to the wildlife art he specialized in. The result was some of the most lifelike dinosaur images achieved up to that time. His evocative, snorkeling *Diplodocus* head was a fitting end of an era in which dinosaurs were portrayed as classic reptiles. In contrast, his athletic *Allosaurus* running down *Camptosaurus* would fit well into a current show of dinosaur art.

Until this time, artists who illustrated dinosaurs relied upon scientists for guidance. It was a young scientist-artist in the late 1960s and early 1970s who made forever obsolete the traditional mode of restoring the archosaurs. In a series of black-and-white sketches, Robert Bakker portrayed sauropods striding across Mesozoic drylands like dinosaurian giraffes, ceratopsids galloping like charging rhinos, and small theropods skittering about hither and yon like their avian relations. Suddenly, what had been a rather plodding genre of art was exciting and dynamic. It became even more so when Sarah Landry drew a feathered dinosaur for Bakker's seminal *Scientific American* article in 1975. Even Burian's last works were influenced by the new style; his final *Brachiosaurus* scene shows one of the two colossi ashore.

Freed from the constraints of the modernist consensus, and with growing public demand for illustrations of the new view of dinosaurs and all the new discoveries, a community of dinosaur artists appeared and began to expand with increasing rapidity in the late 1970s and 1980s. Central to this unprecedented flourishing of dinosaur art were Mark Hallett, Doug Henderson, John Gurchie, Bob Walters, Bill Stout, and yours truly. Now theropods were shown attacking sauropods that had been foolish enough to retreat to the water; duckbills were portrayed as caring for their nestlings; and ceratopsids were deployed in defensive rings around their young.

In 1986, Sylvia Czerkas organized the first major show of dinosaur art and an accompanying symposium volume, *Dinosaurs Past and Present*. Things have only picked up since then. As museums added and upgraded their ever more popular dinosaur exhibits, and books and documentaries on the subject became numerous, the demand for dinosaur art grew. Japan became a major market, in addition to America and Europe; China is becoming one as well. A market for full-size 3-D dinosaurs appeared—to be filled by the big sculptures of Stephen Czerkas, Brian Cooley, the late Dave Thomas, and others. This trend was taken high-tech in the 1980s, when companies started producing full- and near-scale robotic dinosaurs, often set within ancient habitats. Sound in concept and popular with the public, the execution of these machines has not always been what one might hope for. In 1993, *Jurassic Park* was the first blockbuster film to feature modern-view dinosaurs, animated with hi-tech computers and robotics. It was like throwing gasoline on the fire. Today, there are so many artists around the world illustrating dinosaurs that they cannot be listed. Considering that dinosaur paleontology is continuing to thrive and expand, and fascination for dinosaurs grows in an increasingly educated world, the future of dinosaur art in the new century looks promising.

The Art of Charles R. Knight

by Gregory S. Paul

During the first half of the twentieth century, paleontologists typically thought of dinosaurs as small-brained, tail-dragging reptiles that practiced little socialization and parenting. In recent years, it has become increasingly apparent to some researchers that many dinosaurs were quite active and communal. But the earlier view, which Stephen Jay Gould of Harvard University has dubbed the "Postmodern Consensus," held for many decades. Although paleontologists were responsible for this trend, the American artist Charles R. Knight (1874–1953) popularized it. The murals he painted for museums around the country dominated the way people viewed prehistoric life, not only during his professional career—which extended from the turn of the century to the 1940s—but for several decades after his death as well. Indeed, the current generation of dinosaur illustrators, including myself, grew up admiring his renditions. And these images will very likely continue to inspire paleoartists in the years to come.

Knight's influence prevails in large part because he was both a superb artist and a naturalist who possessed a deep understanding of anatomy. He had the ability to apply his vast knowledge of anatomical structure to make prehistoric creatures come alive again. His paintings remain on display at many museums, including the Field Museum of Natural History in Chicago and the Natural History Museum of Los Angeles County, and they form an important part of the new dinosaur halls at the American Museum of Natural History in New York City.

The first published accounts of fossils that today are believed to be from a dinosaur appeared in 1824. Throughout the 1800s, scientists collected numerous teeth and bones from excavations in Europe and the United States. The public naturally clamored for descriptions of the long-gone giants. But the jumbled skeletons the fossil hunters found offered only sketchy information to artists hoping to re-create the prehistoric animals. The most notable effort to satisfy society's curiosity came from Richard Owen, the preeminent paleontologist who coined the name "Dinosauria" in 1841. In 1854, he commissioned full-size dinosaur sculptures—which are still standing today—for the grounds of the Crystal Palace in London.

The only complete skeleton unearthed before the 1880s came from Germany: a small, carnivorous, birdlike animal named *Compsognathus*. The situation improved dramatically during the 1870s and 1880s, when scientists began to excavate the dinosaur-rich sediments in the arid western United States. There they uncovered whole skeletons of sauropods, predaceous allosaurs, and plated stegosaurs from the Jurassic period. Knowledge about the shape and size of dinosaurs quickly started to accumulate. Shortly thereafter, in the 1890s, Knight began painting them.

Despite Knight's good timing, it is somewhat remarkable that he became the most famous dinosaur artist of his time. Knight was a sensitive character prone to phobias. And although he showed early promise—he began drawing animals and landscapes at age five or six—he was very nearsighted. In addition, a severe injury to his right eye during childhood further impaired his vision. All the same, encouraged by the adults around him, including an artistic stepmother and a talented family friend, Knight attended a series of art schools in and around New York City as he grew older. At age 16, he got his first, and only, full-time job, painting nature scenes for church decorations.

Quickly thereafter, Knight moved from Brooklyn—and away from his increasingly jealous stepmother—to Manhattan. He soon launched a successful freelance career as an illustrator for several natural history publications. He enjoyed going to the city's zoos and parks, and he chronicled his trips by making numerous, meticulous sketches of animals, plants, and other objects. The exercise enhanced his work, as did his habit of visiting the American Museum of Natural History. There he honed his knowledge of anatomy by dissecting carcasses. It was also at the museum that Knight found his calling, when a paleontologist there asked him as a favor to create a replica of an extinct mammal.

After an extended trip to Europe—during which he studied art and visited even more zoos—Knight turned his attention to dinosaurs almost exclusively. He went to work for a short while under Edward Drinker Cope, just before the renowned vertebrate paleontologist died. Cope and his rival, Othniel C. Marsh of Yale College, had brought about the first great rush of American interest in dinosaurs during the 1870s.

But Knight formed his most important association again at the American Museum of Natural History, collaborating with the aristocratic paleontologist Henry Fairfield Osborn. As director of the museum, Osborn wanted someone to translate his collections of dry bones into captivating, living images. Such pictures, he thought, could draw crowds and make his museum the leading center of natural science.

Knight quickly won attention for the museum and for himself, fashioning restorations that reflected many of Osborn's early ideas. Osborn proposed, for example, that sauropods may have been terrestrial high browsers, and so, under Osborn's direction, Knight painted just such a sauropod—a brontosaur—rearing up on its hind legs as though in search of foliage. Knight also showed large theropods—the most successful

Duck-billed dinosaurs of the genus *Anatosaurus* were painted by Knight in 1909. He based the composition on two skeletons mounted at the American Museum of Natural History. In the museum's newly renovated dinosaur halls, the mounts and Knight's painting are display side by side.

Fighting carnivores of the genus *Dryptosaurus* are shown here as they were described by the paleontologist Edward Drinker Cope. Knight completed the paintings in 1897, shortly after Cope died. Within a decade, most scientists frowned on the idea that these dinosaurs leaped into the air. Some scientists now think these therapods may have been quite aggressive hunters.

predatory dinosaurs—leaping into the air. Although he was correct to characterize these theropods as agile hunters, most paleontologists of the time rejected that notion.

During the early twentieth century, digs in North America and Asia produced remnants of remarkable dinosaurs from the Late Cretaceous period—among them terrible tyrannosaurs, horned ceratopsians, duck-billed hadrosaurs, and armored ankylosaurs. Knight's paintings from this time—primarily murals for the American Museum of Natural History and for the Field Museum of Natural History—were sophisticated works of art. He typically painted misty scenes, possibly because of his poor long-range vision, filled with finely rendered, highly realistic figures of well-known dinosaurs. These were Knight's most productive years, and his illustrations became the world's most celebrated.

Knight's personal life was also at its zenith during the 1920s. He and his wife, the spirited Annie Hardcastle, were a popular couple on New York's social scene. Annie secured a comfortable life for them, managing all Knight's money matters, from his pocket change to his payments for paintings (he was notoriously absentminded about finances). At age 13, their daughter, Lucy, took charge. Seven years later, she successfully obtained $150,000 from the Field Museum for her father's murals on display there. In the 1930s, Knight augmented his income by giving lectures, and his authority expanded even further. Today, dinosaur restoration is a minor industry, practiced around the world. Necessarily, much of the romance that Knight enjoyed—having been almost alone in the field—has disappeared.

Knight worked in close collaboration with paleontologists. Thus, his art reflects the scientific dogmatism of his times. This dogmatism was by no means absolute, however. For example, in "Life through the Ages"—a catalogue of dinosaurs Knight compiled in 1946—he called dinosaurs "unadaptable and unprogressive" and "slow-moving dunces" that were well suited for extinction in favor of "alert, little warm-blooded" mammals. But on the same page, he noted that one predaceous dinosaur was "lightly constructed for quick action, and fairly sagacious for a reptile." And he did not always draw dinosaurs as "typical" reptiles. In one painting, he depicted a pair of *Triceratops* watching over a youngster. On occasion, he placed social groups of plant-eating dinosaurs in his work. And after the discovery of dinosaur nests in Central Asia, he painted, at Osborn's suggestion, diminutive protoceratopsids guarding their eggs.

The limitations of the time are most apparent in Knight's best-known piece, showing a lone, horned *Triceratops* facing down two tyrannosaurs. Knight did not know that enormous beds of bones would eventually reveal that some horned dinosaurs lived in herds. Moreover, in Knight's picture, little action takes place between the herbivore and its predators. Every foot is planted firmly on the ground. In fact, the "every-foot-on-the-ground"

Horned *Agathumus*, one of Knight's earliest works, was finished under the direction of Cope in 1897 for the American Museum of Natural History. During Cope's day, paleontologists offered many fanciful and unsubstantiated descriptions of dinosaurs. The animal shown sports what would seem by current standards to be an extreme number of adornments.

rule is true of almost all Knight's dinosaur figures. Although he frequently drew mammals, even large ones, walking and running, he almost never depicted dinosaurs doing so. Knight most often colored dinosaurs in rather drab shades of solid dun and green. Dinosaurs may have been such hues, but they probably had color vision much like reptiles and birds, and their scaly skins would have been suitable bases for more intense pigmentation. For these reasons, most of today's artists often apply vivid colors to their dinosaurs.

Knight used his vast knowledge of anatomy to make extinct forms appear so real that his viewers could easily believe he had seen them. This ability no doubt explains why his pictures continue to look plausible today. But this seeming realism was in some ways superficial. Although Knight sketched detailed musculoskeletal studies of living animals, he did not produce similar studies of dinosaurs—in part because skeletons reveal limited information about an animal's musculature. Instead, Knight drew skeletal mounts, made rough sculptures, or composed life restorations freehand—a tradition in which many dinosaur artists have followed.

One particular anatomical convention that Knight practiced perplexed me when I was a budding dinosaur artist in the late 1960s—back in the days before the idea that dinosaurs were energetic had gained any popularity. I knew that dinosaurs were considered to be reptiles, and that lizards and crocodilians have narrow thigh muscles attached to small hips. Consistent with this theory, Knight made his dinosaurs with narrow, reptilelike thighs. Yet looking at skeletons, I thought that dinosaurs seemed to be built more like birds and mammals, with large hips anchoring broad thigh muscles. What was a teenage dino-artist to do? I copied my hero Knight, even though Alfred S. Romer, the esteemed vertebrate paleontologist of Harvard, had correctly depicted big-hipped dinosaurs with broad, birdlike thigh muscles in his classic 1920's studies of the evolution of tetrapod musculature. The paradox was resolved in the 1970s, when the new hypothesis that dinosaurs were "warm-blooded" at last emerged. An animal having broad hips and large thigh muscles would need to have an aerobic system capable of sustaining high levels of activity for extended periods.

Artists are a bit like magicians: We use optical illusions to fool people into thinking they are seeing a version of reality. Because one's bag of optical tricks gets bigger with time, most artists tend to get better with age. Knight's last decade of restorations, however, did not meet his earlier standards. Deteriorating eyesight may have been the culprit. Also, Osborn was long departed, and the Great Depression and World War II had sent dinosaurology into an era of quiescence that did not lift for 30 years. Knight never knew of dinosaur nesting grounds, the mass migration of herds, polar habitats, the shape of *Apatosaurus*'s head, giant meteoritic impacts, or the fact that birds are living dinosaurs. Even so, his re-creations currently set the highest standards for artistic quality—and they continue to motivate those of us who follow in his footsteps.

Dinosaurs Got Hurt Too

by Rebecca R. Hanna

Millions of years ago, during the Jurassic Period, a meat-eating dinosaur named Big Al lived . . . and died. Although it is not known for sure what ended Big Al's life, abnormalities in his skeleton provide clues to events that occurred while he was alive. Bone is a living tissue that is constantly being remodeled by cells. As you read these words, your very own bone cells are hard at work. Because of this, our bones are able to mend after they are broken, and dinosaurs were no different. Scarring on bones can result from injury, infection, or disease, and is referred to as an abnormality, or pathological condition. Because the bone dies with the animal, mending of injuries can occur only while the animal is alive. Therefore, abnormalities reflect life events and provide important clues to things that do not fossilize, such as behavior. In fact, certain dinosaurs may actually be characterized by different patterns of abnormalities (for example, tail injuries in duck-billed dinosaurs), and these patterns may be a reflection of their lifestyle.

Disease in the Fossil Record

Paleopathology, the study of disease in the fossil record, is a term that was coined by Sir Marc Armand Ruffer, who studied maladies in Egyptian mummies in the late 1800s. In paleopathology, disease is defined as any deviation from the normal or healthy state of a bone. A medical doctor named Roy L. Moodie was the first person to thoroughly explore the fossil record for signs of disease. Dr. Moodie, largely responsible for the establishment of the field, published numerous papers between 1916 and 1930 describing prehistoric abnormalities in both vertebrates and invertebrates. His recurrent plea in many of these manuscripts was for paleontologists to include descriptions of abnormalities in their publications, and for medical students to use the fossil record to learn about the history of disease. Regardless, the field of dinosaur paleopathology remained largely untouched until the mid-1980s when paleontologist Larry D. Martin and medical doctor Bruce M. Rothschild joined forces to work in this field. Even so, much work remains to be done.

Diagnosing a specific disease process based on a bone abnormality can be challenging and is not always possible. Bone cells can do only three different things: make bone, destroy bone, and maintain bone. So the response of a bone to a disease is limited to bone production, bone destruction, or a combination of both. Furthermore, we will never have the full picture of dinosaur disease because not all disorders affected bones.

Scarring from an injury typically occurs as a mass of newly formed bone (also called a callus) that serves to reconnect the fractured segments. As healing progresses over time, the callus is remodeled by the bone cells until scarring is barely visible. However, if bone segments are out of alignment when healing occurs, then the callus will always be noticeable. Additionally, infection is sometimes a direct consequence of injury, especially with open wounds or compound fractures that expose bones to outside microbes.

Infection in bone produces a variety of patterns, including large growths (exostoses), holes or openings (abscesses), and/or channels into the bone for drainage (cloacae). The drainage channels form when bacteria infecting the bone are pus-producing (pyogenic). Other diseases affecting dinosaur bones include birth defects, tumors, and osteoarthritis.

By far the most common abnormalities seen in the fossil record are injuries resulting from trauma. Infection is the second most common pathological condition, and, as mentioned above, is sometimes caused by an injury. Although plant-eating dinosaurs were more abundant than meat-eaters, it is the meat-eaters or carnivores that most frequently show signs of injury.

Carnivores

The individual with the most thoroughly studied pathologies is a subadult *Allosaurus*, dubbed Big Al, that sustained injuries to its tail, feet, toes, finger, ribs, and shoulder. This animal also withstood chronic infection in its foot, finger, a rib, and possibly its shoulder. Big Al's earliest injury was to the base of its tail where this allosaur was struck by a force such as a kick or fall. We know the injury occurred early in its life because the mass of bone that formed in response to the fracture, the callus, was reduced in size and smoothed out by bone cells, a feat several years in the making.

The worst problem this allosaur had was the infected bones in its right foot, and associated flesh wounds where pus produced during infection was drained through an opening in the skin. Big Al's middle toe on this foot was so inflamed with infection that it pushed against the adjacent toes. Because of this infection, the allosaur had a limp and was most likely in a great deal of pain.

During pursuit of prey or attempted capture, Big Al hyperextended the toes of its left foot and damaged the second toe claw. The second finger of its right hand, broken by a twisting force and later infected, may have been damaged during prey capture as well. Adding insult to injury, Big Al fractured four right ribs either during a fall or when struck by another animal. While healing, the broken ribs limited some of this allosaur's

Ouch! When dinosaurs fought each other, like these two *Daspletosaurus* tyrannosaurs, the common result was injuries. If a bone were damaged, the defect might be preserved in the fossil skeleton.

activities with their soreness. Big Al's right shoulder blade shows evidence of either an early-stage infection and/or injury that could have been inflicted by another allosaur. As if this weren't enough for a subadult allosaur to withstand, Big Al's hip and vertebrae also have abnormalities of unknown origin.

Did all allosaurs have as many problems as Big Al, or was this dinosaur just particularly clumsy? As it turns out, Big Al appears to possess an unusually high number of abnormalities for a single individual. However, other allosaurs also injured their tails, feet, toes, fingers, ribs, and shoulders, in addition to withstanding infections in their feet, hands, ribs, and shoulders. The interesting thing is that some of these problems are seen in other large carnivores as well (e.g., *Acrocanthosaurus*, *Albertosaurus*, *Ceratosaurus*, *Deinocheirus*, *Gorgosaurus*, *Majungatholus*, *Marshosaurus*, *Tyrannosaurus*).

In contrast to Big Al's set of malformations, other allosaurs also survived fractures, infection, and, in one case, a cancerous tumorlike growth, in their arms that, unlike in *Tyrannosaurus rex*, were useful appendages. An intriguing abnormality occurs in ancestors of *T. rex*—*Albertosaurus* and *Daspletosaurus*—whose upper arm bones were fractured and healed fairly frequently (three out of six, and one out of two, respectively). Only one case of injury to a *T. rex* humerus has been found thus far, and it is attributed to tendon avulsion. Tendons connect bone to muscle, and tendon avulsion occurs when the tendon is literally ripped off the bone. Given that an adult *T. rex* was 40–50 feet long, and its arm bones were about the same size as those on a large human (although substantially more muscular), this animal had extremely small arms in comparison to its overall body size. *T. rex* could not touch its two hands together, or even reach its mouth! The case of tendon avulsion in this one specimen, which happens to be the infamous "Sue" now in Chicago at the Field Museum, may have occurred when the animal's arm got in the way during prey capture or was caught on something during a fall. The frequency of breaks in the upper arm bones of *T. rex*'s ancestors (*Albertosaurus* and *Daspletosaurus*) suggests that their arms were more functional, but maybe not quite strong enough to withstand the pressures of whatever they were using them for (e.g., prey capture).

The *T. rex* nicknamed Sue has other abnormalities that help us infer behavior. In addition to the tendon avulsion, Sue had an injured neck rib, face, tail, and lower leg bone, as well as an infected finger bone. The neck and face injuries were apparently caused by another *T. rex* when the two were grappling with each other mouth to mouth. Sue's skull has scrape marks that match up with another *T. rex*'s teeth and show evidence of bone scarring, indicating that Sue lived for some time after this occurred. The neck rib (also called a cervical rib) may have been broken by this same event. Sue's tail and lower leg bone may have been fractured during a fall or by a blow from another animal. The lower leg bone (fibula) is not a true weight-bearing bone since its counterpart, the shinbone, or tibia, bears most of the animal's mass. Thus, Sue was able to get around while the lower leg healed. Two other occurrences of fractured fibulas occur in the tyrannosaurids *Albertosaurus* and *Gorgosaurus*. Other *T. rex* abnormalities include injured ribs, as well as two possible birth or genetic defects: one in the backbone and one in the tooth row (two different individuals). Tyrannosaurids did sustain injuries and sometimes infection in their feet, although these types of abnormalities did not occur as frequently as in allosaurs.

Small raptorlike dinosaurs, such as *Deinonychus* and *Troodon*, show injuries to the second toe that had a lethal sickle-shaped claw at the end of it. This claw was held off the ground during locomotion, and its large range of motion and size made it a perfect tool for killing and disemboweling prey. *Deinonychus* shows injuries to the ribs that may have been caused by feisty prey. Additionally, signs of muscular strain and fractures in the legs and feet of *Troodon* and *Syntarsus* may imply injuries sustained during prey capture or pursuit. This appears to be a rare occurrence since these animals survived trauma to weight-bearing elements. Other abnormalities seen in *Troodon* include

Massive force generated by *T. rex* in the "puncture and pull" biting technique was sufficient to have created the huge furrows on the surface of the section of a fossil *Triceratops* pelvis. The enormous body of *T. rex* and its powerful neck musculature enabled the "pull" in the "puncture and pull."

abscesses of unknown origin in a skull and backbone.

The lifestyle of a carnivorous dinosaur may have predisposed it to injury during competition with others for carcasses, mates, and territory, or in pursuit of fleeing prey. Carnivore appendages may have been injured when hunting or killing. Additionally, arms, hands, ribs, and tails could be hit, bit, or traumatized in a fall while pursuing prey, competing for a carcass, or protecting territory. Face scars on *T. rex* allow inference of territorial combat between members of the same species in the form of grappling mouth to mouth.

Carnivores also show evidence of infection, which can place high physiological demands on an individual, requiring strong immune and inflammatory responses. Although it is difficult to make generalizations about the nature of an individual's immune response simply from the presence of infection, two extremes exist: The immune response of the animal may have been inadequate because long-lived infection developed when it would not have done so in a healthy individual; or it may have been adequate if localized containment in a particular bone prevented an otherwise fatal infection from spreading throughout the body. Food and fluid consumption must be increased to sustain effective immune and inflammatory responses.

Depending on the location and severity of the problem, carnivores seem to have lived longer with injuries and infections than herbivores. Meat-eaters may have retained enough of a confrontational attitude to avoid predation while healing, as long as their wounds were not too severe. It also appears that large meat-eaters sustained more injuries and infections than their smaller cousins, although this could be a preservational bias of the fossil record.

Herbivores

Evidence of injuries in plant-eaters is relatively uncommon, but not because these are rare animals in the fossil record. In contrast, remains of plant-eaters are far more abundant than those of meat-eaters because, like today's predator-prey ratios, there were more herbivores than carnivores around. However, since a sick or injured herbivore is more easily taken by a predator, the animal may not survive long enough for the bone cells to produce a record of the disease.

In ceratopsian dinosaurs, like *Triceratops*, the most common injury is rib fractures. These injuries mainly occur in the section of ribs between the hips and midflank. Similar injuries occur in adult male American bison, which incur rib traumas when flank-butting.

Horn and frill disruptions occurring in the skulls of ceratopsians (e.g., *Triceratops*, *Torosaurus*, *Pentaceratops*, and *Diceratops*) may be evidence for fighting between members of the same species. Skull, neck, and shoulder pathologies are also noted in *Triceratops*, *Einiosaurus*, and *Pachyrhinosaurus*. All injuries to this region of the body probably resulted from interactions between members of the same species involved in some manner of head- or flank-butting. Ceratopsians also show stress fractures in their toe bones caused by foot-stamping, sudden accelerations, or long-distance migrations.

Hadrosaurs, or duck-billed dinosaurs, commonly received injuries to their tails that were stiffened by a strut-work of ossified (bony) tendons that balanced their bodies over their hind legs. The rigid tail could be moved side to side slightly, but not up and down. One consequence of having such an unyielding tail was injury. A hadrosaur tail could be broken when another animal stepped on it, although this would occur only when the hadrosaur was lying on the ground, because they did not drag their tails. A second possibility is that the tails of females were broken by males during copulation. However, this would cause multiple breakage events, which are not seen in the fossil record. It is also possible that breaks and infection were caused when the tail was bitten by a large carnivore or bashed against an object (e.g., a tree, rock face, or another animal) while the animal was moving.

Injury and subsequent infection also occurred in hadrosaur hands, arms, hips, and feet, albeit less frequently than tail pathologies. Only one dinosaur specimen (a hadrosaur) has been diagnosed with a dental abscess or jaw infection that caused tooth loss. Problems in dinosaur teeth are rarely seen because, unlike humans, dinosaurs continually grew new teeth throughout their lives. A possible birth defect noted in the foot of a nest-bound baby hadrosaur would have put this animal in danger had it survived long enough to leave the nest.

Sauropods, the largest animals to ever walk on land, rarely show signs of disease. An *Apatosaurus* tail with a bite injury that matches up with *Allosaurus* teeth lends support to the idea that allosaurs were active predators, in addition to being scavengers. Another interesting aspect of sauropod tails is the fusion of two to four vertebrae into a solid unit resulting from the ossification of ligaments. This appears to be a regular growth phenomenon rather than a disease process and has been suggested to be a sexually dimorphic character, as it was noted in approximately 50 percent of examined specimens. On rare occasions, sauropods also withstood injuries and infections in ribs, hips, and tails.

A small, bipedal hypsilophodont (*Leaellynasaura*) survived a long-lived infection that resulted in a grotesque malformation in one of its weight-bearing bones. This infection may have contributed to the demise of the animal by reducing its effectiveness at obtaining food and water, or by making it unable to escape a predator.

Although attacks on armored dinosaurs such as ankylosaurs and nodosaurs seem far-fetched given the amount of bony plates covering their bodies, these animals were affected by trauma and infection on rare occasions. An ankylosaur named *Tarchia* had a bony growth in its sinus caused by injury or infection that was connected to the outside of the face by an opening in the skull. *Edmontonia*, a nodosaur, sustained an injury to its upper jaw. These instances of trauma most likely occurred during combat between armored dinosaurs for territory or mates, as it is hard to imagine a carnivore even pursuing such a heavily shielded prey.

Stegosaurus sustained injuries to their tail spikes fairly frequently, with posttraumatic infection occurring in about half of the cases examined. These animals were probably using their tail spikes defensively to ward off predators or other competitors.

Contrary to our expectations, given the enormous size and weight of some dinosaurs, osteoarthritis rarely occurred in dinosaurs. Osteoarthritis is a disease affecting weight-bearing joints that is characterized by destruction of cartilage and formation of bone spurs. Two different *Iguanodon* specimens have abnormalities in their weight-bearing bones caused by osteoarthritis. This affliction is degenerative in animals today and occurs as joints deteriorate.

Herbivores show signs of disease (i.e., injury, infection, etc.) less frequently than carnivores. A sick or injured herbivore represents an easier target for a predator, and, therefore, may not live long enough for bone cells to record the malady. Some of the common injuries seen in plant-eaters include injured ribs in ceratopsian dinosaurs and broken tails in duck-bills. These patterns provide important clues for deducing behavior (e.g., flank-butting in ceratopsians), especially when compared to modern day analogs.

Conclusions

Because abnormalities reflect life events, the study of disease in the fossil record provides insight into dinosaur behavior and lifestyle. In fact, it appears that certain types of dinosaurs may be characterized by a pattern and frequency of injuries and infections in their skeletons. Carnivores participate in activities that predispose them to injury more frequently than herbivores, which may help explain why meat-eaters show signs of injury and infection more frequently than plant-eaters. Additionally, carnivores probably lived longer with injuries than herbivores, depending on the location and severity of the problem. Diseased or injured herbivores were more vulnerable to predation and may not have lived long enough for bone cells to record the pathology. It also appears that large meat-eaters may have sustained more injuries and infections than small carnivores, although this could be a preservational bias of the fossil record. The study of disease in the fossil record not only shows the sorts of problems dinosaurs lived with, but also assists in the reconstruction of dinosaur behavior and lifestyle.

Chapter Three
Relationships and Evolution

Introduction

Because we humans have evolved a unique level of verbal sophistication, we like to name things. Indeed, labeling items is one of the crucial foundations of science because it helps us categorize and organize things before we analyze them. But categorizing is often a surprisingly difficult process. George Olshevsky details the vexations involved in naming dinosaurs, while noting that it is ironically easier than you might think.

Having named dinosaurs, the next question to answer is how were they related to one another? A leading researcher in this field, Thomas Holtz, first explains how the process of determination works, with emphasis on the cladistic techniques that have greatly improved restoring the relationships of fossil creatures. It is widely agreed that all dinosaurs descended from a single ancestor; but are primitive forms such as *Eoraptor* true dinosaurs or just very close relatives? Where do the bizarre therizinosaurs, which have been placed among theropods, prosauropods, and sauropods, really belong? Which dinosaurs are most closely related to birds?

Of course, one the most fascinating results of dinosaur phylogenetics has been the realization that birds really are the direct descendants of advanced theropods, which are similar to *Deinonychus*. This issue remains controversial because of undue media focus on the few remaining dissenters, who are fighting a losing rear-guard action with the discovery of the feathered dinosaurs predicted by the dinosaur-bird link. Philip Currie gives a first-hand account of this remarkable development in dinosaur paleontology. Nowadays, the real controversy is how and why birds evolved from dinosaurs. Kevin Padian and Luis Chiappe present what has become the majority view: that running dinosaurs learned to fly from the ground-up, by using protowings to begin to generate thrust and lift as they ran full-tilt across the Mesozoic landscape. A paleoornithologist, Andrzej Elzanowski, gives a different spin to the origin of dinosaurs from birds by proposing that some small theropods were climbers that learned to fly from the trees-down. This hypothesis is gaining new support from the fossil record as the remains of very small theropods with climbing adaptations are found.

It all used to be so simple—birds were birds because they were feathered, dinosaurs were reptiles because they were scaly. Now we know that some dinosaurs were feathered and that birds are dinosaurs. More complicated, but much more interesting.

Naming the Dinosaurs

by George Olshevsky

It is ridiculously easy to name a dinosaur. First, of course, you must have a *specimen*: a fossil tooth, a bone, a footprint, an egg or nest of eggs, an associated group of bones, an articulated skeleton, or simply a collection of bones from the same place that you believe belongs to the same kind of dinosaur. Next, you must be quite certain that the specimen represents a kind of dinosaur that has not already been named. Then you must coin your name according to rather loosely followed rules of Latin grammar and make sure that the name hasn't already been used by someone else for a different organism. And finally, you must publish the name in a paper that presents the reasons you believe a new name is warranted for your specimen. That's all there is to it.

Well, not quite; and the "not quite" is what this article addresses. I'll start by describing a few of the general principles and rules that govern how animals, including dinosaurs, are scientifically named, and by defining some of the terms that biologists use when discussing the process of *biological nomenclature* (that is, naming organisms). A name created in accordance with these principles and rules is a *scientific* name; all other kinds of names are *vernacular*. A few dinosaur names are vernacular and have no more scientific standing than the word cat, but the vast majority are scientific.

The Principle of Priority is the key rule of biological nomenclature. It states that, if, in the course of time, an organism or group of organisms acquires two or more different published names, the earliest of these names (the *senior synonym*) prevails; the later names are sunk as *junior synonyms*. Like dark matter, junior synonyms exist in a phantom zone of unused names, appearing from time to time only in bibliographies and catalogues as names no longer available to zoology. This is what happened to the name *Brontosaurus*. In 1903, Chicago paleontologist Elmer Samuel Riggs realized that the dinosaur called *Brontosaurus* by Othniel Charles Marsh in 1879 was the same dinosaur as *Apatosaurus*, named by Marsh two years earlier, in 1877. *Brontosaurus* sank as a junior synonym, to await the time when, if ever, somebody shows that Riggs was wrong. Only when the senior synonym for one reason or another proves invalid or is removed from synonymy is the next earliest junior synonym pulled out of the phantom zone to become the organism's new official name.

Always popular, *Apatosaurus* used to be known as *Brontosaurus*. Unfortunately, the first name was coined first, and most researchers think the two names apply to the same kind of sauropod, so *Apatosaurus* has priority over the better known name. Past confusion over the correct nature of the skull has no relationship to the confusion of the names.

The Principle of Priority originated to bring order to the welter of scientific names that emerged in the wake of the 1758 publication by botanist Carolus Linnaeus of his *Systema Naturae*, the work that introduced the system of biological nomenclature and classification in use today (in several editions: 1758 is the year of the tenth, taken as the first year of biological nomenclature, the first year in which a name may be considered as scientifically created). The Principle of Priority became part of the Stricklandian Code presented by Hugh Edwin Strickland to the British Association for the Advancement of Science in 1842, the same year that Richard Owen published the name Dinosauria in a lengthy report on British fossil reptiles to that same association. The Stricklandian Code, the first significant attempt at a systematic, internationally accepted set of rules for naming zoological organisms, was supported by zoologists worldwide, including Charles Darwin, and led, through a series of international conferences toward the end of the nineteenth century, to the 1905 *Rëgles Internationales de la Nomenclature Zoologique*. This was the direct predecessor of the *International Code of Zoological Nomenclature* (ICZN), which (in its fourth edition, January 1, 2000) presently governs the creation of scientific names of all animals, dinosaurs included. (Botany has its own nomenclatural code and conventions, as does microbiology. These codes resemble the ICZN but are independently administered.)

Linnaeus, in his treatises, presented the idea that life on Earth can be organized by similarity of form and anatomy into a hierarchy of groups and subgroups called *taxa* (singular: *taxon*); the method of classifying organisms into this hierarchy is known as *biological taxonomy* or *systematics*. Originally, at the bottom of the Linnaean hierarchy were *species*, and above them *genera* (singular: *genus*), *orders*, *classes*, and *kingdoms*.

There were three kingdoms in Linnaeus's classification: Animal, Plant, and Mineral. Later, *families* were added between genera and orders, *tribes* between genera and families, *cohorts* between orders and classes, and *phyla* (singular: *phylum*) between classes and kingdoms. So, for example, the Animal Kingdom is divided into about three dozen phyla, each phylum is divided into a number of classes, each class into several orders, and so forth, each category at each level conveniently uniting an array of animal species that share certain specified features. In Linnaeus's day, the descriptions and taxon names were written in Latin, the *lingua franca* of eighteenth-century science. Indeed, the name Carolus Linnaeus itself is the Latinization of Carl von Linnè, Linnaeus's Swedish given name. Nowadays, only the taxonomic names themselves must be Latinized—a last vestige of those earlier times.

Until Darwin published his *On the Origin of Species*, naturalists had no idea why the natural world permitted itself to be hierarchically organized this way; the hierarchies seemed divinely ordained. But the advent of the theory of evolution brought with it the idea that organisms in the same species share a close common evolutionary ancestor, thus accounting for their resemblance to one another; and that the species in a genus shared a more remote common ancestor, the genera in a tribe or family shared a still more remote common ancestor, and so on up the hierarchy. The more remote the ancestor, the more time there was for its descendants to lose their old features and to acquire new ones, and thereby to diverge and diversify. Distantly related species in the same phylum or kingdom share only their very deepest and most basic features.

With the coming of cladistic analysis into zoology, scientists realized that Linnaeus's taxonomic ranks are, above the species level at any rate, entirely arbitrary. There is nothing in the "tree of life" that makes any particular common ancestor more important than any other, no reason to select one rather than another as the ancestor of a class or an order or a family. Also, the number of common ancestors of an organism and its relatives far exceeds the number of Linnaean ranks that can plausibly be invented to fit in between a kingdom and its constituent species. Each such ancestor is potentially the progenitor of a whole taxon. So, in the long run, scientific classifications will dispense with phyla, classes, orders, and the rest as relics of a bygone era. Each species will be classified as the end of a long string of progressively more inclusive but otherwise unranked taxa, representing the ever more remote common ancestors of that species with its ever more distant relatives.

But in the meantime, the ICZN codifies the creation of taxa only at the Linnaean rank of family and lower: superfamilies, families, subfamilies, tribes, subtribes, genera, subgenera, species, subspecies, and any other ranks that zoologists might contrive to fit in between them. The names and ranks of the higher taxa (such as Class Mammalia and Subclass Dinosauria) are technically in an anarchic free-for-all, but because they are well known to their respective researchers and because their scope is so much wider than that of the lower ranks, they remain quite stable. This anarchy is not an inconvenience; indeed, it supports the introduction of new taxonomic methods, such as cladistics, at the higher levels.

Linnaeus in his treatises gave each species a two-part name (or *binomen*). The first part, or *generic name*, is a singular Latin noun, the name of the genus to which the species belongs. The second part, the *species epithet* or *trivial name*, modifies the first part, and the two parts together comprise the *species binomen*, the name of the species. The most famous species binomen created by Linnaeus is *Homo sapiens*, for the human species. Generic names and species binomens (or binomina) are always written *italicized* or <u>underscored</u> in publications. Above the genus level, the names of the taxa are Latinized plural nouns with various codified suffixes. They are written in roman type, not italicized or underscored. These conventions, built into the rules of biological nomenclature, are still followed today. Most dinosaurs are known to the general public by just their generic names, even though all dinosaur genera (except one) also have a species epithet for each included species. *Tyrannosaurus rex* is perhaps the only dinosaur species popularly known by its entire species binomen. (In case you're interested, the one dinosaur genus that still lacks a species epithet is *Macrodontophion*, created in 1834 for a single, slender, medium-sized tooth belonging to a meat-eater, found in what is now the Ukraine. Nobody has bothered to redescribe it and create a species epithet for it. It may even prove to be nondinosaurian.)

Now let's examine the three aspects of zoological naming as they apply to dinosaurs. Bear in mind that when you create and publish a zoological name, you are offering a hypothesis to the world. The hypothesis is that the specimen you are naming belongs to a species unlike any previously described, and that it is distinct from its closest, most similar relatives. Every biologist knows that a species is a reproductively isolated population of freely interbreeding organisms, and every paleontologist knows that there is no way to test whether the individuals represented by their fossils could ever have interbred. The only handle we have on the nature of extinct species is how closely two fossils resemble each other. If they look enough alike—if they share enough features—and come from rocks of more or less the same age, then we are free to conclude that they *likely* belonged to the same species. So, be prepared to have your hypothesis questioned and examined by other zoologists, who may dredge up all kinds of reasons that your specimen belongs to a species already described and that the differences you cite are merely the result of individual variation within an already-named species. Testing of hypotheses is a major part of science, and if your new species survives the tests, you will have served science well.

The Specimen

Before you create a scientific name, you must have something to apply the name to. Naming hypothetical creatures, such as "missing links" that are supposed to exist but have not yet been discovered, is not supported under the ICZN. Names of fictitious animals and fictional animals from novels and films also have no scientific standing, and, of course, fraudulent names are frowned upon by the scientific community. Doing paleontology is difficult enough without having to contend with these kinds of problems. So if you want to name a dinosaur, you must have a real specimen of that dinosaur at hand.

The most difficult part of dinosaur paleontology is to go out into the field and exhume a new species of dinosaur. This is, however, one of the things that professional dinosaur paleontologists do, and, as a result, they get to name the lion's share of new dinosaurs. But even if you are an amateur paleontologist and lack the means to mount an expedition or the expertise to locate and prepare dinosaur fossils on your own, there are ways you may find new dinosaurs. One way is to haunt the collections of the major museums, which house specimens already unearthed and prepared by professionals. After spending a few years learning how to tell one dinosaur from another, you may, every so often, turn up a specimen that was inadvertently misidentified as a known dinosaur and actually belongs to a new species. If you enjoy good relations with the collection curator or the museum's head paleontologist, and nobody else is already working on that part of the collection, you may be allowed to describe the specimen and name the new dinosaur. Indeed, many paleontologists would appreciate having an interested and competent worker do this for them.

Another way is to patronize the libraries of universities and museums and read the early papers of dinosaur science. In the old days, the methods and philosophy of naming fossil organisms were less precise than they are today, and as a result some of the dinosaurs slipped through the cracks. Knowledge of dinosaurs has expanded considerably during the twentieth century, and there is room in paleontology for workers willing to revise earlier works in the light of this new information. Early dinosaur literature is immense, and there are by no means enough people to sift through it all looking for misnamed dinosaurs and misidentified specimens. This is where the dinosaur hobbyist can make a real contribution to the science. And a specimen already thoroughly described in the nineteenth century need not be redescribed; you may simply cite the original description in the paper when you give it its new name.

Creating a Name

So let us assume you have found a dinosaur worthy of naming (or renaming). If your dinosaur represents both a new genus and a new species, then you will have to create the generic name and the species epithet. But if it represents a new species within an already established genus, then half your work has been done for you: The generic name of the species will be the name of the established genus, and all you'll need to do is create a species epithet to distinguish your species from the others in the genus. On the other hand, if you have found a species in, say, a library search that has been classified in the wrong genus, then again half your work has been done for you. In this case, you retain the species epithet and create a new generic name for it. And last but not least, there is one other case to consider, namely, that you have found a species placed in an incorrect genus that belongs in another already-established genus. In this case, you will probably not need to create either a species epithet or a generic name; you will simply transfer the species to its correct genus and replace the old generic name with the new one. Your species binomen becomes a *new combination* (of generic name and species epithet) rather than a new species name. Only if, after doing this transfer, you find that there is already a species in the correct genus with the same species epithet, will you have to create a new species epithet for your dinosaur to avoid what taxonomists call *secondary preoccupation*. This occurred when the dinosaur *Protiguanodon mongoliensis* was moved to the genus *Psittacosaurus*, which already had a species named *Psittacosaurus mongoliensis* in it. The redescriber created the new species binomen *Psittacosaurus protiguanodonensis* for the second *Psittacosaurus mongoliensis*. (Now, fortunately, the latter species is considered a junior synonym of the former, and there is presently no need for a separate species called *Psittacosaurus protiguanodonensis*.)

The ICZN specifies that a generic name must be a singular Latin noun in nominative case and that the species epithet must be linguistically constructed as a modifer of that noun. But dinosaurs, and indeed most of the organisms that have been named by zoologists, were unknown to the Roman civilization, for which Latin was the spoken tongue. There simply are no Latin words for them. So it is necessary to make up new Latin words, especially for dinosaur generic names. The species epithets, however, can be ordinary Latin adjectives, such as *parvus* (meaning "small") or *immanis* (meaning "gigantic") or *fragilis* (meaning "delicate" or "easily broken"). In the eighteenth and nineteenth centuries and early years of the twentieth century, knowledge of the classical languages (Latin and Greek) was a prerequisite to being a cultured, erudite individual such as a scientist or a naturalist. Therefore, those languages are the source of most of the scientific names in use today. But, nowadays, roots from all kinds of local languages are coming into use in dinosaur names, with Chinese dinosaurs frequently having Chinese-rooted names, Mongolian dinosaurs having Mongolian-rooted names, and so on. Check the etymology of a recently formed dinosaur name, and you could well learn its country of origin.

The ICZN supports any pronounceable combination of the 26 letters of the Roman alphabet as a valid generic name or species epithet. The key requirement is that the new name be different from all generic names previously published in zoology. If, for example, somebody has already used a name such as *Protognathus* for a beetle (which happened in 1950), then one cannot thereafter name a dinosaur *Protognathus* (which happened in 1988). This is known as *primary preoccupation*; the earlier *Protognathus* preoccupies the later *Protognathus*, and another name must be created for the genus with the preoccupied name (in this case, *Protognathosaurus*). It is not always easy to demonstrate that the name one wishes to bestow on a new genus has not already been used for some other animal (it is like trying to prove that all cows are not green), but many good museum libraries contain reference books and nomenclators in which to look up your name to see whether it has already been used. The more references you don't find it in, the better are the chances that your name is not preoccupied. In particular, you may wish to concoct a pronounceable nonsense combination of letters, such as *Bliwox* or *Oxofob*, for your new dinosaur. Since there is an excellent chance that nobody else has already used this combination of letters, your name will be safe from preoccupation. Indeed, there is a dinosaur with just such a species epithet, an armored dinosaur called *Tianchisaurus nedegoapeferima*. The species epithet is a nonsense word constructed from the first one, two, or three letters of the last names of the principal actors in the motion picture *Jurassic Park*. For this honor, the producers of the movie contributed a sum of money to the namer's organization to further paleontological research in China.

But most scientists stick with tradition and try to create Latinized names that are euphonious and mnemonic, names that will evoke an image in the mind of the reader of the creature they are naming. As even children's dinosaur books make abundantly clear, every dinosaur name has a meaning. Sometimes this meaning is obvious: The Greco-Latin name of the celebrated predator *Tyrannosaurus rex* can readily be translated as "tyrant-lizard king," a fitting appellation for an animal as tall as a house that could devour a carcass in quarter-ton chunks. In other instances, the meaning is more obscure. For example, the name *Coelophysis bauri*, for a medium-sized, lightly built predator unearthed in New Mexico, roughly translates as "Baur's hollow form." This peculiar phrase refers to narrow tunnels inside the dinosaur's neck vertebrae. "Baur," however, is Georg Hermann Carl Ludwig Baur, a German anatomist who worked at the Yale Peabody Museum during the later years of the nineteenth century. He aided Edward Drinker Cope, the dinosaur's describer and one of America's most famous classical paleontologists, in an astringent campaign to discredit Yale's director, Othniel Charles Marsh, Cope's bitter rival. Cope rewarded him by putting his name in the species epithet. Many paleontologists consider Cope's names among the best-formed and most pleasing Latin names ever created in zoology.

Sometimes, scientists follow tradition all too well. The first scientifically named dinosaur genus was *Megalosaurus*, published by the Reverend William Buckland in 1824. The

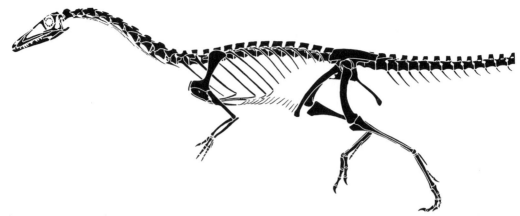

Skeletal restoration of *Coelophysis* by Gregory S. Paul. The name *Coelophysis* was first applied to some bones that are too fragmentary to identify other than belonging to an early theropod. When a quarry full of complete skeletons was discovered nearby at Ghost Ranch the old name was applied to them as well. Recently some researchers argued that because *Coelophysis* was nondefinitive, another name should be used for the Ghost Ranch skeletons. But the International Commission of Zoological Nomenclature ruled that *Coelophysis* should remain the name for the Ghost Ranch materials, a decision that has not made all happy.

name had actually been invented a few years earlier by another reverend, William D. Conybeare, but Buckland published it with a formal description, so the genus is generally credited to him. Using the Greco-Latin root *saurus* (it is written *sauros* in Anglicized Greek, and it means "lizard'") for extinct-reptile names was already something of a tradition in early nineteenth-century Britain, and *Megalosaurus* (meaning "huge lizard'") fit right in with such other names as *Ichthyosaurus* and *Plesiosaurus* (both genera of extinct nondinosaurian aquatic reptiles). Hundreds of dinosaur generic names published after 1824 also use the root *saurus*, all part of the dinosaur-naming tradition that started with *Megalosaurus*.

The second dinosaur to be scientifically named was originally slated to be named *Iguanosaurus* by its discoverer, Gideon Algernon Mantell, until the Reverend Conybeare gently reminded him that an iguana is already a *saurus*, or lizard, and suggested that a better name would be *Iguanodon* ("iguana tooth'"), because the specimens that Mantell had were mainly teeth resembling those of a hypothetical gigantic iguana. The ending *odon* is Latinized Greek for "tooth." Mantell published the name *Iguanodon* in 1825, and the name has generated its own retinue of scores of copycat *odon* dinosaur-genera names, such as *Deinodon*, *Troodon*, *Trachodon*, and so forth. After *saurus*, it is the second most popular ending for a dinosaur name.

In creating names for dinosaur genera, you will be torn between being thoroughly original and following any tradition that has been established for the dinosaurs of the group to which your genus belongs. If your dinosaur is a ceratopian, or horned dinosaur, you may want to use the popular ending *ceratops*, meaning "horned face." If "it" is a predator, you may want to use *raptor*, meaning "thief" or "robber," which has become popular since the film *Jurassic Park*; *venator*, meaning "hunter"; or *nychus*, meaning "claw." If it's an ostrich dinosaur, the customary ending is *mimus*, meaning "mimic," together with a prefix derived from some kind of bird, as in *Gallimimus*, or "chicken mimic." And if it's an armored dinosaur, a common ending is *pelta*, meaning "little shield," a pretty good description of the bony plates imbedded in the skin of ankylosaurs. Roland W. Brown's *Composition of Scientific Words* contains thousands and thousands of Greco-Latin roots and their meanings that can be used for naming dinosaurs.

Many dinosaur names honor the paleontologists who worked with their specimens. Sometimes the name is incorporated into the generic name, along with another root word, as in *Lambeosaurus*, or "Lambe's lizard," honoring Canadian paleontologist Lawrence Morris Lambe for his work on duck-billed dinosaurs. And sometimes the person's name becomes practically the entire name, as in *Janenschia*, an African dinosaur named for German paleontologist Werner Janensch, whose lifelong work was on the dinosaurs and other fossil vertebrates of Africa. Nor do dinosaur names need to be austere and serious; the fish-eater *Irritator* acquired its name because the describers were irritated to discover that the specimen they were working with had been faked up with automobile body filler to make it more attractive for sale to fossil collectors. *Elosaurus*, meaning "marsh lizard," was created as a pun on the surname of Marsh, who had died a few years before.

A generic name is a Latin noun, and the species epithet modifies this noun, so that a species binomen is technically a noun phrase. As such, it is required by the ICZN to be grammatically correct (as far as possible). There are four kinds of species epithets commonly used within species binomens: adjectives, nouns in apposition, genitive nouns, and locative nouns or place names. If your species epithet is an adjective, then it must agree in gender with your noun. Masculine genera require adjectives with a masculine declension, feminine genera require a feminine declension, and neuter genera require a neuter declension. If a species moves from a genus with one gender to a genus with a different gender, it may be necessary to respell the species epithet to agree with its new genus. The masculine dinosaur species *Thecodontosaurus diagnosticus* became *Efraasia diagnostica* when it was moved to the feminine genus *Efraasia*.

A noun in apposition is simply another noun attached to the generic name. In English, for example, "barber" is a noun in apposition to "Floyd" in the phrase "Floyd the barber." The best-known dinosaur species epithet of this kind is *rex*, which means "king," in the names *Tyrannosaurus rex* and *Othnielia rex*. Nouns in apposition have their own genders and are not respelled when a species moves to a new genus. The nonsense epithet *nedegoapeferima* noted earlier is a noun in apposition to the generic name *Tianchisaurus*.

Frequently, a scientist may choose to honor his or her coworkers, field personnel, finders of specimens, or other people who have assisted or supported his or her research by using their names in species epithets. The customary way to do this is to Latinize the honoree's name and make a genitive noun out of it by giving it the appropriate ending. If the honoree is only one male, the genitive ending is the masculine singular *-i*; if only one female, the ending is the feminine singular *-ae*; if more than one person, among whom at least one is male, the ending is the masculine plural *-orum*; and if more than one person, all of whom are female, the ending is the feminine plural *-arum*. Recall the aforementioned *Coelophysis bauri*, with which Cope honored his assistant Baur. These endings apply to any genitive noun in a species epithet, not just a person's name. *Iguanodon exogyrarum* was named after the numerous shells of the genus *Exogyra* found in the rocks along with the specimen.

Place names, usually the localities or provinces where the dinosaur specimen was unearthed, often wind up in species epithets, and sometimes in generic names as well. In a species epithet, a place name is Latinized if it was once part of the Roman Empire or is derived from such a place name; otherwise, it is just used bodily with a suffix such as *-ensis*, *-ianus*, or *-icus*. These convert the place name into an adjective, so the rules about gender agreement with the generic name must be followed. Chinese dinosaur paleontologists are particularly fond of using localities in their dinosaur generic and species names, as in *Mamenchisaurus hochuanensis*, named after the Mamenchi Ferry station on the Yangtze River, near where the first *Mamenchisaurus* skeleton was unearthed, and Hochuan County in China, where the second skeleton—the one representing the species *Mamenchisaurus hochuanensis*—was found.

All in all, dinosaurs have been named for the shapes of their skulls, their bodies, their teeth and claws, their vertebrae, their ribs and other bones, their ankles and feet, their armor plates, and their horns, frills, and other protuberances. They have been named for their supposed habits and lifestyles, the kinds of food they may have consumed, and the sorts of habitats they may have occupied. They have been named after gods, devils, mythological monsters, astronomical constellations, industrial corporations, weapons, birds and other animals, famous paleontologists, wives of paleontologists, children of paleontologists, obscure museum directors, goat herders, scientific foundations, families of ranchers, and extinct European tribes. They have been named after continents, rivers, mountains, prominent landmarks, geological formations, states, large cities, small villages, and long-lost countries. They have even been named after one another. To study dinosaur names is truly to explore the range of human experience.

Publishing the Name

Now that you have your specimen or specimens and have chosen a name for your new dinosaur, you must publish the name in a place where other workers will read about it and, you hope, begin to use it. To publish your dinosaur name, you might submit your paper to a paleontological or zoological journal, but this is not, and never has been, necessary. All you need to do to publish your name is produce your own journal or newsletter with your paper in it, in an edition available for sale or for free to the public in a suitable quantity (a good round number is 100) of identical copies. Dinosaurs are occasionally named or renamed in dissertations by students as part of earning their academic degrees, but the ICZN considers dissertations to be too limited in distribution to qualify as nomenclatural publications. Likewise, abstracts of talks distributed at conferences, handwritten notes, computer printouts, society newsletters, and Internet Web sites and other electronic media are not proper publications in zoological nomenclature. If your paper is worded so as to disclaim being a nomenclatural act, which you might want to do in a preliminary report on a new species before actually describing it, then your paper disqualifies itself from being the formal publication of the new name.

Of course, if you want the scientific community to accept and use your new dinosaur name right away, publication in a proper journal, where your paper will be peer-reviewed, is best. Scientists believe that peer review establishes that you are not some kind of crank, that your reasons for creating a new name are valid, and, generally, that your paper is worth spending precious time reading. Peer-reviewed journals, however, are a feature of contemporary science. During the nineteenth century, when some of our best-known dinosaurs were described and named, such journals for all practical purposes did not exist, and the validity of a name was certified by the reputation and prestige of the namer. Cope and Marsh, the premier American dinosaurologists of the nineteenth century, each had what was virtually his own personal journal in which to publish: Cope's was *The American Naturalist*, and Marsh's was *The American Journal of Science*. These journals had been set up to get Cope's and Marsh's letters and papers describing new fossils swiftly into print, to establish priority for their discoveries.

When your paper appears in print, your dinosaur specimen becomes the *type specimen* of your new species, if you are describing a new species. If you are simply transferring one species into another genus, the type specimen of the species goes along with it into the new genus. If you are naming a new species, the ICZN mandates that you must explicitly designate a type specimen for your species to be valid. Without a type specimen, a species name is a *nomen nudum*, or "naked name," lacking scientific standing— just another vernacular name. This is the usual designation for dinosaur names announced in newspaper articles or magazines before their formal descriptions are published. It is not good practice to telegraph your new dinosaur name like this; ideally, its first published appearance should also be its formal description. But there is no real harm done to your dinosaur name if you do. Most of the millions of named species of organisms—beetles and butterflies, shells and crustaceans, trilobites and ammonites— have a type specimen of some kind residing in an institution of some kind, where it may be examined and compared at any time with other specimens that are being classified and named. This is one of the grand functions of university and museum collections.

Just as a species is anchored to a type specimen, so must every new genus have an explicitly designated type species, and every new family have a type genus. That's as far as the ICZN goes. If you are creating a new genus as well as a new species, then you must explicitly designate your species (or some other species) as the type species of the genus. If you are creating a new genus for an already-established species, you must designate that species as the type species of your new genus. Without a type species, a genus is a *nomen nudum* with no scientific standing (unless it was created before the rules about types went into effect; this is how *Macrodontophion* has managed to survive without a named type species). Numerous complications can arise with respect to the various kinds of type specimens and their respective species, genera, and families, and a good deal of the ICZN is devoted to rules, conventions, and recommendations about what to do under such special circumstances. It can take years to become familiar with the ICZN and how to apply it correctly. Fortunately, as noted at the beginning of this article, most instances of name creation are simple and straightfoward: find a specimen, create the name, publish the name. You don't have to become an ICZN expert to do this.

Naming a dinosaur, or any organism, confers a kind of immortality on the namer. The name—if it survives the tests of validity, if you have done your homework and followed the rules of zoological nomenclature, if your description is well written and well illustrated—will be used for as long as zoology persists as a science; and when your name is referred to in scientific papers, your paper will usually be cited as well. This is one of the reasons that scientists apply themselves to the naming process so diligently and assiduously.

Classification and Evolution of the Dinosaur Groups

by Thomas R. Holtz, Jr.

Recovering the Family Tree of Dinosaurs

In 1842, Sir Richard Owen coined the term Dinosauria, the "fearfully great lizards," and began the study of the classification of the dinosaurs. Owen recognized that among all of the extinct Mesozoic era reptiles found in Great Britain, three forms, carnivorous *Megalosaurus*, plant-eating *Iguanodon*, and armored *Hylaeosaurus*, differed from all the others and from all modern reptiles in a number of skeletal features. These features included extra vertebrae in the hips, specialized double-headed ribs, and limb bones constructed so that the legs were held directly underneath the body.

Owen was not an advocate of evolutionary theory or "transmutationism," as it was then called. However, as with many natural historians of his era, he constructed systems of classification based on the presence of one or more specialized feature found among the creatures he was examining. Owen's colleague Charles Darwin discovered the underlying reason for the shared presence of specializations in various groups of organisms. This reason was the process of evolution, or "descent with modification," to use the phrase Darwin preferred. Within an ancestral population, a trait or suite of traits gives the individuals with this variation an advantage in response to some environmental factor. For example, hindlimbs that were more upright, and fusion of back or tail vertebrae with those of the hips, would allow the animal to run faster and more continuously than its sprawling-legged, smaller-hipped kin. If those creatures with the new trait survive, they will pass this characteristic on to their descendants. The descendant groups may diverge into different modes of life (some as predators, for example, and others as herbivores), but all will pass on these traits unless they are further modified by evolution. Thus, ancestral stems of the "tree of life" diverged into

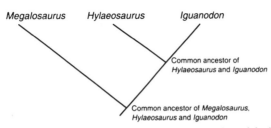

Cladogram showing the relationship between the original three members of Sir Richard Owen's Dinosauria. The cladogram shows that the lineage leading to *Megalosaurus* diverged prior to the split between the lines leading to *Hylaesaurus* and to *Iguanodon*. In other words, *Iguanodon* and *Hylaeosaurus* share a common ancestor that was not shared with *Megalosaurus*.

various different branches that acquired their own new features but retained some of the traits of their ancestors. Darwin suggested that descent from common ancestors should be the criterion on which classification is based; in other words, the goal of classification would be the description of the branching pattern of this "tree of life."

Darwin recognized that a single feature could develop independently in different lineages of organisms, a condition called convergence. Thus, he reasoned, an approach to forming a classification based on the pattern of descent with modification should take into account the aggregate of many different characters, to find those that show concordance in their distribution among the organisms in question.

Although Darwin suggested making descent from common ancestors the criterion of biological classification in his classic 1859 work *On the Origin of Species*, it would be nearly a century before a methodology was developed to accommodate this model. Willi Hennig, an East German entomologist, developed an analytical technique called "phylogenetic systematics" (or "cladistics") by which different potential patterns of descent with modification could be evaluated. Taxonomists between Darwin and Hennig had constructed various forms of "family trees," linking earlier fossils with later forms on the assumption that these particular early specimens were part of the population directly ancestral to the more recent forms. In contrast, Hennig realized that, given the chancy nature of preservation of organisms as fossils, the probability of the populations directly ancestral to later forms would be fossilized, preserved, and uncovered by paleontologists would be extremely small. Instead of searching for the direct lines of ancestor and descendant, Hennig developed an approach that searched for patterns of divergence through time.

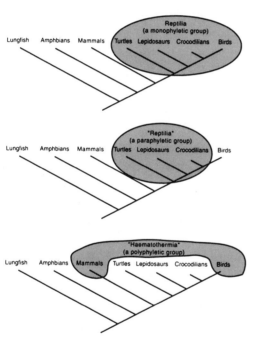

Three different types of groups have been recognized by taxonomists through the years. Monophyletic groups, or clades, represent whole branches of the tree of life; an ancestor and all of its descendants. Paraphyletic groups represent the descendants of an immediate common ancestor, but exclude one or more descendant of that ancestor (perhaps because it is more greatly transformed). Polyphyletic groups are unnatural assemblages of forms with no immediate common ancestor. Modern systematists use only monophyletic groups in their classifications.

Hennig's system portrays the branching pattern of recency of common ancestry on a figure called a cladogram. In these figures, those organisms on branches connected to a common node are considered to share a more recent common ancestor with organisms whose branches do not connect at that node. For example, *Hylaeosaurus* and *Iguanodon* share a more recent common ancestor with each other than either does with *Megalosaurus*. (To put it another way, the lineages leading to *Hylaeosaurus* and *Iguanodon* branched from each other more recently than the lineage containing both these forms diverged from *Megalosaurus*).

Hennig (as had others before him) recognized that there were three potential sorts of similarities between creatures: shared derived characters, which represent evolutionary novelties developed in the common ancestor of the organisms being studied; shared primitive characters, which represent characters found in that common ancestor but also found more widely distributed among the relatives of that ancestor; and convergences.

Biologists had long recognized that convergences were not useful in recovering the pattern of ancestry and descent because these represent independent evolutionary events. For example, modern turtles and modern birds both have toothless beaks, an unusual feature among vertebrates. However, this shared feature does not indicate that turtles and birds both come from a common ancestor that was also similarly toothless; in fact, primitive extinct turtles and primitive extinct birds have teeth. The toothless beaks of turtles and birds thus represent an evolutionary convergence.

Shared primitive characters were similarly not useful in reconstructing the pattern of descent. For example, humans and opossums both have five fingers on their hands, while horses have only one. Clearly human and opossum hands are more similar to each other than human and horse hands (or opossum and horse hands). However, this similarity does not represent some evolutionary novelty acquired by an ancestor of humans and opossums that is not shared by horses. Instead, having a five-fingered hand is a primitive feature found in the common ancestor of humans, opossums, and horses—and duck-billed platypi and crocodiles and dinosaurs and turtles and frogs and salamanders—indeed, of all the modern groups of land-dwelling vertebrates! The shared presence of five fingers is simply the retention of a primitive feature; and, instead, humans and horses are more closely related to each other than either is to opossums (both have placentas, for example). The single finger of the horse hand is the result of an evolutionary novelty in the history of the horse lineage long after it diverged from a common ancestor with primates.

Hennig observed that only shared derived characters (which he termed synapomorphies), which represent new specializations among a subset of the organisms studied, show the branching pattern of evolution. For example, compared to other groups of extinct reptiles, the presence of extra hip vertebrae and upright hindlimbs in dinosaurs represent shared derived characters in the common ancestor of Dinosauria.

If scientists had a total knowledge of the anatomy of all the species of dinosaurs that ever lived, reconstructing the phylogeny (family tree) of Dinosauria would be trivially easy, as we could clearly see which similarities between some groups are convergences and which are shared derived features. However, as is common in the natural sciences, we have only a small fraction of the total pattern. Many dinosaur species are known only from single fragmentary fossils, and there are whole sections of the history of dinosaurs (for example, Middle Jurassic North American and Late Jurassic South American dinosaurs) that are totally unknown because we do not have fossils of land animals of those ages from these regions. Given that our knowledge represents only a partial sample of the whole, how can we reconstruct the family tree?

Hennig's system of cladistics offers a way of choosing between the different possible branching patterns of the family tree. Since we cannot definitely tell a convergence from an inherited shared derived character without knowing the family tree (which is itself that which we are trying to reconstruct), we consider all nonprimitive similarities in the groups we are looking at to be potential shared derived characters. We then compare all the different possible configurations of the branching patterns of common ancestry (that is, all the different cladograms possible for the organisms in question) by counting up the number of evolutionary changes from the primitive to the derived condition necessary to describe that tree. The tree (or trees) with the fewest number of evolutionary changes required is preferred. Because the number of possible cladograms can be quite

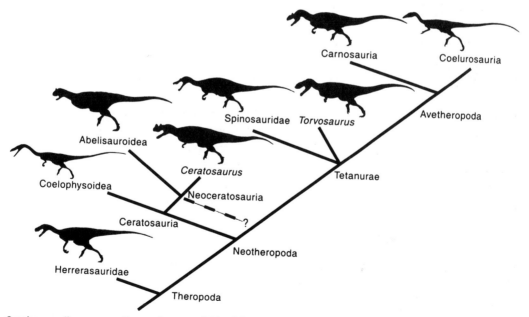

Carnivorous dinosaurs, or theropods, were all bipedal running forms. It is uncertain at present whether the neoceratosaurs shared a more recent common ancestor with the gracile coelophysoids or the more advanced tetanurines.

143

high (for example, 16 different possible trees for four groups, 105 for five, and more than 34 million for 10 groups), computer analyses are required to evaluate these clado-grams.

The rise of cladistic methodology and of computers and software capable of evaluating the different possible trees has moved reconstructing the phylogeny of dinosaurs from an art to a science. However, much work remains to be done. Because different researchers may evaluate the anatomical features in different ways, the cladograms that their studies produce can be different. Furthermore, because many species are known only from partial fossils, the missing data may result in numerous different possible cladograms that explain the observed data equally well; potential future discoveries of more complete fossils will help in resolving these uncertainties. Finally, some groups of dinosaurs are known at present only from highly specialized descendants. Because their more primitive ancestors are not yet known, it is uncertain whether some of the features that are similar between them and other specialized dinosaurs are convergences or these features are instead inherited from a common ancestor.

Because of these uncertainties, there are differences between the cladograms produced by different paleontologists, but there are also many broad agreements. In most of the rest of this chapter, we will examine both the consensus and some of the controversies surrounding the phylogeny of Dinosauria. First, however, we will briefly examine the place of dinosaurs among the reptiles.

Dinosaurs in the Family Tree of Reptiles

In the 1970s through the mid-1980s, there was some debate among paleontologists over whether dinosaurs should be considered reptiles. That debate did not concern a differ-ence of opinion as to the position of dinosaurs in the family tree of vertebrates. It instead centered on the debate over dinosaurs' physiology: Were dinosaurs cold-blooded, like "reptiles" (as the term was used then) or were they warm-blooded, like their descen-dants the birds? To most paleontologists today, dinosaurs are considered a type of rep-tile, and birds are considered a type of reptile. This shift has occurred because of the way biologists use their formal taxonomic names.

Scientists have, over the decades, proposed many different groupings of organisms over geologic time. These groups can be described as being one of three different types:

- Monophyletic groups, also called clades (branches), are groups composed of a direct common ancestor and all its descendants. "Reptilia" in the modern usage of the term (that is, turtles, lepidosaurs, crocodilians, and birds) is a monophyletic group. Most scientists today use only monophyletic groups in their classifications.

- Paraphyletic groups are those with a direct common ancestor that exclude one or more of their descendants. Reptilia in the traditional usage of the term (that is, turtles, lepidosaurs, and crocodilians) is paraphyletic, as it excludes birds (which are more closely related to crocodilians than crocodilians are to other reptiles).

- Polyphyletic groups are those with no direct common ancestor that is also part of that group. Polyphyletic groups often turned out to be based on convergences. For example, a group Haematothermia, or "warmbloods," composed of birds and mammals but not turtles, lepidosaurs, and crocodilians, represents a polyphyletic grouping.

Once scientists accepted that monophyletic groups would be the only type used in taxonomy, the debate whether dinosaurs were reptiles was over. The name Reptilia now applies to a particular branch of the family tree of the vertebrates, not to some general "grade" of development (that is, cold-blooded terrestrial vertebrates with a shelled egg). Since dinosaurs are part of that branch, whether they were cold-blooded or warm-blooded is not a consideration in their classification: they are reptiles.

Paleontologists have reached a general consensus on the basic structure of the family tree of terrestrial vertebrates. Among living vertebrates, mammals, turtles, squamates (snakes and other lizards) and the New Zealand tuatara, crocodilians, and birds together form a clade with the shared derived feature of an amniotic egg. (Only the echidna and platypus among living mammals still lay such an egg; in marsupial and placental mammals, the tissues associated with the egg are internalized). All these sorts of terrestrial vertebrates are called collectively amniotes.

Amniota is divided into two branches. Synapsids (mammals and extinct protomammals such as sail-backed *Dimetrodon*) are one lineage; the sister group (closest relative) to the synapsids is Reptilia. Synapsids show an enlarged skull opening for an increased volume of their jaw muscles; later synapsids show increasingly specialized feeding structures, such as teeth differentiated into incisors, canines, premolars, and molars. Synapsids were the dominant group of vertebrate on land during the Pennsylvanian (323–290 million years ago) and Permian (290–251 million years ago) periods, but they began to decline in importance during the Triassic (251–200 million years ago). After the great extinction at the end of the Cretaceous, the mammals (derived synapsids) once again rose to dominate.

Reptilia, in monophyletic terms, are all descendants of the most recent common ancestors of turtles, lepidosaurs (tuataras and lizards, including snakes), and archosaurs. This differs from the classical paraphyletic use of the term for any amniote that was unlucky enough not to have evolved into a mammal or a bird. Thus, in traditional taxonomy, a primitive synapsid was considered a reptile, but today it is not regarded as a reptile at all: synapsids split off from the lineage leading to reptiles prior to the origin of Reptilia. Additionally, although birds were not considered reptiles in the traditional sense, we now recognize that they are simply one (extremely diverse and highly specialized) branch of the reptile radiation.

Although there is some debate over the structure of the reptile family tree (primarily over the placement of turtles), many researchers agree that there are two main branches in Reptilia: Anapsida and Diapsida. Anapsida, a group containing various Permian and Triassic period reptiles and (most likely) modern and extinct turtles, diverged from the other reptiles not long after the synapsid-reptile split. Anapsids have various specializations of the skull and limbs. They have never been a particularly dominant group of animals in ecological terms.

Diapsids have been (and remain) far more diverse than anapsids. Diapsids are characterized by several important evolutionary novelties, among which are a pair of specialized openings behind the eye-socket for attachment of enlarged jaw muscles, and hindlimbs that are typically longer than the forelimbs. Many living quadrupedal diapsids (lizards and crocodilians) will actually run primarily on their long hindlegs when scared or threatened.

The diapsid radiation produced many diverse groups. The lepidosaurs are a group containing the tuatara of New Zealand and squamates, the lizards (a clade that also contains the extinct marine mosasaurs and the snakes). Another group of diapsids were the extinct euryapsids, a clade of aquatic and fully marine reptiles of the Mesozoic. While many lineages of euryapsids retained legs with individual fingers and toes, two clades (the superficially dolphinlike ichthyosaurs and the plesiosaurs) independently evolved paddle-shaped limbs.

Yet another group of diapsids were the archosaurs. This clade was characterized by a pair of extra openings in the skull, the antorbital fenestra between the eye socket and the nostril and the mandibular fenestra in the back of the lower jaw. The two main branches of the archosaurs are the Pseudosuchia and the Ornithodira. Pseudosuchians include the modern crocodilians and their extinct ancestors. During the Triassic period, pseudosuchians were quite diverse: armored herbivores, huge carnivores, superficially crocodile-like aquatic predators, and other forms dominated the landscape in which the first dinosaurs appeared.

Ornithodirans share the common evolutionary novelties of a narrow compressed foot and a neck capable of folding into an S-shape. The pterosaurs, or flying reptiles of the Mesozoic, are one major branch of the ornithodirans. Pterosaurs had extremely long fourth (ring) fingers on their hands, which supported a wing membrane.

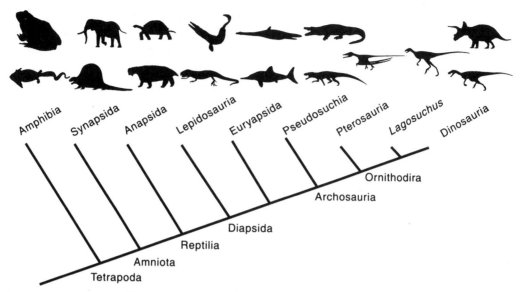

Phylogenetic relationships of the terrestrial vertebrates, living and extinct. Amphibians must return to the water in order to lay their eggs, while amniotes have shelled (or internalized) eggs that they lay on the land. Synapsids include mammals and their extinct relatives, such as fin-backed *Dimetroden*. Anapsides include various extinct forms of the late Paleozoic and early Mesozoic Eras; turtles are probably the only living group of anapsids. Lepidosaurs include tuataras and aquamates, lizards, snakes and the marine mosasaurs and various other extinct groups. Euryasids are all extinct: the clade of aquatic reptiles included various groups that lived near shore, as well as the fully marine plesiosaurs and icthyosaurs. Pseudosuchians were once highly diverse; crocodilians are the only surviving clade of pseudosuchian. Pterosaurs, the flying reptiles, include the short-tailed pterodactyloids and various primitive long-tailed lineages. *Lagosuchus* is a primitive ornithodiran close to the ancestry of the dinosaurs. Dinosauria itself represents one branch of ornithodiran archosaurian diapsid reptiles.

The other major branch of ornithodirans was characterized by very long lower portions of the leg (tibiae and metatarsi). The femur, or thighbones, of these forms was bent so that the head faced inward: this allowed these reptiles to hold their legs directly underneath the body. Early members of this branch, such as *Lagosuchus*, were typically quite small (10 inches or so long). Very likely, they were excellent runners, a useful trait in a world populated by much larger predatory pseudosuchians. It is from within this group that dinosaurs evolved.

The Root of the Dinosaur Family Tree

Dinosaurs, therefore, were ornithodiran archosaurian diapsid reptilian amniotes, in terms of their position on the family tree of the vertebrates. The ancestral dinosaurs retained the various specializations that had been acquired during the history of the terrestrial vertebrates (such as shelled eggs, antorbital fenestrae, and so forth) as well as evolving their own unique features.

The dinosaur hand was quite specialized compared to that of other reptiles. The joint between the thumb and the first metacarpal (palm bone) that supports it was offset, so that when a dinosaur would close its hand, the thumb folded toward the palm. This would allow for a better grip. Also, the fourth (ring) and fifth (pinky) fingers of the hands of primitive dinosaurs were reduced in size compared to those of other reptiles. These features indicate that the earliest dinosaurs did not use their hands in locomotion most of the time, and the skeletons of primitive dinosaurs confirm this. The first dinosaurs were all bipedal. Their hindlimbs, like those of their closest relatives, were held directly underneath the body. The inside of the hip socket, which is a shelf of bone in typical amniotes, was just a cartilaginous sheet in dinosaurs, perhaps related to an increased range or rate of fore-and-aft motion of the leg. Related to the change in the hip bones, one or more vertebrae are added to the hip region of dinosaurs, probably to strengthen this region for more rigorous walking and running than in other reptiles.

The most primitive members of each of the dinosaur groups were around 3–5 feet long, suggesting that the common ancestor of all dinosaurs was much smaller than many of its descendants. Although this common ancestor to all dinosaurs has not yet been recovered, a genus that probably closely resembles it is known. This is *Eoraptor*, a 3–foot long primitive dinosaur from the Late Triassic Ischigualasto Formation of Argentina. *Eoraptor* itself cannot be the actual ancestor of all dinosaurs because it occurs at the same time as primitive representatives of all three major branches of the dinosaur radiation (theropods, sauropodomorphs, and ornithischians). However, anatomically, *Eoraptor* seems to have more closely approximated the generalized common ancestor of all dinosaurs than any other known genus. Its discoverers and describers (Paul Sereno, Fernando Novas, and colleagues from Argentina and the United States) consider it to be a basal member of the theropod lineage; other paleontologists remain skeptical and suggest it might have branched off prior to the divergence of the sauropodomorph and theropod lineages. *Eoraptor* has teeth that are bladelike, with fine serrations running up and down the front and back sides. Such teeth were widely distributed among archosauriforms, and suggest a primarily carnivorous diet for this primitive form.

Most paleontologists agree that there are three primary clades within Dinosauria. These are Theropoda (bipedal, primarily carnivorous forms), Sauropodomorpha (long-necked herbivores), and Ornithischia (a diverse clade of beaked herbivores). The consensus of most studies of dinosaur relationships agree that the basalmost split within Dinosauria is between Ornithischia and a clade called Saurischia, composed of Theropoda and Sauropodomorpha. Derived features uniting the saurischians included an elongated neck and long neck ribs, an enlarged thumb claw, and a hand in which the second (index) finger is the longest. However, a few authors have advocated an alternative (not-as-yet supported by an explicit numerical cladistic analysis) in which sauropodomorphs and ornithischians form a clade (called "Phytodinosauria") to the exclusion of theropods. Sauropodomorphs and ornithischians did share a leaf-shaped tooth form with large denticles rather than fine serrations; such teeth, also exhibited in some modern herbivorous lizards and some advanced theropods, is most likely associated with a diet of plants.

Whether it is the sister group to theropods and sauropods or just to the Sauropodomorpha, the monophyly of Ornithischia has not been seriously questioned. All ornithischians share a unique bone, the predentary, which caps the front end of the lower jaw. The predentary supported a horny beak and may have allowed for slight rotation outward of the sides of the lower jaw to facilitate chopping up plant matter. The lower jaws of ornithischians also have enlarged coronoid processes to serve as the anchor for enlarged jaw muscles.

The most primitive ornithischian known, *Pisanosaurus*, was a contemporary of *Eoraptor*. Although the only known skeleton is incomplete, the fragmentary pelvis suggests that this 3–foot long dinosaur lacked the backward-pointing pubis found in all later ornithischians. This repositioning of the pubis expands the volume of the abdominal cavity, allowing for a greater length of intestines for the digestion of plant material. The Early Jurassic southern African form, *Lesothosaurus*, appears to be the most primitive ornithischian with a backward-pointing pubis currently known. Most workers agree that it is outside a clade composed of all remaining ornithischians, although some consider it to be a basal member of the ornithopod lineage.

The remaining ornithischians fall into three different clades: the armored Thyreophora, the ridge-headed Marginocephalia, and the primarily bipedal Ornithopoda. All these dinosaurs had a tooth row in the upper jaw that was inset from the side, suggesting the presence of a muscular cheek (*Lesothosaurus* did not have this inset tooth row).

The oldest well-dated dinosaur fossils come from formations in Argentina and Brazil of the Late Triassic epoch, although sauropodomorph remains are known from a Madagascar unit that is tentatively assigned a Middle Triassic date. During the early part of the Late Triassic, dinosaurs remained an important but subdominant component in a terrestrial ecosystem shared with diverse pseudosuchian archosaurs and advanced synapsids. Toward the end of the Late Triassic, the first large herbivorous dinosaurs (prosauropods) emerged, but the smaller carnivorous and smaller herbivorous ecological guilds were shared with nondinosaurian amniotes. Only after the great extinctions during the end of the Late Triassic did dinosaurs become the major group of large terrestrial vertebrates, dominating the land until the Cretaceous-Tertiary extinction event.

The Theropods: Dinosaurs Red in Tooth and Claw

The Theropoda ("beast-footed ones") in some ways retained more of the primitive features of Dinosauria than did their relatives. For example, most Mesozoic theropod genera retained bladelike serrated teeth, indicating a carnivorous diet; all known theropods walked bipedally; many theropods retained grasping hands with trenchant claws; and so forth. However, some particular theropod groups were among the most highly specialized of all dinosaurs.

The most primitive theropods known are the herrerasaurids of the Late Triassic. Best known for *Herrerasaurus* of Argentina, this clade is also known from Brazil and North America, and quite likely extended over most of the unified Triassic supercontinent of Pangaea. Herrerasaurids shared with later theropods a specialized hinge in the lower jaw; this intramandibular joint, between the tooth-bearing dentary and the various postdentary bones, probably served to absorb stresses associated with feeding on large living prey. Although herrerasaurids were overshadowed by larger predatory archosaurs in their environment, they represented the first model in a basic design of carnivore that dominated the terrestrial environment for 165 million years: fully bipedal, striding carnivores with powerful grasping hands.

There is some debate over the precise phylogenetic position of the herrerasaurids. Most recent analyses (such as those of Paul Sereno of the University of Chicago) support its position as a primitive member of Theropoda. However, herrerasaurids lack some features found in sauropodomorphs and definite theropods, suggesting that these are early primitive forms that diverged from other saurischians prior to the split between Sauropodomorpha and Theropoda.

Definite theropods all shared numerous important specializations. While the ancestral dinosaurs and herrerasaurids walked with all four toes on the hindfeet touching the ground, advanced carnivorous dinosaurs (the Neotheropoda) walked on only the middle three toes (digits 2–4). The inside toe, digit 1, is greatly reduced. Neotheropods also had extra openings in the snout, associated with a soft-tissue air sac system whose function is not fully understood, but that may have served in part as a heat exchange mechanism. New information suggests that all neotheropods may have had their clavicles (collarbones) fused into furculae, or wishbones; previously, this structure was known only for tetanurines. In modern birds, wishbones serve as an elastic energy storage mechanism during flight; in extinct neotheropods, this structure may have served as a brace for the chest against the forces generated by prey grasped in the forelimbs.

Most analyses divide Neotheropoda into two clades, Ceratosauria and Tetanurae. Ceratosaurs are characterized by numerous specializations of the vertebrae and by fusion of various hip and hindlimb bones in the adults. Among the ceratosaurs are two clades. Coelophysoids are best known from fossils from the Late Triassic and Early Jurassic, and were generally gracile in build. Coelophysoids shared a specialized kink along the tooth row of the upper jaw: similar wave-shaped tooth margins in other reptiles (such as some modern crocodilians) are believed to be an adaptation to assist in handling struggling small prey. Some coelophysoids, such as *Segisaurus* and *Procompsognathus* were only 3 feet long; others, such as *Dilophosaurus* and *Gojirasaurus* were 20 feet or more and massed 800–900 pounds: these latter were among the first large dinosaurian carnivores. The mid-sized (10–14 feet) coelophysoids *Coelophysis* and *Syntarsus* are both known from sites in which many dozens or hundreds of skeletons were buried simultaneously, suggesting that at least some members of this clade were gregarious in some circumstances. The light build of coelophysoids suggests that they were relatively fast.

The remaining ceratosaurs comprise the clade Neoceratosauria. Most neoceratosaurs were robustly built, with deep skulls. The Late Jurassic *Elaphrosaurus*, which may be a primitive neoceratosaur or a late-surviving coelophysoid, had a gracile build, and limb proportions comparable to some of the ornithomimids (the ostrichlike dinosaurs described below). The best-studied Jurassic neoceratosaur is *Ceratosaurus* of the Late Jurassic Morrison Formation of the American West. This genus shared numerous derived features with the Cretaceous abelisauroid ceratosaurs, such as the addition of several extra hip vertebrae. The abelisauroids are best known from the Cretaceous of the southern supercontinent Gondwana (and in particular from South America, India, and Madagascar), although some possible abelisauroid material is known from the Late Cretaceous of Europe. Abelisauroidea contains the small noasaurids, which exhibit some convergences with various coelurosaur groups (in particular a retractable sickle-shaped claw on pedal digit 2) and the larger abelisaurids. The latter were the dominant predatory group during the Late Cretaceous of South America, India, and Madagascar. Abelisaurids converge on tyrannosaurid coelurosaurs in a number of features, in particular in fusion of the nasal bones in some groups and in the reduction of the forelimbs in *Carnotaurus* (although it is more extreme in this ceratosaur than in any known tyrannosaur).

Although most published phylogenetic analyses place coelophysoids and neoceratosaurs as sister groups, an alternative hypothesis is suggested by some preliminary studies. In this alternative, neoceratosaurs (as a clade or as a paraphyletic grade) are more closely related to tetanurines than they are to coelophysoids. This is supported by a number of tetanurine-like features of the skull and hindlimb found in neoceratosaurs but lacking in coelophysoids, and it is more consistent with the general historical distribution of these clades; that is, coelophysoids are most common in the Late Triassic and Early Jurassic, while neoceratosaurs and tetanurines are best known after the Middle Jurassic. (However, recent discovery of a Late Triassic tetanurine in Argentina has closed this stratigraphic "gap"). Additionally, union of neoceratosaurs and tetanurines outside of coelophysoids would require that the numerous vertebral and hindlimb features shared by the traditional "ceratosaurs" either would have evolved convergently or have been ancestral to neotheropods and lost in tetanurines.

Tetanurae ("stiff tails") derives its name from the elongation of the prongs of the tail vertebrae, which result in a more rigid structure in the rear half of the tail than in other theropods. This adaptation appears to be associated with increased agility; such a tail can function better as a dynamic stabilizer, allowing the dinosaur to make tighter turns. Similar adaptations arise in hypsilophodont ornithopods and (outside of the dinosaurs) in some long-tailed pterosaurs and some fast-running lizards. Tetanurines also possessed proportionately larger hands than found in ceratosaurs, and their teeth tended to be concentrated in the front of their snouts (while those of other dinosaurs continued further back in the jaws).

Most tetanurines belong to the specialized clade Avetheropoda ("bird theropods"); however, there are several tetanurines that lie outside this group but whose relationship to each other remains uncertain. These "megalosaurs" may comprise a clade (the Spinosauroidea), as suggested by Sereno; alternatively, they may form a paraphyletic grade with respect to avetheropods. Several of the "megalosaurs" were relatively bulky, and many have had short, powerfully built arms with large thumb claws. Otherwise, these are relatively unspecialized large forms such as *Torvosaurus*, *Piatnitzkysaurus*, and *Megalosaurus* itself.

One group of megalosaurs, the Spinosauridae of the Cretaceous, was a highly specialized group. Spinosaurids had an extremely long snout with large conical teeth (unlike the typical bladelike teeth of most theropods). The teeth at the end of their snouts were larger than the rest, and formed a "rosette." This skull and tooth morphology is similar to that of many fish-eating reptiles of today (such as some crocodilians) and of the past (such as the parasuchians of the Triassic), suggesting that spinosaurids may have had a diet of large fish. Indeed, spinosaurids are known from fossil assemblages that also contain abundant large fish. However, the rest of their skeleton does not reveal any particular aquatic adaptations. One suggestion is that spinosaurids may have been able to feed on fish from lakes or streams (and thus sample food otherwise inaccessible to typical large theropods), but were capable of crossing overland to other lakes and streams fairly easily (and thus have access to greater range than any individual large crocodilian). However, like crocodilians today, spinosaurids would almost certainly have been capable of taking down large terrestrial prey as well. Indeed, the original specimen of the English spinosaurid *Baryonyx* contains both fish scales and bones of a young *Iguanodon* in its gut contents.

Avetheropods shared numerous derived features. Among these was a specialized wristbone, the semilunate carpal block, which allowed a greater range of motion for the hand in the plane of the forearm (perhaps to allow the large hand to be folded closer to the body when not in use). Avetheropods also had an additional opening in the snout

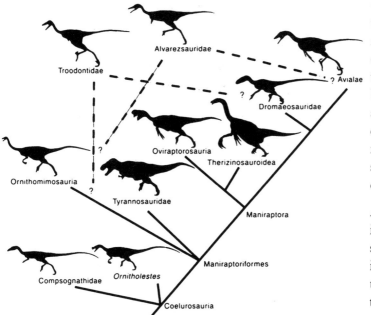

Coelusaur phylogeny has been the subject of a number of studies, but no comprehensive consensus yet exists. This is in part due to the highly specialized skeletons of the maniraptoriform coelusaurs, including the giant short-armed tyrannosauroids and the herbivorous long-necked therizinosauroids. The phylogenetic positions of the bird-like troodontids and alverezsaurids are particularly problematic.

(the maxillary fenestra) that is not found in herrerasaurids, ceratosaurs, or "megalosaur"-grade tetanurines. The chambers in the vertebrae of avetheropods were very complex, suggesting a more complex air sac system than in other dinosaurs.

Avetheropoda is divided into two clades, Carnosauria and Coelurosauria. These names were used for much of the twentieth century as divisions between large and small theropods, respectively. Their modern use does not represent divisions based on size, but instead on phylogeny. Carnosauria comprises the well-known Late Jurassic theropod *Allosaurus* and its allies; Coelurosauria represents a highly diverse clade sharing many derived features.

The carnosaurs were characterized by highly pneumatic snouts as well as specializations in the dorsal vertebrae. Many carnosaurs had crests of some form along the nasal or lacrimal bones: these are especially spectacular in the forward-facing crest of the Early Jurassic *Cryolophosaurus* of Antarctica and the huge hollow central crest of *Monolophosaurus* of the Middle Jurassic of China. Although the more advanced carnosaurs, such as *Allosaurus* and *Acrocanthosaurus*, had only three fingers, primitive forms such as *Sinraptor* retained the fourth metacarpal—this suggests that the loss of the fourth metacarpal occurred independently in advanced carnosaurs and coelurosaurs. At present, the largest known theropods are the carcharodontosaur carnosaurs *Carcharodontosaurus* of the early Late Cretaceous of northern Africa, and the contemporary Argentine form *Giganotosaurus*: the latter exceeds the largest known specimens of *Tyrannosaurus* by about 10–15 percent. Although carnosaurs seem to have been the most common large theropods from the Middle Jurassic through the Early Cretaceous, none is known beyond the first stage of the Late Cretaceous. Instead, the top predators of typical Late Cretaceous dinosaurian faunas of North America and Asia were tyrannosaurid coelurosaurs, while those found in Gondwana and possibly Europe were abelisaurid ceratosaurs.

Coelurosaurs differed from other theropods in a number of features. The brain cavity of coelurosaurs, where known, seems to have been larger than noncoelurosaurian theropods of the same size. Coelurosaur hands tended to be more slender than those of their relatives. The chevrons (bones beneath the vertebrae in the tail) were pointed both forward and backward, forming an even more stiffened tail than the basal tetanurine condition. Additionally, recent discoveries of fossils with preserved skin structures from the Yixian Formation of China indicate that filamentous integumentary elements (almost certainly homologous with the feathers of birds) were present on both primitive (compsognathid) and advanced (oviraptorosaur, therizinosauroid, and dromaeosaurid) coelurosaurs. This suggests that all coelurosaurs may have had a covering of "protofeathers" ancestrally.

A number of Jurassic and Cretaceous coelurosaurs do not show any particular special affinity with the more advanced clades. Many of these primitive forms, such as the compsognathids (such as *Sinosauropteryx* of the Yixian Formation), *Ornitholestes*, and *Coelurus*, were less than 6 feet long; others, however, such as *Deltadromeus* and *Dryptosaurus*, were comparable in size to the largest neoceratosaurs and typical carnosaurs.

The remaining groups of coelurosaurs, collectively, the maniraptoriformes, represent some of the most specialized of the dinosaurs. This group contains the ornithomimosaurs, tyrannosaurids, oviraptorosaurs, therizinosauroids, troodontids, dromaeosaurids, alvarezsaurids, and birds. Although there is agreement that all these groups are more closely related to one another than they are to other clades of theropod, paleontologists dispute the particular relationships among the maniraptoriforme. As a group, this clade is characterized by reduced number of tail vertebrae, by slender hands, by the development of a secondary palate in the snout, and by various modifications of the braincase. Although some maniraptoriforme clearly retained the primitive carnivorous habit (tyrannosaurids and dromaeosaurids, for instance), others show considerable divergence in jaw and tooth form from the ancestral condition, suggesting that these groups evolved diets including food other than vertebrate flesh.

The ornithomimosaurs (bird mimics) show some convergences in form with modern ostriches and other flightless birds. They had a small skull (with 220 tiny teeth in the primitive, Early Cretaceous *Pelecanimimus*, but toothless in typical Late Cretaceous forms) at the end of a long neck and a compact body with extremely long legs. Unlike modern ostriches, however, the ornithomimosaurs had slender bony tails and long arms ending in three fingers. The hands of ornithomimosaurs were distinctive, in that the fingers were of equal length and formed a hooking-and-clamping structure similar to that seen in modern sloths. However, the long-legged ornithomimosaurs seem unlikely to have been habitual tree dwellers; instead, these hooking-and-clamping hands may have been used for grasping branches to gain access to leaves and fruit. The long hindlimbs of ornithomimosaurs ended in long slender feet; in most of these forms, this foot was further modified by a pinched middle metatarsal bone. Functional analysis of this structure in the mid-1990s suggested that this adaptation (termed an "arctometatarsus") may have served as a form of shock absorber, consistent with the interpretation of ornithomimosaurs as one of the fastest-running groups of dinosaurs.

Another clade of coelurosaur that is characterized by an arctometatarsus is the tyrannosaurids. Most famous for the last and largest member of the clade, *Tyrannosaurus rex*, tyrannosaurids at first glance seem very far removed from the ostrichlike ornithomimosaurs. However, the hindlimb anatomy of tyrant dinosaurs is nearly identical to that of ornithomimosaurs; indeed, their proportions plot along the same growth curve (so that young tyrannosaurs have the same long gracile legs as do adult ornithomimosaure). These observations suggest that tyrannosaurids were, like the ostrich dinosaurs, swift-running animals for their size; however, at larger size it may be that tyrant dinosaurs relied less on speed and more on strength in acquiring their food. Additional derived features in the pelvis and skull suggest in my own analyses that tyrannosaurids were closer to ornithomimosaurs than to other coelurosaur groups. That said, additional data suggest other alternative placements (tyrannosaurids, for example, as primitive maniraptorans; or tyrannosaurids as the sister group to the rest of the maniraptoriforme) in analyses by Paul Sereno and Peter Makovicky of the American Museum of Natural History, and Hans Sues of the Royal Ontario Museum. Regardless of their precise position, all observations indicate that tyrannosaurids are derived coelurosaurs, and developed their giant form independent of true carnosaurs such as *Allosaurus* and *Giganotosaurus*. Tyrannosaurids are additionally characterized by incisor-like nipping teeth in the premaxilla, by extremely powerful jaws and lateral teeth (which, unlike those of typical theropods, are relatively thick side to side), and extremely reduced forelimbs with only two claws. Their anatomy suggests that tyrant dinosaurs were predators specialized for swift running and for dispatching and dismembering their prey with their jaws alone. While later tyrannosaurids were 40–50 feet long and massed up to 7 tons, the earliest known fossils (including teeth from the Early Cretaceous of Japan) suggest that tyrant dinosaurs began as relatively moderate-sized animals 9–10 feet long or so, and only acquired their giant size after the extinction of the carnosaurs.

The remaining groups of maniraptoriforme comprise the Maniraptora (hand grabbers). These forms typically had very long arms and enlarged sterna (breastbones). The oviraptorosaurs, including feathered *Caudipteryx* of the Yixian Formation, had short boxy skulls; in advanced forms, these jaws are entirely toothless. The diet of oviraptorosaurs is a matter of some debate—everything from shellfish to vegetation to eggs has been suggested, although the only definite gut contents known for a representative of this group are small vertebrates. Many oviraptorosaur species are characterized by ornate crests. Oviraptorosaur specimens from the desert deposits of Mongolia and China indicate that these dinosaurs brooded their nests in the manner of modern birds (similar evidence exists for troodontids). Some oviraptorosaurs had an arctometatarsus, suggesting that they were swift moving; others had a more primitive unpinched foot.

Therizinosauroids were for a long time a problematic group in terms of their phylogenetic position. They seemingly combined elements of advanced coelurosaurs, prosauropods, and ornithischians. However, discovery of primitive specimens of therizinosauroid, as well as reanalysis of the advanced forms, indicates that they represent a clade of maniraptoran coelurosaur, most likely the sister group of the oviraptorosaurs (although Sereno suggests that they are more closely allied with the ornithomimosaurs). Therizinosauroids had teeth similar in size and shape to those of prosauropods and ornithischians, suggesting that they, too, were primarily herbivorous. Also like the ornithischians, the pubis of therizinosauroids was directed backward, greatly increasing the gut volume. In all but the most primitive members of the clade, the foot is extremely short and broad, suggesting that these were slow-moving theropods.

Troodontids were a highly derived clade of coelurosaur; indeed, they are so highly derived that there is no definite phylogenetic position for them that doesn't require considerable convergences. As in dromaeosaurs and birds, the tail was extremely mobile at the base, and the foot had an enlarged, retractable sicklelike claw at the end of the second toe. Like in therizinosauroids (and primitive ornithomimosaurs), the teeth were small, pinched at the base, and had large denticles rather than fine serrations. Like ornithomimosaurs, troodontids had an arctometatarsus and a specialized inflated pocket in their braincase. Additionally, unlike most other maniraptoran groups, relative to the length of the femur, the arms of troodontids actually quite short were instead, comparable in proportion to primitive tetanurines. The diet and habit of troodontids is likewise uncertain; some have suggested that they were raptorial predators in the mode of dromaeosaurids, while others propose a more omnivorous diet that included eggs, insects, soft-bodied invertebrates, and vegetation, as well as meat. Their long, gracile hindlimbs indicate that they were among the fastest dinosaurs. Work by Dale Russell and colleagues in the 1970s and early 1980s demonstrated that troodontids had extremely large brains compared to typical large dinosaurs; however, as the braincases of other maniraptorans were then unknown in detail, it is not certain whether troodontids had atypically large brains or that all maniraptorans might share such an enlargement.

Dromaeosaurids may have reached the attention of the general public in the form of *Velociraptor* (the main villains of *Jurassic Park*), but their importance to dinosaur paleontology came much earlier. The description of the first relatively complete dromaeosaurid fossils, that of Early Cretaceous *Deinonychus*, by John Ostrom of Yale University in 1969, was a pivotal event in the study of the Dinosauria. The recognition of the long, grasping forelimbs, the tail that was highly mobile at the hips but stiffened by long bony rods throughout most of the length, and the enormous retractable sickle-claw on the second toe of the foot suggested that dromaeosaurs were extremely agile and active hunters. This challenged the then-prevailing notion that dinosaurs were primarily slow and sluggish animals. Furthermore, comparison of the skeletons of dromaeosaurids and of the most primitive bird, Late Jurassic *Archaeopteryx*, indicated numerous anatomical similarities and suggested to Ostrom that birds were the direct descendants of derived coelurosaurs close to the dromaeosaurids. Indeed, subsequent phylogenetic analyses have consistently found that the dromaeosaurids are among the closest groups of carnivorous dinosaurs to the bird clade. Dromaeosaurids share with birds a backward-pointing pubis. Unlike in therizinosauroids and ornithischians (for which the pubis was redirected to make space for larger plant-digesting intestines), this bone was probably reoriented in response to changes in the musculature of the pelvis, tail, and hindlimb associated with knee-driven rather than hip-driven walking. The modified tail anatomy of dromaeosaurids and early birds allowed it to be more efficiently used as a dynamic stabilizer, so that these dinosaurs could turn more quickly while running.

The vast preponderance of data demonstrates that birds are part of the maniraptoran coelurosaur dinosaur radiation. Indeed, the difficulty in distinguishing early birds from other dinosaurs is exemplified by the problematic phylogenetic status of the Cretaceous clade Alvarezsauridae. The alvarezsaurids (which had long slender limbs, backward-pointing pubes, and short powerful arms ending in an enormous thumb claw) do not unambiguously fall in any particular part of the coelurosaur tree. Some skeletal evidence suggests that they were closer to ornithomimosaurs than to all other groups. On the other hand, additional features (including many aspects of the skull and vertebrae) suggest that these were highly specialized early birds. Resolution of this position awaits additional evidence, particularly in the form of ancestral alvarezsaurids.

The avian status of other dinosaur fossils, such as *Archaeopteryx*, *Confuciusornis*, and *Hesperornis*, is unquestioned by most researchers. Discoveries during the last two decades have greatly increased our knowledge of the radiation of birds during the Cretaceous period, including tree-dwelling perchers, fish-eating shorebirds, and flightless running forms. The majority of Mesozoic bird groups retained teeth. Only the toothless neornithine birds, however, survived the great extinction event at the end of the Cretaceous. Thus, this clade alone of all the Dinosauria persists to the present.

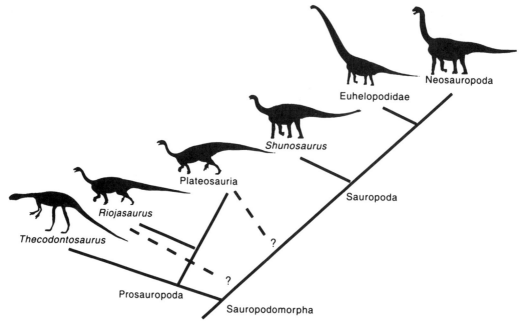

Relationships of the primitive sauropodomorphs are not well resolved; some analyses support a monophyletic clade leading to the true sauropods.

The Sauropodomorphs: Giants on the Earth

Although many people are familiar with *Tyrannosaurus*, *Stegosaurus*, and *Triceratops*, the creatures that most immediately come to mind when the word "dinosaur" is mentioned are the giant sauropods. Up to 100 feet or more in length, and massing 100 tons, the largest of these dinosaurs (*Argentinosaurus*, *Argyrosaurus*, and the reported but since lost remains of *Amphicoelias fragillimus*) were by far the most enormous creatures to walk the surface of the Earth, exceeded in size only by the blue whale (*Balaenoptera musculus*) among all animals.

However, these giant dinosaurs were descendants of much smaller forms. The Sauropodomorpha as a whole contains the true sauropods and an assemblage of more primitive forms of the Late Triassic and Early Jurassic. Some studies indicate that these smaller, earlier forms comprise their own clade Prosauropoda, while others suggest that they, instead, represent a paraphyletic series of transitional clades between the ancestral dinosaurs and Sauropoda. (Perhaps both models represent an aspect of the true phylogeny, with a monophyletic Prosauropoda and a handful of primitive forms outside the prosauropod-sauropod clade).

All sauropodomorphs had small skulls and long necks. Prosauropods and primitive sauropods had leaf-shaped teeth with large denticles, similar to those of ornithischians, suggesting a herbivorous diet. Many of the primitive prosauropods (such as Late Triassic *Saturnalia* of Brazil and Early Jurassic *Thecodontosaurus* of Great Britain) were small bipedal forms 7–10 feet long. However, larger prosauropods such as *Massospondylus*, *Plateosaurus*, *Riojasaurus*, and *Melanorosaurus* show adaptations of the forelimb, suggesting that they spent a fair amount of time on all fours, although they retained grasping ability in the hand. In these larger forms (some of which reached 35 feet or more in length), the long necks and the ability to walk bipedally gave these dinosaurs access to leaves high up in trees, a resource untapped by other large herbivores (which were either small, fully quadrupedal, or both) during the Early Mesozoic. By the end of the Triassic, prosauropods were among the most common large herbivorous vertebrates.

This large size had additional benefits. It would serve as a defense against smaller predators and allow a larger gut volume for digesting vegetation. The combination of these selective forces may, indeed, have been responsible for the origin of the true sauropods. Sauropods are all characterized by enormous size (the smallest adult sauropods are still larger than most other dinosaurs) and, consequently, the necessary quadrupedal bearing on columnar limbs, as well as the addition of at least one extra hip vertebra. The hind-feet of sauropods are broader and more massive than those of prosauropods. *Shunosaurus* of the Middle Jurassic of China is among the best known of the early sauropods.

Sauropod evolution seems to be characterized by different specialized methods of approaching the ecological niche of tree-feeding. Several Chinese sauropods (*Euhelopus*, *Omeisaurus*, and *Mamenchisaurus*) of the Middle Jurassic through the Early Cretaceous of China are characterized by extremely long necks that incorporate three or more extra vertebrae than found in typical groups. The analyses of Paul Upchurch of Cambridge University support a monophyletic grouping of these three taxa as the clade Euhelopodidae; those of Jeff Wilson (now at the University of Michigan) and Paul Sereno, instead, suggest that the "euhelopodids" are a polyphyletic group (if so, a remarkable example of convergence limited to a particular geologic span and geographic setting). Another primitive form is the recently described 70–foot-long *Jobaria* of the Early Cretaceous of Niger, known from some of the most complete skeletons of any sauropod. This giant, like its sister group the Neosauropoda, had a modified hand in which all the long bones (metacarpals) formed a pillarlike structure.

The Neosauropoda comprise the greatest part of the sauropod diversity of the Late Jurassic and Cretaceous. In addition to their columnar hands, neosauropods had broad hips and teeth lacking denticles along the edges. The two main clades of neosauropods are Diplodocoidea and Macronaria. Diplodocoids had long sloping skulls and nostril openings above their eye sockets. The teeth of diplodocoids were pencil-shaped and

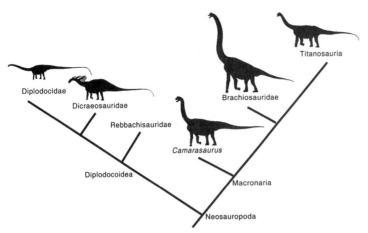

The giant neosauropods include the largest land animals in Earth's history. Different skull shapes, tooth forms, neck lengths, and limb proportions suggests differing approaches to browsing in trees in each clade.

apparently used for scraping leaves and needles off of tree branches. The ends of the tails of diplodocoids were slender and, at least in the case of *Diplodocus* and *Apatosaurus* (formerly called *Brontosaurus*), extremely long; some paleontologists have suggested that they were used as whips in defense against attacking theropods. Diplodocoids also had forelimbs reduced in size compared to the hindlimbs, suggesting to some that these dinosaurs habitually reared on their hindlegs in order to feed even higher in trees. Diplodocoids include the extremely long-necked diplodocids such as *Diplodocus*, and *Apatosaurus*; short-necked dicraeosaurids, with elongate neural spines of the vertebrae culminating in those of Early Cretaceous *Amargasaurus* of Argentina; and the recently recognized rebbachisaurids of the Cretaceous of South America and Africa. This latter clade seems to have evolved highly specialized broad snouts with a dental battery of continuously replacing teeth comparable to those of hadrosaurids and ceratopsian ornithischians. Although poorly known at present, rebbachisaurids reveal that (contrary to the interpretation of some workers) sauropods were not declining in the Cretaceous but were, instead, diversifying in their adaptations during this period.

Macronarian sauropods include the primitive Jurassic form *Camarasaurus* and the specialized brachiosaurids and titanosaurs. Macronaria ("big noses") derive their name from the greatly enlarged size of their nostrils (which exceeds the size of their eye sockets); however, this trait is also found in *Jobaria* and is probably the ancestral condition for Neosauropoda as a whole. An elongate hand and additional features of the skull, vertebrae, and pelvis unite the macronarians relative to other sauropods. Also, *Camarasaurus*, brachiosaurids, and some primitive titanosaurs share robust, spatulate teeth, indicating a more powerful bite than in diplodocoids.

Brachiosaurids were a Jurassic and Early Cretaceous clade of slender-necked macronarians. They differed from many other sauropod clades in possessing elongate neural spines just behind the shoulder region, perhaps associated with muscular or ligamentous support of their long necks. As traditionally interpreted, these are the only sauropods (indeed, the only quadrupedal dinosaurs) in which the forelimbs were longer than the hindlimbs, giving them a superficially giraffelike profile; however, there is some question as to the precise proportion between the fore- and hindlimbs. Brachiosaurids are among the largest and tallest dinosaurs; the recently discovered *Sauroposeidon* of the Early Cretaceous of Oklahoma appears to have been 80 feet or so long, half of which was neck.

Although they have been known since the 1800s, titanosaurs have only recently become relatively well understood. Although the name is apt for some forms (*Argentinosaurus* and *Argyrosaurus* from the Late Cretaceous of Argentina, for example, are among the largest dinosaurs known from material currently in collections), titanosaurs are not merely characterized by size. Instead, they represent a major radiation of primarily Cretaceous taxa found worldwide. New discoveries suggest that their skulls were similar to those of brachiosaurids, but in advanced forms they had pencillike teeth as in diplodocoids. Studies by Jeff Wilson and Matt Carrano (now of the State University of New York, Stony Brook) suggest, based on evidence such as a more flexible spine, wider stance, and flared hips, that titanosaurs had more sophisticated locomotory behavior than more primitive sauropod clades, perhaps including a greater period of time spent in a bipedal stance.

The Thyreophorans: The Best Offense Is a Good Defense

The armored dinosaurs comprise a clade named Thyreophora ("shield bearers"). The most primitive of these, *Scutellosaurus* of the Early Jurassic of the American West, generally resembled the basal ornithischian *Lesothosaurus*. However, *Scutellosaurus* possessed a series of keeled armored scutes along its neck, trunk, and tail. This armor might have served as protection against smaller predators but would have been less effective against larger carnivores such as the contemporaneous *Dilophosaurus*. Thyreophorans from later in the Early Jurassic, such as *Emausaurus* of Germany and *Scelidosaurus* of Great Britain (and possibly the United States and China)

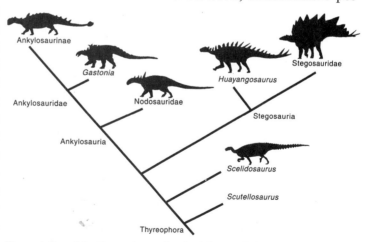

The evolution of the thyreophorans involved the acquistion and modification of armored scutes. In stegosaurs this dermal armor was modified into various forms of plates and spikes, including the thagomizer (tail-spike club). Ankylosaur evolution involved the development of heavy armored scutes over most of the body, including the eyelids.

were larger and more heavily armored; consequently, their forelimbs suggest that they spent most or all of their time on all fours.

The thyreophorans of the Middle Jurassic through the Cretaceous are divided into two clades, the plated stegosaurs and the intensely armored ankylosaurs. In both taxa, the hands and feet are very broad to support their large mass. A large spine is present on the shoulder of most stegosaurs and at least some ankylosaurs; this may be a synapomorphy of the advanced thyreophorans, or it might have developed convergently in Stegosauria and some ankylosaur clades.

Stegosauria is best known from the North American Late Jurassic *Stegosaurus*, but the greatest diversity of this clade is found in China. In stegosaurs, the scutes closest to the midline of the spine were tall, forming flat plates, conical spines, or shapes intermediate between these. The tails of stegosaurs were adorned with pairs of long backward-and-sideway-facing spines: this specialized tail club arrangement has been termed a "thagomizer" after a *Far Side* cartoon by Gary Larson. Additional armor was present in at least some stegosaurs as shoulder spikes, hip scutes, and a mass of small rounded knobs along the base of the throat. In early stegosaurs, such as Middle Jurassic *Huayangosaurus*, the forelimb was not much shorter than the hindlimb, and the vertebrae were relatively unmodified; in later forms (such as African *Kentrosaurus*, Chinese *Tuojiangosaurus*, and *Stegosaurus* itself), the forelimbs were considerably shorter than the hindlimbs, and the vertebrae were stretched vertically. Stegosaurs were present in the Early Cretaceous in Europe and Asia, but this clade does not seem to have survived into the Late Cretaceous.

Ankylosauria is characterized by armor fused to the roof of the skull and by two U-shaped rings of fused armor scutes in the neck and shoulder region, as well as a mosaic of some large and many smaller scutes over the rest of the back and parts of the throat, belly, and limbs. In many ankylosaurs, the broad hip region was covered by a mass of fused scutes; in at least some forms, there was even an armor plate on each eyelid. Relationships among the lineages within Ankylosauria are the subject of some debate at present. The club-tailed ankylosaurids of the Cretaceous of Asia and western North America clearly form a clade; however, the various other forms traditionally called "nodosaurs" almost certainly represent a paraphyletic grade. Nodosaurids proper seem to comprise primarily a Cretaceous North American clade, while various other primitive ankylosaurs from across the globe lie closer to the club-tailed ankylosaurids (for example, the Early Cretaceous *Gastonia* of North America and *Polacanthus* of Europe), while others (for example, *Minmi* of Australia) may have diverged from other ankylosaur lineages prior to the nodosaurid-ankylosaurid split.

The Ornithopods: Beaks, Bills, and Crests

Although most paleontologists consider the Early Jurassic southern African *Lesothosaurus* to be a primitive ornithischian outside of the more advanced clades, this dinosaur and other unspecialized forms have been traditionally interpreted as primitive members of the Ornithopoda. Ornithopods did retain many primitive features lost or modified in thyreophorans and marginocephalians; for example, most primitive ornithopods were small bipedal forms lacking armor, horns, or crests.

However, true ornithopods differed from other ornithischians in having a tooth row for the premaxilla (the front bone in the upper jaw) that is lower than that in the maxilla (the rest of the upper jaw). This, coupled with the general ornithischian condition of a jaw joint placed lower than the level of tooth-to-tooth contact, gave the early ornithopods a mechanical advantage in biting. Much of the evolution of the ornithopods involved the elaboration of their feeding mechanisms.

Heterodontosaurids were the earliest radiation of ornithopods (although some workers have considered them more closely related to the marginocephalians than to hypsilophodontians and iguanodontians). They are best known from Early Jurassic *Heterodontosaurus* of southern Africa, but they include dinosaurs from around the Early Jurassic world and persisted until the Late Jurassic. Although the hands of other early ornithischians are poorly known at present, those of heterodontosaurids resembled those of primitive theropods and prosauropods in being large, grasping structures with trenchant claws; this might reflect retention of the ancestral condition for Dinosauria or convergence with primitive Saurischia. The teeth of heterodontosaurids were highly specialized. There is a single large canine-like tooth in the lower jaw that fits into a corresponding gap in the

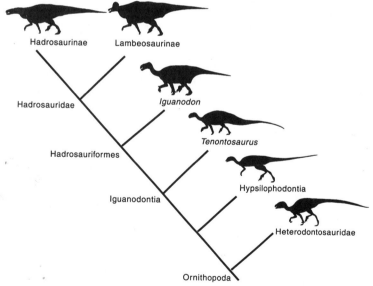

The history of the ornithopods, as with other dinosaur clades, involved an increase in size; in this case from 3–5 foot-long forms like the heterodontosaurids. Additionally, elaborate jaw joints and the development of a grinding dental battery indicates more sophisticated chewing ability in the advanced ornithopods, especially the duckbilled hadosaurids.

upper jaw; this has been interpreted as having a social display function rather than indicating a carnivorous diet. The rest of the teeth have a characteristic chisel shape, probably for slicing and grinding harder food than in the diet of typical ornithischians.

Many of the smaller ornithopods of the Jurassic and Cretaceous are considered members of the clade Hypsilophodontia; however, preliminary work by various paleontologists suggests that this assemblage might well represent a paraphyletic grade between primitive ornithischians and iguanodontians. Typical hypsilophodontians (such as *Yandusaurus* of the Middle Jurassic of China, *Hypsilophodon* of the Early Cretaceous of Europe and North America, and *Parksosaurus* of the Late Cretaceous of North America) were 3–6 feet long, with relatively short skulls, and tails with rod-like bony extensions somewhat like those in dromaeosaurid theropods. These were probably fast running but otherwise unspecialized ornithopods. Some latest Cretaceous forms, such as *Thescelosaurus* of western North America, had elongate snouts and reached lengths of 12 feet or so, suggesting the beginnings of a possible radiation of large advanced hypsilophodonts cut short by the Cretaceous-Tertiary extinction event.

The remaining ornithopods comprise the iguanodontians. Although some, such as *Gasparinasaura* and *Dryosaurus*, were comparable in size to hypsilophodontians, most iguanodonts were much larger—20–40 feet long. In iguanodontians, the teeth are lost from the premaxillary bone in the front of the upper jaw, and the predentary bone on the lower jaw develops a scalloped surface. *Camptosaurus* and *Tenontosaurus* of North America and *Muttaburrasaurus* of Australia represent variations on the primitive iguanodontian theme. The opening of the nostril in iguanodontians is enlarged (especially so in the advanced hadrosauriforms), as it is independently in carnosaurian theropods, macronarian sauropods, and horned dinosaurs. Larry Witmer of Ohio State University and Scott Sampson of the University of Utah have suggested this convergent evolution of enlarged nostril regions was associated with soft tissue structures associated with high rates of respiration and thermoregulation. Except for the smallest forms, iguanodontians seem to have spent a substantial amount of time on all fours. Consequently, the forelimbs were more heavily built, and the wristbones were fused to support their greater weight. However, the hindlimbs were still much larger than the forelimbs, and so these dinosaurs almost certainly could walk or run bipedally.

The more advanced iguanodontians, the hadrosauriforms, include *Iguanodon*, *Altirhinus*, and the duck-billed hadrosaurs. In typical dinosaurs (as with most reptiles), there was little chewing ability: food might be pulped or sliced by vertical movement of the lower jaws, but there was no lateral motion to grind the food. Hadrosauriforms were characterized by a joint in the upper jaw that would allow the maxillae to swing outward when the jaws closed, giving them a unique mode of chewing. To deal with the wear generated by this active grinding motion, the teeth of hadrosauriforms were more numerous and more densely packed than in primitive ornithopods, with 20 or more sets of teeth per row and no space between each set. Hadrosauriforms were larger than other ornithopods (and included the largest ornithischians). The long bones of the hands were long, and the three middle fingers transformed into short hooves, indicating that hadrosauriforms spent a great amount of time walking on all fours. However, they also had an opposable fifth finger (homologous to the human pinky), which suggests that they used their hands to grasp and manipulate food.

The hadrosauriforms include a series of Early Cretaceous forms, such as *Iguanodon* of Europe, Asia, and North America, *Altirhinus* and *Probactrosaurus* of Mongolia, and *Ouranosaurus* of Africa, in which the thumb was transformed into a solid spike, possibly for defense. A similar condition is found in the North American *Protohadros* and *Eolambia*, both from the very beginning of the Late Cretaceous (about 90 million years ago). These latter forms, however, also have dental batteries (tooth rows that form a continuous grinding surface). *Protohadros* and *Eolambia* represent the closest known relatives to the hadrosaurids of the Late Cretaceous. Commonly called "duck-bill dinosaurs" due to the expanded end of their snouts, hadrosaurids are one of the best-studied Late Cretaceous dinosaur groups. They are known from complete growth series from unhatched eggs to adults, and many species are known from complete or nearly complete skeletons. In hadrosaurids, the thumb was lost. The hadrosaurids comprise two major clades. In the hadrosaurines, the snouts were extremely broad, and the nostril openings were quite large. In lambeosaurines, the nasal passages within the skull formed elaborate crests, which may have served as both vocal (air blown through different length and shaped tubes would produce different sounds) and visual signals. Each species of lambeosaurine had a unique crest shape, thus potentially giving each species a distinctive sound and appearance. Furthermore, the crests of specimens interpreted as females and males of each species are slightly different, while hatchlings and juveniles lacked crests, suggesting relatively complex social behaviors within their communities.

The Marginocephalians: Boneheads and Horn-Faces

The last major group of ornithischians is characterized by a shelf of bone that overhangs the back of the skull, leading Paul Sereno to name them the Marginocephalia, or "ridge heads." Except for problematic *Stenopelix* of the Early Cretaceous of Germany (the only definite marginocephalian known from outside of Asia and western North America), the dinosaurs within this group divide into two separate clades: the dome-headed pachycephalosaurs and the deep-beaked ceratopsians.

The bipedal pachycephalosaurs (thick-skulled lizards) derive their name from the thickened layer of bone over their braincases. In the most advanced forms, this layer formed a thick dome. Most paleontologists interpret this dome as an adaptation for shoving

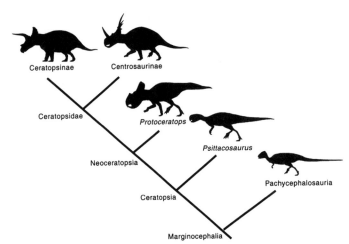

Marginocephalians are characterized by specializations inferred to serve in social interactions, including the thickening skulls of pachycephalosaurs (for use in head-butting combat) and the frills of neoceratopsians (which probably served as both anchors for powerful jaw muscles and as large visual signals). The horns of ceratopsids likely served both as a defense against tyrannosaurids and as weapons for intraspecific combat: broken horn cores and punctures in frills suggest vigorous head-pushing matches between large horned dinosaurs.

matches, and possibly even more vigorous clashes comparable to those seen in bighorn sheep today. At present, pachycephalosaurs are known only from the Late Cretaceous, where they represent a minor part of the herbivorous dinosaurian fauna of Mongolia, China, Canada, and the western United States.

Until recently, ceratopsians were known only from the Cretaceous, but the discovery and analysis of *Chaoyangsaurus* of the Middle Jurassic of China has greatly extended their stratigraphic range. As with other ceratopsians, this form has an extra bone (the rostral) at the front tip of the upper snout. This forms part of the deep beak found in all ceratopsians. The primitive ceratopsian *Psittacosaurus*, one of the most common dinosaurs of the Early Cretaceous of Asia, retained the ancestral bipedal habit, as do the more primitive members of the Neoceratopsia (such as *Archaeoceratops* and *Microceratops*). The neoceratopsians are further specialized in that the shelf on the back of their skull is elongated into a frill, which serves as the attachment surface for the jaw muscles. Additionally, neoceratopsians have densely packed rows of teeth, forming a continuous cutting surface, and enlarged jaw muscle attachment surfaces on the lower jaw. Together, the evidence suggests that the neoceratopsians were specialized feeders on relatively durable plant material, although the particular food in their diet is as yet unresolved.

In the more advanced neoceratopsians, such as *Protoceratops*, the skull had become so large that the dinosaurs were forced into quadrupedality. While most of the primitive ceratopsians were only 13 feet long, the most advanced forms were much larger. *Turanoceratops* of Asia and *Zuniceratops* of New Mexico represent the most primitive of the horned ceratopsians. In these forms, there was a single small horn on the nose and one over each eye. In the giant ceratopsids, a strictly North American clade, these horns were greatly elaborated. The ceratopsids are among the most common Late Cretaceous herbivorous dinosaurs of western North America; evidence from quarries in which the total number of specimens of a single species number in the dozens or hundreds indicates that at least some ceratopsids lived in herds. Within the Ceratopsidae are two major clades: the shorter and deeper-snouted centrosaurines, in which the nose horn dominated over the brow horns, and the longer-snouted ceratopsines (sometimes called "chasmosaurines'"), including the world famous *Triceratops*, which had longer brow horns than nose horns.

The Future of Dinosaur History

The general phylogenetic relationships among the major groups of dinosaurs have been the subject of many analyses for the last 15 years, whereas the relationships of the individual species within each group have yet to be examined in as much detail. This will change in the near future, as studies are ongoing to estimate the phylogeny of almost all the major dinosaurian clades.

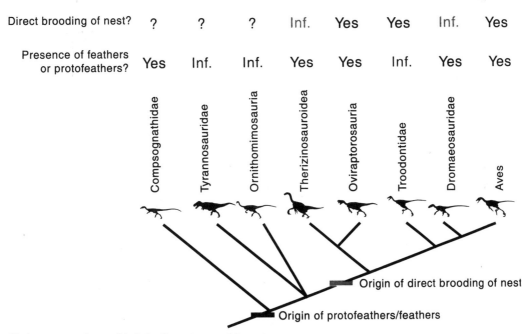

Cladograms can be used to infer the presence of "protofeathers" in dinosaurs whose integument is not known, or brooding behavior in groups for which nests have not yet been discovered. Assuming a single origin for the protofeather structures (preserved in fossils from the Early Cretaceous Yixian Formation of China), the most recent common ancestor of all coelurosaurs would have had this fuzzy body covering. This implies that tyrannosaurids, ornithomimosaurs, and troodontids (whose skin is not yet known) would be the descendants of feathered animals and most likely fuzzy themselves. Similarly, discoveries of oviraptorosaur and troodontid brooding nests in the same manner as birds implies that the common ancestor of all maniraptorans was a brooder, and thus its descendants would have inherited this condition. However, because information on more primitive theropod nests is currently lacking, it cannot be resolved if brooding behavior would have been present earlier in theropod history.

More than simply recovering the family tree of the Dinosauria, the new phylogenetic analyses allow for predictions of dinosaurian forms as yet undiscovered. For example, skeletal evidence supports the hypothesis that pachycephalosaurs and ceratopsians are sister taxa—though the latter are known from as early as the Middle Jurassic, while the former do not show up in the fossil record until the end of the Early Cretaceous. Because the divergence between bone-headed and horned dinosaurs must have occurred prior to the appearance of a known ceratopsian, there is a gap of about 40 million years in the record of the pachycephalosaurs. Rock units of the Middle and Late Jurassic epochs in Asia almost certainly contain remains of early pachycephalosaurs, but these primitive domeheads have not yet been found.

Additionally, understanding the phylogenetic relationships among the dinosaurs allows for predicting the presence of some anatomical or behavioral adaptations that might not typically be preserved. For example, as mentioned previously, various specimens of coelurosaur from the Yixian Formation of China are preserved with the remains of integumentary structures that are almost certainly homologous with the feathers of birds. The most primitive of these dinosaurs is the compsognathid *Sinosauropteryx*. Assuming that there was a single origin of this structure, then all dinosaurs that descend from the most recent common ancestor of birds and *Sinosauropteryx* would have had this specialization (unless through evolution they subsequently lost it). Strange as it might seem, the evolutionary relationships among the dinosaurs suggest that the ostrichlike ornithomimosaurs and even the tyrannosaurids had a covering of protofeathers. Similarly, specimens of oviraptorosaurs and troodontids have been discovered showing that these maniraptorans brooded their nests in the manner of birds. Although no dromaeosaurid nests have yet been discovered, the most parsimonious explanation of the data supports nest brooding as ancestral to all maniraptoran lineages, and suggests that *Velociraptor*, *Deinonychus*, and their kin also kept their eggs warm and protected by sitting on them. Such predictions would seem unlikely if not for our growing knowledge of the phylogeny of Dinosauria.

The Flying Dinosaurs

by Andrzej Elzanowski

The idea of birds as flying dinosaurs is almost as old as Darwin's theory of evolution, and its vagaries are discussed in this chapter in the essay by Kevin Padian and Luis M. Chiappe. After a century of domination of individual prominent authorities, such as Thomas Huxley and then Gerhard Heilmann, a new era in the study of avian origins was launched in the seventies by John Ostrom: that of collective work by a small but dynamic international community of scientists. A collective endeavor in science obviously spawns more controversy, but it also provides a more reliable overall picture than any single authority.

Birds Are Flying Theropods

As in human genealogy, modern phylogenetics answers the question about the origins of a group by identifying its closest relatives. That birds are specialized archosaurs is a solid piece of knowledge, one based on traditional phylogenetics and acquired long before the advent of today's cladistic methodology. All birds express the archosaurian trend to bipedalism. Other archosaurian features—such as the antorbital opening in front of the eye orbit and peculiar secondary ossifications called interdental plates that alternate with teeth on the inner (lingual) side of the jaws—are present in *Archaeopteryx* but transformed and obliterated in modern birds.

The perfect consensus about avian relationships ends at the level of archosaurs, but the majority of experts see enough evidence to recognize birds as theropods and, therefore, as dinosaurs. A close relationship of birds and theropods is attested by computer-based reconstructions of genealogy that involve statistical evaluation of how reliable each of the reconstructed groupings is. Some of the characters used for these reconstructions are robust, that is, well pronounced and unique or rare. But even unique characters may evolve independently and in groups that are derived from similar ancestors. The frequent phenomenon of parallelism and convergence, especially in the evolution of related lineages, makes it especially important to base phylogenetic reconstructions on characters that evolved under different selective pressures. That is why it is so important that birds reveal their theropod ancestry in both cranial and postcranial characters.

Birds are identifiable as dinosaurs (or, in a traditional wording, as their descendants) by such technical details as the presence of a hinge "mesotarsal" joint between the proximal and distal tarsals, and the opening in the inner wall of the hip joint cavity. Within the dinosaurs, birds reveal their theropod ancestry in a mix of cranial and postcranial, all of which are extremely unlikely to have resulted from parallelism. In the skull of *Archaeopteryx*, the theropod ancestry is marked by the details of maxillary bone (which has an ascending ramus with additional openings), and at least one bone of the root of the mouth, the ectopterygoid, which has a lateral hook. In the postcranial skeleton, the theropods and primitive birds share the presence of five sacral vertebrae, loss of fingers 4 and 5, and a functionally three-toed foot with the toe 1 raised above the ground and a vestigial metatarsal 5.

The characters shared with all the theropods not only demonstrate that birds and theropods are close relatives; the majority of students now accept that birds originated from within the maniraptors (in a broad sense of the term, maniraptoriformes) and thus from within the theropods, rather than being a "sister group" of all the theropods. The origins from within the theropods are demonstrated by the presence of advanced characters uniquely shared by the birds and a subgroup of the theropods. There are sets of unique characters that unite birds with a series of ever less encompassing subgroups of the theropods (tetanurans, coelurosaurs, maniraptors, etc.), starting with the most primitive theropods. For example, *Archaeopteryx* shares with all other coelurosaurs (or even tetanurans) a semi-lunate wristbone that facilitates sweeping movements of the hand, and with the maniraptors long grasping forelimbs, the ulna bowed to the back, and metacarpal 2 more slender than metacarpal II. A successive appearance of the *Archaeopteryx* characters at the consecutive nodes of the tetanuran phylogeny provides a guarantee of the origin of birds from within that group. Birds are well nested in the theropod phylogeny. In contrast, critics of the theropod origins of birds often invoke isolated appearances of birdlike characters in single groups of the archosaurs, which a priori are likely to represent convergence rather than similarity attributable to common origin.

A sister-group relationship of birds and theropods would be thinkable only if the coelurosaurs were the most primitive theropods, and thus the similarities of birds and coelurosaurs could be primitive. But this is impossible, as we now know what the most primitive theropods (*Herrerasaurus, Eoraptor*) looked like and that the maniraptors are highly derived within the nonavian theropods. What is far from being clear, however, is which group of the maniraptors is closest to birds, an issue that defines the present cutting edge in the exploration of bird origins.

Several cladistic reconstructions based on the postcranial skeleton support the closest relationship between birds and dromaeosaurids. The dromaeosaurids and *Archaeopteryx* share some detailed similarities such as the pubic foot (a terminal expansion of the pubic bones) projecting only posteriorly. But most of these similarities are restricted to the postcranial skeleton, which is prone to convergence in response to similar demands of locomotion. A warning to this effect comes from the skeleton of *Velociraptor* with the pubis directed stronger to the rear than it is in *Archaeopteryx*, which is an obvious case of convergence with more advanced birds.

In contrast, the whole slew of unique cranial similarities uniting birds with other maniraptors, especially the oviraptorosaurs, ornithomimosaurs, and therizinosauroids, is unlikely to be convergent and thus likely to mark their close relationship, because otherwise the skulls and jaw functions differ dramatically between these groups. In terms of the jaw function, the oviraptorids are comparable to the dicynodonts and the therizinosauroids (a group of mammals' relatives), and may be convergent with the ornithischians in having cheeks. Most prominent and consistent among the characters uniquely shared by birds, oviraptorosaurs, ornithomimosaurs, and therizinosauroids are the bony root of the mouth and the lower jaw. Of course, some of the characters, such as the toothless beak that appears in the oviraptorosaurs, advanced ornithomimosaurs, and at least two lineages of advanced birds, are evidently convergent; the reduction of teeth is a recurrent theme of vertebrate evolution.

Identifying the closest avian relatives among the theropods proved more difficult than most authors admit. There are good, biologically meaningful reasons for these difficulties. One reason is that the most birdlike coelurosaurs and maniraptors are large, late Cretaceous survivors of the true avian ancestors (that probably lived in the Jurassic), and differ from them in morphology. Large animals are not simply magnified copies of small animals, but they differ in proportions (because of growth rate differences between various parts), locomotion (primarily because of inertia), habitat range (which, among other things, depends on locomotion), feeding and food, and in other respects. All these differences may lead to qualitative differences in morphology.

Another reason the identification of the closest avian relative remains controversial is that the phylogeny of nonavian maniraptors is unsettled, and its reconstructions are remarkably variable, which suggests major difficulties in distinguishing between common ancestral and convergent characters. A possibility has to be kept in mind that at least some maniraptors currently considered nonavian could in fact be secondarily flightless, even if they branched off before *Archaeopteryx*. (This means that their lineage may have evolved flight early, then for whatever survival adaptive reasons, evolved back to a flightless morphology.) And we now know that *Archaeopteryx* does not represent *the* most primitive stage in the evolution of avian flight. Gregory Paul once proposed that many or the most birdlike theropods were secondarily flightless. The idea seems to be gaining support. It may account for the presence in various maniraptors of a very birdlike shoulder girdle and wishbone (furcula), the presence of a shortened tail, and of reduced flight and tail feathers in *Caudipteryx*. If flightlessness had evolved at a stage of avian evolution close to *Archaeopteryx,* it would be extremely difficult to distinguish it from the primary flightlessness.

A few scientists, led by Alan Feduccia of the University of North Carolina at Chapel Hill, continue to disagree with the theropod origins of birds and argue that birds are derived from the dinosaurian ancestors (once collectively called thecodonts) but not from the dinosaurs themselves. Because the primitive theropods are close to the ancestry of all dinosaurs, this view would be agreeable with birds being the sister group of the theropods, but stands in strong opposition to the prevailing view that the origin of birds is nested *within* the theropods. The two main persisting lines of criticism for the theropod origin of birds are, one, the contradictory identification of fingers by embryology and paleontology and, two, an observation that the known coelurosaurs are too large and lived too late in geological time to have possibly given rise to the birds. Also invoked against the close relationships of theropods and birds have been differences between their respiratory systems, but this argument is simply spurious. Modern birds have a truly unique respiratory system that includes extensive air sacs and allows for largely unidirectional flow of the air. Neither the theropods nor any other vertebrate group have it. The avian respiratory system evolved largely de novo within birds, and thus cannot provide evidence of avian relationships.

Homology (common ancestry) versus development of the only three fingers left in the avian hand is indeed the most puzzling issue of the origins of birds. In limbs used as props to support the body, frequently reduced are the fingers from both the inner and outer sides of the hand or foot: first to disappear are finger 1 (thumb or big toe) and finger 5. But this clearly wasn't the case in the evolution of the theropod hand. In all dinosaurs, there is a tendency toward the reduction of fingers 4 and 5, which are vestigial in most primitive theropods (such as *Herrerasaurus*); and finger 4 persists in many theropods, but only three fingers are left in the tetanuran theropods. Because the hand of *Archaeopteryx* shows a detailed, evidently homologous similarity to that of other coelurosaurs, there can be no doubt that its three fingers represent digits 1, 2, and 3 and the numbers of bones in each finger agree. However, all studies of the embryonic development of the manus that were performed repeatedly by top experts in the field identify a pattern of embryonic mesenchyme condensations and cartilages that leads to digits 2, 3, and 4. The situation is a genuine case of conflicting evidence, which calls for a special explanation. The entrenched opponents of the theropod origins of birds use it as their most important argument, but in the absence of an alternative ancestral group of archosaurs with manual digits 2, 3, and 4, this is only negative evidence that is no match to the overwhelming positive evidence for the theropod ancestry of birds. A change of developmental pathways is the only plausible explanation for the development of avian hand. Over the long stretches of evolution, changes of developmental pathways are known to have taken place; thus, development cannot be used as an ultimate proof of

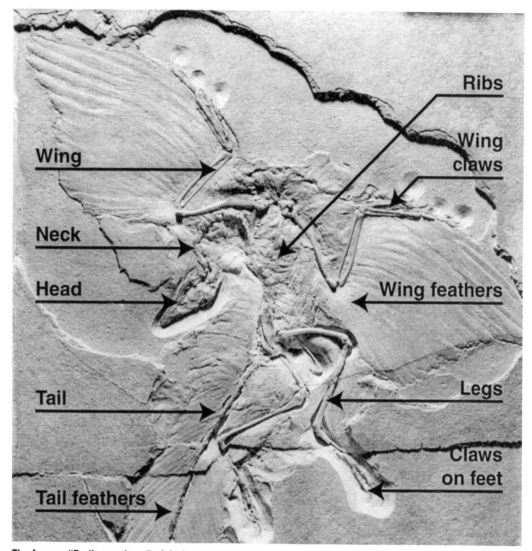

The famous "Berlin specimen" of Archaeopteryx shows that this little dinosaur was clearly a bird, or, conversely, that this little bird was clearly a dinosaur.

homology. For example, the skeleton of the posterior (but not anterior) dorsal and anal fins of actinistian fishes (such as the famous *Latimeria*) is built according to a plan found in the paired fins, which does not mean that any of the dorsal fins is homologous with a paired fin. It does mean that similar or identical genes can be expressed in different morphological contexts to produce similar but nonhomologous results.

What must have happened in the evolution of birds is a shift of embryonic condensations that give rise to fingers in relation to the expression domains of genes that shape them. Such shifts are now known to occur in vertebrate evolution. Each finger is shaped by a specific set of genes, which are activated upon a contact with an external activator substance. Which genes are activated depends on the timing and positioning of a condensation, not its origin, that is, homology. Guenter Wagner and Jacques Gauthier propose that condensations 2, 3, and 4 moved into the positions of those for fingers 1, 2, and 3, and thus their later development switched to that of fingers 1, 2, and 3. It is equally possible, however, that condensations 1, 2, and 3 moved into positions 2, 3, and 4, which caused an embryonic deviation of the early development (a phenomenon known as caenogenesis), but their late development remained unchanged.

The last common ancestor of birds and known maniraptors lived sometime back in the Jurassic (or conceivably, but much less probably, in the Late Triassic) and has yet to be found or identified. The protobirds were not bigger and possibly even smaller than *Archaeopteryx*, and the known maniraptors are much heavier when grown up than *Archaeopteryx* and other primitive birds. This indicates that concomitant to the origin of birds was a substantial reduction of body size. There are two reasons it must have been so—that is, small size as a precondition for the origin flight. First, the smaller an animal, the higher its surface-to-volume ratio; thus, small vertebrates need less power per unit of body mass to generate lift than large vertebrates. Aerodynamically, an optimal body mass for the origin of vertebrate flight is in the range of ⅓ to 3 ounces, which is much less than the body mass 5 to 6 ounces of the smallest *Archaeopteryx* specimen. Second, small vertebrates are more agile and versatile in locomotion, especially in climbing, which takes them to the heights that facilitate the initiation of flight. Small mammals and lizards climb trees without pronounced climbing adaptations, and juvenile crocodiles are known to climb river banks much better than the grown-up individuals.

If the protobirds (a term for ancestral birds) were small, then their absence in the fossil record isn't that surprising. Jurassic deposits with terrestrial vertebrates are scarce and small; delicate skeletons are seldom preserved as fossils, especially with abundant reptilian scavengers that swallow the whole carcass and digest the skeleton, as do today's crocodilians and monitors. This is because food passes through the alimentary tract of cold-blooded animals much longer than in mammals and birds, leaving enough time for a complete dissolution and digestion of bone.

Lifestyle of the Early Birds

Avian flight is a costly adaptation that interferes with other functions of the forelimbs and requires a separate, second locomotor apparatus, which has to be built up in ontogeny and maintained through life. The main flight muscles make up 15–25 percent of the body mass. In contrast to bats and pterosaurs, flight does not replace, but functions in addition to, terrestrial locomotion (or swimming) in all primitive birds as well as the majority of modern birds. Such a costly adaptation could have evolved only in response to predation, since only the survival traits that lead to reproduction are maximized at any cost, while other functions are optimized in terms of lowest cost/benefit ratios. Most exposed to predation are the juveniles. In conjunction with their small size and locomotor flexibility, they are prime candidates for the pioneers of avian flight.

For bipedal animals, climbing up is impossible without the help of forelimbs (or the beak that replaces them, in the parrots) and climbing down is much more difficult than climbing up (even more difficult than in quadrupeds). This explains the peculiar structure of the forelimb in the earliest birds, which is a functional chimera of a wing and a grasping organ with powerful claws. The forelimb was used both for climbing up, then as a parachute to return to the ground. The avian flight may have developed as a means of returning to the ground. However, the evolutionary sequence that led to the avian flight cannot be demonstrated without a reasonably complete fossil record of the earliest birds, which isn't available and may never be. The closest we can currently get in the fossil record to the origin of flight is *Archaeopteryx*, which retains its position as the most primitive of all known unquestionable birds. *Archaeopteryx* is beyond the initial stage of the evolution of avian flight but seems to be primitive enough to cast light on the lifestyle and locomotion of the avian ancestors. That is why most current reconstructions of *Archaeopteryx* are heavily biased toward authors' preconceptions about the origin of birds and their flight.

What makes a modern avian wing adapted for flight are its plumage and unique skeletal and muscular anatomy. *Archaeopteryx* had modern-looking flight feathers, but their arrangement may have been more different from that of modern birds than is commonly assumed in popular life reconstructions. For example, we don't know which feathers, if any, filled out the space between the forearm and the body.

In the shoulder girdle of modern birds, the big keeled sternum is located way below the shoulder joint and provides the main insertion area for powerful flight muscles. Because the sternum is deep, and the coracoid bone that attaches to the front of the sternum is strutlike and slender, the distance between the origins and the insertion of the flight muscles is large, and their fibers are long, which allows for a great amplitude of wing beats (nearly 180 degrees in some birds). Another consequence of the shape of modern avian coracoid is the formation of an opening between the ends of three bones—coracoid, scapula, and clavicle (hence the opening is called triosseal). An insertion tendon of one of the flight muscles (supracoracoideus) that originates from the sternum passes through this opening and is thereby deflected to attach on the upper surface of the arm bone. As a result, the muscle elevates the arm and raises its front margin (rotates it nose up, a movement called supination) instead of lowering it as it did in the ancestors. Supination of the wing counteracts the opposite nose-down rotation called pronation, which happens automatically during the power stroke because the flight feathers extend behind the bones (that is, behind the rotation axis of the wing). Supination is necessary for some flight maneuvers, especially for breaking at landing.

The hand skeleton of modern birds acts as a single module (except for finger 1, which supports the allula feather). Distal wristbones and metacarpals are fused into a single bone (carpometacarpus), which has a pulley (carpal trochlea) at its proximal end that guides the movement of two prominent wristbones, one U-shaped (called cuneiform or "ulnare") and another polyhedral (called scapholunar or radial), which in turn guide the movement of the entire hand relative to the forearm. The finger bones are flattened and articulate very tightly with one another and the carpometacarpus. These adaptations limit the hand movements to well-defined planes and help control them precisely, coordinate the flexion and extension of the forearm and manus, and withstand the torsional and shearing stresses of the power stroke.

Archaeopteryx has none of the modern avian adaptations of the wing skeleton. The pectoral girdle is small, the coracoid is short, and the rib cage below the shoulder joint is shallow, which indicates that the wing depressors must have been much smaller and probably weaker, their fibers much shorter, thus the amplitude of wing beats much more limited than in modern birds. The hand skeleton isn't consolidated, and its movements aren't guided as they are in modern birds, which reveals two major constraints on flight compared with modern birds. First, the wings of *Archaeopteryx*, as well as its tail, lacked the maneuverability used by modern birds for landing, takeoffs, and flight between obstacles (such as tree branches). Second, *Archaeopteryx* was poorly if at all adapted for flight at low speeds, which is aerodynamically complex, as it requires controlled changes of the pitch of the entire wing relative to the body and of the plane of the manus relative to the wing. The minimum power speed (at which the least work has to be done) for the *Archaeopteryx* was around 18 mph.

Inspired by the origin of birds from the theropods, which are terrestrial as adults, there have been several recent attempts to show that *Archaeopteryx* was capable of a ground-up takeoff, but none of them was successful. *Archaeopteryx* must have been able to somehow climb to an elevated point for its forelimbs to be used as wings. As all maniraptorans and other primitive birds (such as *Confuciusornis*), *Archaeopteryx* had powerful manual claws, whose most likely function was in climbing. However, aside from phylogenetic preconceptions about the origins of birds from terrestrial adult theropods, a genuine objection has been raised that the use of wings for climbing interferes with their use for flight and that flight and climbing adaptations do not go together in one limb. This conflict is refuted by immature but grown-up hoatzins, which continue for a certain time their juvenile habit of using wings to clamber around among the branches, resulting in damaged flight feathers. The other side of the coin is that they do it and survive some damage to the wings. The unique relationship between climbing and flying in *Archaeopteryx* and *Confuciusornis*, with no parallels in today's world, can be understood only in the context of their paleobiological reconstruction.

The sharp-pointed teeth of *Archaeopteryx* indicate that it was insectivorous, at least in the broadest sense of this term, which includes feeding on a range of small animals, but it was not adapted for the pursuit of small, mobile animals among the branches. Such a pursuit requires high maneuverability that does not seem possible for a bipedal animal without the maneuverability of wings or, as in some mammals, a prehensile tail. Only the passerines and a few other avian taxa evolved enough maneuverability to forage effectively in the trees. Today's little specialized arboreal birds, such as cuckoos, feed on abundant and sluggish caterpillars and the similar larvae of sawflies and related hymenopterans. One can safely conclude that *Archaeopteryx* foraged primarily or exclusively on the ground, be it the floor of araucaria groves or a beach. Its foraging in the tree crowns is extremely unlikely, even if trees were present in its range.

A combination of ground foraging, inability to take off on the spot, and the presence of swift predators such as *Compsognathus* in the habitat of *Archaeopteryx* calls for an escape strategy. *Archaeopteryx* certainly could not outrun *Compsognathus*, which was larger, and its best chance was to run to the nearest launching site to take off. The simplest launching site is a break in topography (e.g., a cliff); but there is no reason to believe that *Archaeopteryx* was limited to rugged landscapes, because it had powerful manual claws that were very well adapted for climbing, similar in shape to the claws of modern climbing birds, especially rock climbers. In the Solnhofen area, *Archaeopteryx* may have climbed a variety of elevated objects such as rocks, arborescent bushes, or conifer stem succulents (up to 10 feet tall). This scenario requires the presence of one or a few launching sites in the feeding range. *Archaeopteryx* was certainly able to remember the location of even a single launching "perch" and use it in emergency, as most of today's lizards (probably with less cognitive abilities) remember their single refuges, where they swiftly escape.

Only occasional, emergency use of the finger claws for climbing to safety is much more plausible than its regular use for clambering around in search of food. Just climbing vertical or steep surfaces may not have caused much damage to the wing feathers; and limited damage was tolerable because, in terms of Darwinian fitness, the survival was worth any price as long as there were prospects for reproduction.

Archaeopteryx was certainly at ease on the ground, as attested by its strong legs, with the tibia longer than the femur, and a slightly elevated, short hallux that did not interfere with terrestrial locomotion. However, not a single feature of *Archaeopteryx* represents a pronounced specialization for running. All modern groundbirds have the hindlimb, pelvis, and vertebral column much better adapted for fast ground locomotion than *Archaeopteryx*

Another category of groundbirds comprise the *escape runners*, such as many galliform birds (e.g., domestic fowl) and tinamous (South American flying relatives of the ratites), which run only in emergency to escape a predator. In terms of proportions between the leg segments (called intramembral ratios), *Archaeopteryx* shows the best match with the chicken family (phasianid) galliforms and the tinamous, which are typical escape runners. However, it also comes close to some the arboreal and arboreo-terrestrial foragers among the cuckoos. Interestingly, the leg proportions of *Archaeopteryx* also overlap with those of the primitive birds *Confuciusornis*, *Cathayornis*, and *Sinornis*, which were initially proclaimed to be arboreal; but this proved unfounded for *Confuciusuronis* and requires further scrutiny for the other genera. The comparisons of leg proportions are in perfect agreement with the paleobiological reconstruction of *Archaeopteryx* as an emergency runner that ran to take off to a nearby launching site. In addition, they make it clear is that *Archaeopteryx* was not a fast-running forager.

An integrative, compromise model of *Archaeopteryx* and the origins of avian flight is emerging after a century of seemingly unresolvable contradiction between the arboreal and cursorial groundbreaking models. In agreement with the cursorial hypothesis, the primitive birds were terrestrial and ran for life, but only to climb to the safety of the heights. In agreement with the arboreal theory, the avian flight started from the heights down, as in all other flying vertebrates. Paradoxically, passive flight may have been initiated as a means of returning to the ground from the safety of heights; but once developed and perfected into active flight, it extended the escape sequence and became part of it—the development of active flight allowed a protobird to use even low objects, which do not provide safety themselves, just for the takeoff and then to fly away from the predator. *Archaeopteryx* may represent this stage, if, as most scientists believe, it was capable of active flight.

A consensus is in sight that, as first proposed by Stefan Peters, active flight evolved from the steering movements of the wings—that is, as a means of controlling and extending the passive flight (parachuting and gliding) from the heights down. A disagreement persists, however, as to what made the protobirds climb to the heights and then fly back to the ground. Joseph Garner, Graham Taylor, and Adrian Thomas of Oxford University have just put forward the Pouncing Proavis model, according to which birds evolved from predators that specialize in ambush from elevated sites. However, this model runs into major problems because it explains the origin of flight as a feeding adaptation. A feeding adaptation cannot evolve unless it brings energetic gains, and this seems extremely unlikely for the origins of avian flight, which involved the high energetic cost of building up the flight-and-climbing apparatus in addition to strong legs. Pouncing is used today by advanced skillful fliers (such as the rollers, owls, and some passerines such as the shrikes) and poor walkers. It makes it hard to see what clumsy fliers would gain by poorly controlled pouncing, climbing back, and waiting rather than using their strong legs to pursue the prey on the ground. As Anders Hedenström put it, "We do not know why a cursorial dinosaur became a sit-and-wait ambush predator."

Favored here is the "Escape Theory of the Origin of Avian Flight," which proposes that it was initiated as part of the juvenile defensive sequence that included running, climbing to the safety of heights, then taking off. The protobirds started using flight to parachute back to the ground, then, later on in evolution, away from the predator. The Pouncing Proavis Model and the Escape Theory agree as to the initial cycle of climbing and flying back to the ground, but differ as to the biological roles of these activities.

The Escape Theory explains the size gap between adult theropods and avian ancestors as a result of paedomorphic origin of avian flight. Paedomorphosis (formerly known as neoteny), which is the evolutionary retention of juvenile features of the ancestors, was once proposed by Tonni Thulborn to explain the evolutionary origin of feathers for thermal insulation in the juveniles, which lose their body heat faster than the adults. While this remains a good possibility, the paedomorphic origin of birds looks even more likely in the context of the origin of flight. A reduction of body size, which was necessary for the initiation of avian flight, is obviously conducive to paedomorphosis. In addition, the juvenile maniraptors must have been exposed to a high predator pressure. The same predator pressure on the juveniles that led to the beginning of avian flight also favored the earliest possible use of the limbs and, thus, fast locomotor maturation, which nearly inevitably leads to paedomorphosis. There are several reasons to believe that birds not only are flying dinosaurs, but flying baby dinosaurs.

Cretaceous Diversity

Archaeopteryx stands apart as the most primitive bird, and its closest relatives are yet to be discovered. The second most primitive bird is *Rahonavis* from the Late Cretaceous of Madagascar. This exotic name was given to an array of skeletal fragments, which present a startling mosaic of primitive and a few highly advanced characters. *Rahonavis* combines the *Archaeopteryx*-like pelvis (with the pubis directed straight down) and a long tail of free vertebrae with advanced saddle-shaped (heterocoelous) vertebrae and several bulges on the forearm's ulna, which appear to be quill knobs (for the attachments of secondary flight feathers). *Rahonavis* apparently was a ground predator, with the second toe raised above the ground and with a strong claw, an adaptation that evolved once more in the South American seriemas (today represented by two genera, *Cariama* and *Chunga*). Cathy Forster of the State University of New York at Stony

Brook and her colleagues, who discovered *Rahonavis*, claimed its relationships to *Archaeopteryx*; but all real similarities may be primitive for birds and thus may not indicate a close cladistic (genealogical) relationship. Whether or not a relative of *Archaeopteryx*, *Rahonavis* was a late survivor of the early stage of avian evolution.

All the remaining true birds, called Ornithurae (bird-tailed), have the tail much shortened, and its terminal vertebrae fuse into a pygostyle, which support the fan of tail feathers. The modern avian tail helps steer in flight and break before landing. Another major ornithuran invention is a slender strutlike coracoid, a correspondingly deep thorax, and a well-developed sternal plate (much larger than in *Archaeopteryx*), all of which indicates larger flight muscles with longer fibers. The most primitive ornithuran is *Confuciusornis*, a crow-sized bird found in thousands of specimens in the now famous Early Cretaceous Yixian Formation in the province of Liaoning in northeastern China. Despite all improvements in the wing architecture and the presence of the pygostyle (which is remarkably big in all primitive ornithurans), *Confuciusornis* retained a functional grasping hand with powerful claws, as in *Archaeopteryx*. This suggests a similar lifestyle—that is, primarily terrestrial feeding—the use of hands for climbing, and the inability of take off from the ground. Alan Feduccia and a few others believe that *Confuciusornis* was arboreal, but there is little evidence for this belief. However, the Pouncing Proavis Model is somewhat more likely to be true for *Confuciusornis* with its more advanced flight than it is for *Archaeopteryx*. Next to the oviraptorosaurs, which are either flightless birds or close to avian ancestry, *Confuciusornis* represents the first avian lineage that lost teeth and replaced them completely with a horny beak. As are many other Liaoning dinosaurs, *Confuciusornis* specimens are preserved with feathers, which reveal a striking sexual dimorphism, as some of them have two elongate tail feathers. These have been conjectured to be males, but the other alternative is at least equally likely in view of what can be inferred about the reproduction of an enantiornithine bird, *Gobipteryx*.

The most significant remodeling of the flight apparatus in avian evolution was underway after the *Confuciusornis* lineage branched off, marking the advent of enantiornithines, the "opposite birds," and euornithines, the "proper birds," which include all modern birds. The two groups share an essentially modern wing architecture designed exclusively for flying. This adaptation indicates a radical improvement in flight maneuverability and, thus, for the first time, an unquestionable ability to take off from the ground. Most probably correlated with flight improvements was the first appearance of sparrow-sized birds in both the enantiornithines and euornithines, marking the second dramatic reduction in size in the history of birds and their flight (the first accompanied the origin of flight). A specialized perching foot and arboreal habits developed in each group.

Paradoxically, the loss of flight is one of the most frequent and predictable motifs of avian evolution with some maniraptors. The costly flight apparatus undergoes reduction in terrestrial or waterbirds that do not depend on flight for feeding whenever they manage to escape from predator pressure (either by the increase of body size or geographic isolation). This was probably happening from the earliest stages of avian evolution with some maniraptors like *Caudipteryx* (whether they branched off after or before *Archaeopteryx*), which are first candidates for the distinction of the oldest flightless birds. But, for now, the first unquestionable flightless bird is *Patagopteryx* from the Late Cretaceous of Patagonia, South America. *Patagopteryx* is the size of a domestic fowl, has wings too short to fly, and shows several peculiarities of the skeleton. Its origins are probably close to the split between the enantiornithines and euornithines.

The oldest known representative of the euornithines is *Liaoningornis*, a sparrow-sized bird from the Yixian Formation of Liaoning, China, the same formation that yielded the much more primitive *Confuciusornis*. The euornithines are diagnosed on such details as the coracoid with a well-defined socket for the scapula and a process (called procoracoid) that supports it, and the sternum with well-developed grooves for the coracoids and a deep keel for the attachment of flight muscles. If it weren't for bulges on the ulna of *Rahonavis*, the presence of quill knobs would be limited to the euornithines, indicating an improvement in the attachment and control of the secondary flight feathers. Most diagnostic details of the Euornithes reveal further improvements of the flight and respiratory mechanics, which depend on the controlled mobility of the chest and shoulder girdle elements.

Very little is known about the radiation of primitive euornithines except for a few predatory waterbirds that lived on fish and squid. Best known among them are the genera *Ichthyornis*, which includes good fliers comparable to the terns, and *Hesperornis*, with its possible relatives, which were flightless, foot-propelled divers, ecologically comparable to today's loons, grebes, and penguins. *Ichthyornis* and *Hesperornis* have been known as textbook examples of Cretaceous birds for over a century, but their relationships to each other, as well as to other birds, remain enigmatic. The postcranial skeleton of *Hesperornis* is too transformed to preserve homologous details shared with the known birds, and the skull of *Ichthyornis* is too fragmentary for conclusive comparisons with the relatively well-known skull of *Hesperornis*. However, the jaws of the two genera show striking similarities, which may reveal their close relationship despite dramatic differences in their locomotor systems.

The presence of teeth in *Ichthyornis* and *Hesperornis* demonstrates that a toothless beak evolved in birds at least twice, once in the enatiornithines (as in *Gobipteryx*) and once in the advanced euornithines. Toothlessness correlated with extensive fusions of jaw bones (especially between the premaxilla and maxilla) is characteristic of modern birds called the neornithines (Neornithes). The resulting great strength of the neornithine beak culminated in its use as a chisel used for wood in the woodpeckers and for hard seeds in many other birds.

Most avian groups met their demise with other dinosaurs. Only a handful of neornithine lineages are known to have crossed the Cretaceous/Tertiary (K/T), but the list is growing. Because of the fragmentary record of Cretaceous neornithines, their identification is one of the most contentious areas in today's avian paleontology.

Feathered Dinosaurs

by Philip J. Currie

Over the last 30 years, we have seen the development of many new ideas concerning these fantastic animals. Creative thinking by paleontologists in turn has led to many new approaches to studying the past. Like detectives trying to solve crimes of long ago, paleontologists have sifted through the clues and applied increasingly sophisticated technologies in the analysis. Nothing is better for a paleontologist than solving some ancient mystery, especially if you learn that your hunches were correct all along. Few paleontological solutions have been as satisfying as the discovery of feathered dinosaurs.

With the development and spread in the 1970s of the idea that dinosaurs might have been warm-blooded animals that were the direct ancestors of birds, paleontologists and paleoartists started to flirt with feathers on dinosaur reconstructions. After all, they reasoned, if dinosaurs were warm-blooded, then the smaller ones would have needed some kind of insulation to help stabilize their body temperatures. Furthermore, if they were ancestral to birds, it would make sense that the first feathers would have appeared on dinosaurs as a form of insulation. Feathers could not have developed in birds at the moment that they sensed a need to fly. The feathers had to have been on the ancestors first, then adapted into a flight mechanism. Although we normally think of birds as being any animals covered with feathers, their form of powered flight was actually the novelty that set birds apart from all other animals.

Warm-blooded dinosaurs and the dinosaurian origin of birds were two of the most hotly argued controversies at the end of the twentieth century. When first proposed, there were far more people opposed to these hypotheses than there were in support of it. Now things have changed, largely because of some remarkable discoveries made in northeastern China.

When I left the Gobi Desert of Mongolia in September 1996, I went to Beijing to spend a few days working with my long-time colleague Dong Zhiming on some dinosaurs we had collected in northwestern China. Shortly after arriving, he showed me a Chinese newspaper report of a beautiful little dinosaur skeleton found in Liaoning, a province northeast of the Chinese capital city. The report suggested that the specimen was actually a bird because it was covered with feathers. I expressed interest in seeing the specimen, mostly because small meat-eating dinosaurs are extremely rare discoveries.

Although I was predisposed to the idea of dinosaurs having feathers, the chances of finding a specimen with them preserved are so poor that I discounted the newspaper report completely—probably just dendrites or an ancient fungal growth, I thought.

Dong made arrangements to see the specimen a few days later at the National Geological Museum of China. Surprisingly, when we arrived and met the museum director, Ji Qiang, we found ourselves surrounded by members of the Chinese and Japanese press, who were there to record our reactions! I have no idea if I gave them the reaction they wanted, but I experienced and expressed some of the most amazing feelings of my professional career. The specimen was beautiful, exposed from the tip of the nose to the tip of the tail on a small slab of rock. But that is not what caught my attention: it was the rim of structures that surrounded almost the entire body. They were real and they belonged! I knew instantly that the first "feathered" dinosaur had indeed been discovered.

The next few months were tumultuous for me. The news spread around the world, slowly at first but with ever-increasing pace. The specimen, as it turned out, had been split down the middle into left and right halves, and the other half had gone to another museum in Nanjing. I went back to China to see this specimen, and I was able to see a second skeleton of a larger individual of the same animal. When it had been found, the discoverer had actually chiseled through many of the "feathers" because he hadn't expected them to be there! The first scientific paper, written in Chinese by Ji Qiang and Ji Shuan named this animal *Sinosauropteryx prima* ("first Chinese dragon feather"). Controversy erupted about whether these really were feathers on *Sinosauropteryx*, which included "scientific" papers written by people who had never even seen the specimens. A third specimen of *Sinosauropteryx* was found, and a delegation of American and German scientists went to see what the fuss was about. By now, controversies on the ancestry of birds and warm-bloodedness in dinosaurs erupted on a scale that surpassed even public and scientific interest in dinosaur extinction!

Protarchaeopteryx was described in 1997, again by Ji and Ji. Whereas *Sinosauropteryx* was a cat-sized animal with short arms and an extremely long tail, *Protarchaeopteryx* had long arms and a relatively short tail. In addition to having downy, featherlike structures covering its body like *Sinosauropteryx*, *Protarchaeopteryx* had long quill-like feathers on the end of the tail. This time there could be no doubt concerning the identification as feathers. I went back to China, and this time I went up to the locality in Liaoning that was producing all of these wonderful fossils. In the meantime, two more feathered dinosaurs were discovered. We thought at first these were also *Protarchaeopteryx*, but they were different by having long feathers behind the arms as well. While one of the new fossils was being prepared, we realized that it represented a third species of feathered dinosaur. This one we called *Caudipteryx*, which means "tail

Caudipteryx, as reconstructed by Gregory Paul is extraordinarily bird-like.

feather." Related to the Mongolian *Oviraptor*, *Caudipteryx* was a turkey-sized animal with long legs that suggest it was a good runner. The feathers behind its arms look like a rudimentary wing, but they and the arms themselves are much too short to have provided any lift. It is more likely that they, and the long feathers on the end of the tail, were used for display.

Dinosaurs were highly visual animals that evolved a fantastic array of ornamentation (crests, frills, horns, spikes, etc.) to attract mates, warn potential rivals, and otherwise enhance their interactive behavior. Once dinosaurs had acquired feathers for insulation, what could be more natural than to adapt them into display structures: they are lightweight, strong, colorful, and can be shed and replaced. Of course, display may not have been the only function for these longer, stiffened feathers. Specimens of the related *Ovi-*

Caudipteryx, as restored to life by artist Joe Tucciarone with its feathery outer covering.

raptor have been found on nests of eggs, and their positions suggest that long feathers on the backs of the arms might have helped protect and warm the eggs.

Two more feathered dinosaurs were described in the final year of the twentieth century. *Beipiaosaurus* is a much larger dinosaur than the other feathered forms. It has long, stiff, featherlike structures on the backs of its arms. Approximately the same size as a large man, *Beipiaosaurus* probably had a relatively small head with leaflike teeth, a long neck, long arms, and a relatively short tail. The other feathered dinosaur is *Sinornithosaurus millenii* ("Chinese bird-reptile of the millennium"). This dog-sized animal had sharp serrated teeth and raptorial claws, which is not surprising considering it is closely related to *Velociraptor*, an animal that has become famous for its ferocity thanks to the movie *Jurassic Park*.

There are now five species of "feathered" dinosaurs from northeastern China. More will be described within coming years because the locality is extremely rich and is being excavated at an unprecedented scale. The five species are all theropods, or meat-eating dinosaurs, but represent five different lineages that are as different from one another as cats, dogs, bears, and weasels are among modern mammalian carnivores. *Sinosauropteryx* is a compsognathid theropod, closely related to the European *Compsognathus*, which was known as the smallest dinosaur for more than a century. *Protarchaeopteryx* is still only known from a single specimen, and it might be the theropod most closely related to what is generally thought of as the earliest bird, *Archaeopteryx*. *Caudipteryx* is an ovi-raptorosaur, while *Beipiaosaurus* is a therizinosaur. Finally, *Sinornithosaurus* is a dromaeosaurid theropod. The fact that these dinosaurs represent such a diverse assemblage of theropods strongly suggests that many, if not most, of the meat-eating dinosaurs were probably feathered. Consider this for example: *Tyrannosaurus* is on the same branch of the family tree of dinosaurs as all the feathered forms. If the more primitive *Sinosauropteryx* has some form of featherlike structures, and all of its closest relatives had feathers, is it possible that even the mighty *Tyrannosaurus rex* had feathers somewhere on its body at some stage in its life? Perhaps it did, although we can be pretty sure that such a large animal would not have needed them for insulation as an adult. Pebbly skin impressions are preserved for the related tyrannosaurs *Gorgosaurus* and *Daspletosaurus*, and there are no indications of feathers; still, it is not impossible that the newborn chicks might have had some sort of downy insulation, or that the adults used feathers as attractive crests or fans for display.

The presence of feathers on dinosaurs does not prove that birds came from dinosaurs. There is much stronger evidence in the skeleton to suggest that birds and dinosaurs are more closely related to each other than either is to any other type of animal. However, feathers are such complex structures that the discovery of feathered dinosaurs has done far more to convince people that birds are living representatives of the

Artist Luis Rey's restoration of the *Sinosauropteryx* in death, with a vignette of it in life.

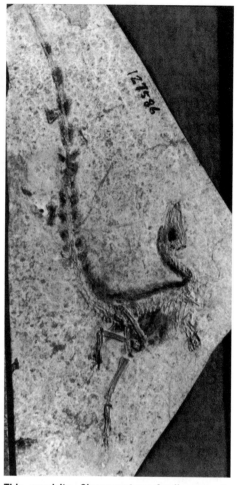

This exquisite *Sinosauropteryx* fossil seems to show the diminutive dinosaur was covered with hair or some kind of proto-feathers.

Dinosauria than the obscure details of wrist and ankle anatomy have. Still, not everyone is convinced, and some of these people are more vocal than the majority of paleontologists who accept that birds descended from dinosaurs. The detractors' arguments are plagued by lack of a convincing alternative for bird ancestry and by a certain amount of circular reasoning. For example, they have argued long and hard that the structure of the ankle is very different in theropod dinosaurs and birds. Now they claim that *Caudipteryx* is a bird because it has feathers. Yet *Caudipteryx* has the same ankle structure as theropod dinosaurs such as *Velociraptor* and *Tyrannosaurus*. Clearly, one of their two arguments has to be wrong because they are contradictory.

Ultimately, it does not matter whether we classify *Caudipteryx* and the others as dinosaurs or birds. Classifications are human concepts that help us organize nature so we can better understand it. Most scientists agree *Archaeopteryx* is the dividing line

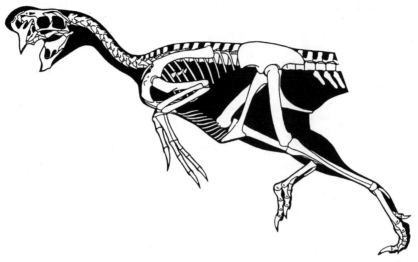

Gregory Paul's skeletal reconstruction of *Oviraptor* shows this dinosaur, too, had many bird-like features.

between dinosaurs and birds. Related animals more derived or advanced than *Archaeopteryx* are birds. But species that are more primitive are not. Using this concept, birds are animals that either fly or, in the case of penguins and ostriches, have direct ancestors that flew. Feathers separate birds from flying insects, flying reptiles, and bats, but they are not what make a bird unique. We could redefine birds as feathered animals, for example. But if we did, we would have to reclassify all feathered dinosaurs as birds. And to do that, we would also have to classify all of their direct descendants as birds. To most people, classifying *Tyrannosaurus rex* as a bird is not as logical as emphasizing flight, rather than feathers, in the definition of what a bird is. The fact that we are having these arguments and that we are having trouble classifying many of the new fossils we are discovering effectively shows how close dinosaurs and birds are to each other. As we draw toward consensus on the ancestry of birds, attention is shifting to equally interesting problems—the evolution of feathers and the origin of flight.

Rey's vision of an *Oviraptor* nesting ground has a distinctly avian feel to it, right down to the feathers and the brooding mother on the clutch of eggs.

The Origin of Birds and Their Flight

by Kevin Padian and Luis M. Chiappe

Until recently, the origin of birds was one of the great mysteries of biology. Birds are dramatically different from all other living creatures. Feathers, toothless beaks, hollow bones, perching feet, wishbones, deep breastbones, and stumplike tailbones are only part of the combination of skeletal features that no other living animal has in common with them. How birds evolved feathers and flight was even more imponderable.

In the past 20 years, however, new fossil discoveries and new research methods have enabled paleontologists to determine that birds descend from ground-dwelling, meat-eating dinosaurs of the group known as theropods. The work has also offered a picture of how the earliest birds took to the air.

Scientists have speculated on the evolutionary history of birds since shortly after Charles Darwin set out his theory of evolution in *On the Origin of Species*. In 1860, the year after the publication of Darwin's treatise, a solitary feather of a bird was found in Bavarian limestone deposits dating to about 150 million years ago (just before the Jurassic period gave way to the Cretaceous). The next year, a skeleton of an animal that had birdlike wings and feathers—but a very unbirdlike long, bony tail and toothed jaw—turned up in the same region. These finds became the first two specimens of the blue-jay—size *Archaeopteryx lithographica*, the most archaic, or basal, known member of the birds (see "*Archaeopteryx*," by Peter Wellnhofer; *Scientific American*, May 1990).

Archaeopteryx's skeletal anatomy provides clear evidence that birds descend from a dinosaurian ancestor, but in 1861, scientists were not yet in a position to make that connection. A few years later, though, Thomas Henry Huxley, Darwin's staunch defender, became the first person to connect birds to dinosaurs. Comparing the hindlimbs of *Megalosaurus*, a giant theropod, with those of the ostrich, he noted 35 features that the two groups shared but that did not occur as a suite in any other animal. He concluded that birds and theropods could be closely related, although whether he thought birds were cousins of theropods or were descended from them is not known.

Huxley presented his results to the Geological Society of London in 1870, but paleontologist Harry Govier Seeley contested Huxley's assertion of kinship between theropods and birds. Seeley suggested that the hindlimbs of the ostrich and *Megalosaurus* might look similar just because both animals were large and bipedal and used their hindlimbs in similar ways. Besides, dinosaurs were even larger than ostriches, and none of them could fly; how, then, could flying birds have evolved from a dinosaur?

The mystery of the origin of birds gained renewed attention about half a century later. In 1916, Gerhard Heilmann, a Danish medical doctor with a penchant for paleontology, published a brilliant book that, in 1926, was translated into English as *The Origin of Birds*. Heilmann showed that birds were anatomically more similar to theropod dinosaurs than to any other fossil group but for one inescapable discrepancy: theropods apparently lacked clavicles, the two collarbones that are fused into a wishbone in birds. Because other reptiles had clavicles, Heilmann inferred that theropods had lost them. To him, this loss meant birds could not have evolved from theropods, because he was convinced (mistakenly, as it turns out) that a feature lost during evolution could not be regained. Birds, he asserted, must have evolved from a more archaic reptilian group that had clavicles. Like Seeley before him, Heilmann concluded that the similarities between birds and dinosaurs must simply reflect the fact that both groups were bipedal.

Heilmann's conclusions influenced thinking for a long time, even though new information told a different story. Two separate findings indicated that theropods did, in fact, have clavicles. In 1924, a published anatomical drawing of the bizarre, parrot-headed theropod *Oviraptor* clearly showed a wishbone, but the structure was misidentified. Then, in 1936, Charles Camp of the University of California at Berkeley found the remains of a small Early Jurassic theropod, complete with clavicles, Heilmann's fatal objection had been overcome, although few scientists recognized it. Recent studies have found clavicles in a broad spectrum of the theropods related to birds.

Finally, a century after Huxley's disputed presentation to the Geological Society of London, John H. Ostrom of Yale University revived the idea that birds were related to theropod dinosaurs, and he proposed explicitly that birds were their direct descendants. In the late 1960s, Ostrom had described the skeletal anatomy of the theropod *Deinonychus*, a vicious, sickle-clawed predator about the size of an adolescent human, which roamed in Montana some 115 million years ago (in the Early Cretaceous). In a series of papers published during the next decade, Ostrom went on to identify a collection of features that birds, including *Archaeopteryx*, shared with *Deinonychus* and other theropods but not with other reptiles. On the basis of these findings, he concluded that birds are descended directly from small theropod dinosaurs.

As Ostrom was assembling his evidence for the theropod origin of birds, a new method of deciphering the relations among organisms was taking hold in natural history museums in New York City, Paris, and elsewhere. This method—called phylogenetic systematics, or, more commonly, cladistics—has since become the standard for comparative biology, and its use has strongly validated Ostrom's conclusions.

Traditional methods for grouping organisms look at the similarities and differences among the animals and might exclude a species from a group solely because the species has a trait not found in other members of the group. In contrast, cladistics groups organisms based exclusively on certain kinds of shared traits that are particularly informative.

This method begins with the Darwinian precept that evolution proceeds when a new heritable trait emerges in some organism and is passed genetically to its descendants. The precept indicates that two groups of animals sharing a set of such new, or "derived," traits are more closely related to each other than they are to groups that display only the original traits, not the derived ones. By identifying shared derived traits, practitioners of cladistics can determine the relations among the organisms they study.

The results of such analyses, which generally examine many traits, can be represented in the form of a cladogram, a treelike diagram depicting the order in which new characteristics, and new creatures, evolved. Each branching point, or node, reflects the emergence of an ancestor that founded a group having derived characteristics not present in groups that evolved earlier. This ancestor and all its descendants constitute a "clade," or closely related group.

Ostrom did not apply cladistic methods to determine that birds evolved from small theropod dinosaurs; in the 1970s, the approach was just coming into use. But about a decade later, Jacques A. Gauthier, then at the University of California at Berkeley, did an extensive cladistic analysis of birds, dinosaurs, and their reptilian relatives. Gauthier put Ostrom's comparisons and many other features into a cladistic framework and confirmed that birds evolved from small theropod dinosaurs. Indeed, some of the closest relatives of birds include the sickle-clawed maniraptoran *Deinonychus* that Ostrom had so vividly described.

Today, a cladogram for the lineage leading from theropods to birds shows that the clade labeled Aves (birds) consists of the ancestor of *Archaeopteryx* and all other descendants of that ancestor. This clade is a subgroup of a broader clade consisting of so-called maniraptoran theropods—itself a subgroup of the tetanuran theropods that descended from the most basal theropods. Those archaic theropods in turn evolved from nontheropod dinosaurs. The cladogram shows that birds are not only *descended* from dinosaurs, they *are* dinosaurs (and reptiles)—just as humans are mammals, even though people are as different from other mammals as birds are from other reptiles.

Early Evolutionary Steps to Birds

Gauthier's studies and ones conducted more recently demonstrate that many features traditionally considered "birdlike" actually appeared before the advent of birds, in their preavian theropod ancestors. Many of those properties undoubtedly helped their original possessors to survive as terrestrial dinosaurs; these same traits and others were eventually used directly or were transformed to support flight and an arboreal way of life. The short length of this essay does not allow us to catalogue the many dozens of details that combine to support the hypothesis that birds evolved from small theropod dinosaurs, so we will concentrate mainly on those related to the origin of flight.

The birdlike characteristics of the theropods that evolved prior to birds did not appear all at once, and some were present before the theropods themselves emerged—in the earliest dinosaurs. For instance, the immediate reptilian ancestor of dinosaurs was already bipedal and upright in its stance (that is, it basically walked like a bird), and it was small and carnivorous. Its hands, in common with those of early birds, were free for grasping (although the hand still had five digits, not the three found in all but the most basal theropods and in birds). Also, the second finger was longest—not the third, as in other reptiles.

Further, in the ancestors of dinosaurs, the ankle joint had already become hingelike, and the metatarsals, or footbones, had become elongated. The metatarsals were held off the ground, so the immediate relatives of dinosaurs, and dinosaurs themselves, walked on their toes and put one foot in front of the other, instead of sprawling. Many of the changes in the feet are thought to have increased stride length and running speed, a property that would one day help avian theropods to fly.

The earliest theropods had hollow bones and cavities in the skull; these adjustments lightened the skeleton. They also had a long neck and held their back horizontally, as birds do today. In the hand, digits 4 and 5 (the equivalent of the pinky and its neighbor) were already reduced in the first dinosaurs; the fifth finger was virtually gone. Soon it was completely lost, and the fourth was reduced to a nubbin. Those reduced fingers disappeared altogether in tetanuran theropods, and the remaining three (I, II, III) became fused together sometime after *Archaeopteryx* evolved.

In the first theropods, the hindlimbs became more birdlike as well. They were long; the thigh was shorter than the shin, and the fibula, the bone to the side of the shinbone, was reduced. (In birds today, the toothpicklike bone in the drumstick is all that is left of the fibula.) These dinosaurs walked on the three middle toes—the same ones modern birds use. The fifth toe was shortened and tapered, with no joints, and the first toe included a shortened metatarsal (with a small joint and a claw) that projected from the side of the second toe. The first toe was held higher than the others and had no apparent function, but it was later put to good use in birds. By the time *Archaeopteryx* appeared, that toe had rotated to lie behind the others. In later birds, it descended to become opposable to the others and eventually formed an important part of the perching foot.

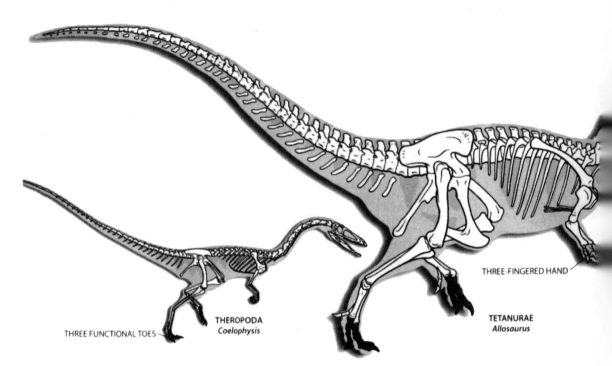

THREE FUNCTIONAL TOES

THEROPODA
Coelophysis

THREE-FINGERED HAND

TETANURAE
Allosaurus

Representative theropods in the lineage leading to birds (*Aves)* display some of the features that helped investigators establish the dinosaurian origin of birds—including in the order of their evolution, three functional toes, a three-fingered hand and a half-moon-shaped wrist bone. *Archaeopteryx,* the oldest known bird, also shows some

More Changes

Through the course of theropod evolution, more features once thought of as strictly avian emerged. For instance, major changes occurred in the forelimb and shoulder girdle; these adjustments at first helped theropods to capture prey and later promoted flight. Notably, during theropod evolution, the arms became progressively longer, except in such giant carnivores as *Carnotaurus, Allosaurus,* and *Tyrannosaurus,* in which the forelimbs were relatively small. The forelimb was about half the length of the hindlimb in very early theropods. By the time *Archaeopteryx* appeared, the forelimb was longer than the hindlimb, and it grew still more in later birds. This lengthening in the birds allowed a stronger flight stroke.

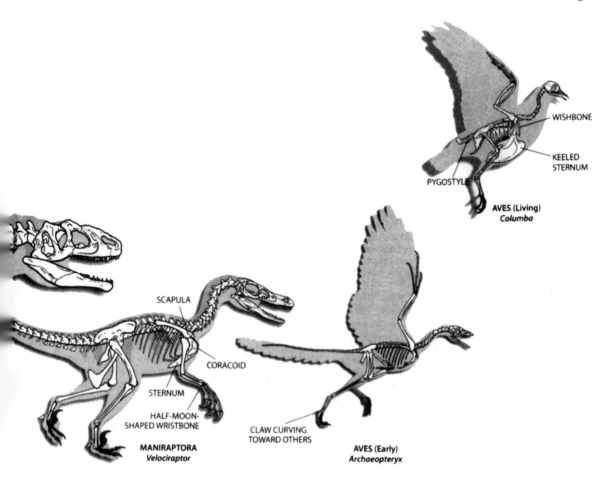

WISHBONE

KEELED STERNUM

PYGOSTYLE

AVES (Living)
Columba

SCAPULA

CORACOID

STERNUM

HALF-MOON-
SHAPED WRISTBONE

CLAW CURVING
TOWARD OTHERS

MANIRAPTORA
Velociraptor

AVES (Early)
Archaeopteryx

new traits, such as a claw on the back toes that curves toward the claws on the other toes. As later birds evolved, many features underwent change. Notably, the fingers fused together, the simple tail became a pygostyle composed of fused vertebrae, and the back toe dropped enabling birds' feet to grasp tree limbs firmly.

The hand became longer, too, accounting for a progressively greater proportion of the forelimb, and the wrist underwent dramatic revision in shape. Basal theropods possessed a flat wristbone (distal carpal) that overlapped the bases of the first and second palm bones (metacarpals) and fingers. In maniraptorans, though, this bone assumed a half-moon shape along the surface that contacted the armbones. The half-moon, or semilunate, shape was very important because it allowed these animals to flex the wrist sideways in addition to up and down. They could thus fold the long hand, almost as living birds do. The longer hand could then be rotated and whipped forward suddenly to snatch prey.

In the shoulder girdle of early theropods, the scapula (shoulder blade) was long and straplike; the coracoid (which along with the scapula forms the shoulder joint) was rounded, and two separate, five-shaped clavicles connected the shoulder to the sternum, or breastbone. The scapula soon became longer and narrower; the coracoid also thinned and elongated, stretching toward the breastbone. The clavicles fused at the midline and broadened to form a boomerang-shaped wishbone. The sternum, which consisted originally of cartilage, calcified into two fused bony plates in tetanurans. Together, these changes strengthened the skeleton; later this strengthening was used to reinforce the flight apparatus and support the flight muscles. The new wishbone, for instance, probably became an anchor for the muscles that moved the forelimbs, at first during foraging and then during flight.

In the pelvis, more vertebrae were added to the hip girdle, and the pubic bone (the pelvic bone that is attached in front of and below the hip socket) changed its orientation. In the first theropods, as in most other reptiles, the pubis pointed down and forward, but then it began to point straight down or backward. Ultimately, in birds more advanced than *Archaeopteryx*, it became parallel to the ischium, the pelvic bone that extends backward from below the hip socket. The benefits derived from these changes, if any, remain unknown, but the fact that these features are unique to birds and other maniraptorans shows their common origin.

Finally, the tail gradually became shorter and stiffer throughout theropod history, serving more and more as a balancing organ during running, somewhat as it does in today's roadrunners. Steven M. Gatesy of Brown University has demonstrated that this transition in tail structure paralleled another change in function: the tail became less and less an anchor for the leg muscles. The pelvis took over that function, and in maniraptorans, the muscle that once drew back the leg now mainly controlled the tail. In birds that followed *Archaeopteryx*, these muscles would be used to adjust the feathered tail as needed in flight.

In summary, a great many skeletal features that were once thought of as uniquely avian innovations—such as light, hollow bones, long arms, three-fingered hands with a long second finger, a wishbone, a backward-pointing pelvis, and long hindlimbs with a three-toed foot—were already present in theropods before the evolution of birds. Those features generally served different uses than they did in birds and were only later co-opted for flight and other characteristically avian functions, eventually including life in the trees.

Evidence for the dinosaurian origin of birds is not confined to the skeleton. Recent discoveries of nesting sites in Mongolia and Montana reveal that some reproductive behaviors of birds originated in nonavian dinosaurs. These theropods did not deposit a large clutch of eggs all at once, as most other reptiles do. Instead, they filled a nest more gradually, laying one or two eggs at a time, perhaps over several days, as birds do. Recently, skeletons of the Cretaceous theropod *Oviraptor* have been found atop nests of eggs; the dinosaurs were apparently buried while protecting the eggs in very birdlike fashion. This find is ironic because *Oviraptor*, whose name means "egg stealer," was first thought to have been raiding the eggs of other dinosaurs rather than protecting them. Even the structure of the eggshell in theropods shows features otherwise seen only in bird eggs. The shells consist of two layers of calcite, one prismatic (crystalline) and one spongy (more irregular and porous).

Ovipator, a maniraptoran theropod that evolved before birds, sat in its nest to protect its eggs, just as the ostrich and other birds do today. In other words, such brooding originated before birds did.

As one supposedly uniquely avian trait after another has been identified in nonavian dinosaurs, feathers have continued to stand out as a prominent feature belonging to birds alone. Some intriguing evidence, however, hints that even feathers might have predated the emergence of birds. In 1996 and 1997, Ji Qiang and Ji Shu'an of the National Geological Museum of China published reports on two fossil animals found in Liaoning Province that date to late in the Jurassic or early in the Cretaceous. One, a turkey-sized dinosaur named *Sinosauropteryx*, has fringed, filamentous structures along its backbone and on its body surface. These structures of the skin, or integument, may have been precursors to feathers. But the animal is far from a bird. It has short arms and other skeletal properties, indicating that it may be related to the theropod *Compsognathus*, which is not especially close to birds or other maniraptorans.

The second creature, *Protarchaeopteryx*, apparently has short, true feathers on its body and longer feathers attached to its tail. Preliminary observations suggest that the animal is a maniraptoran theropod. Whether it is also a bird will depend on a fuller description of its anatomy. Nevertheless, the Chinese finds imply that, at the least, the structures that gave rise to feathers probably appeared before birds did and almost certainly before birds began to fly. Whether their original function was for insulation, display, or something else cannot yet be determined.

The Beginning of Bird Flight

The origin of birds and the origin of flight are two distinct, albeit related, problems. Feathers were present for other functions before flight evolved, and *Archaeopteryx* was probably not the very first flying theropod, although at present we have no fossils of earlier flying precursors. What can we say about how flight began in bird ancestors? Traditionally, two opposing scenarios have been put forward. The "arboreal" hypothesis holds that bird ancestors began to fly by climbing trees and gliding down from branches with the help of incipient feathers. The height of trees provides a good starting place for

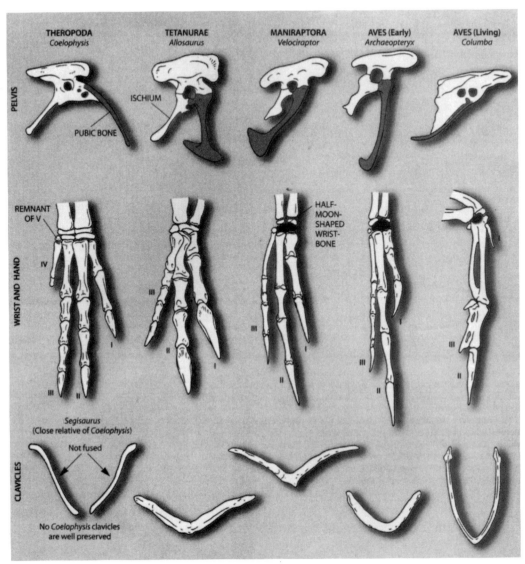

Comparisons of anatomical structures not only helped to link birds to therapods, they also revealed some of the ways those features changed as dinosaurs became more bird-like and birds became more modern. In the pelvis (side view), the pelvic bone initially pointed forward (toward the right), but later it shifted to be vertical or pointed backward. In the hand (top view), the relative proportions of the bones remained quite constant through the early birds, but the wrist changed. In the maniraptoran wrist, a disk-like bone took on the half-moon shape that ultimately promoted flapping flight in birds. The wide, boomerang-shaped wishbone (fused clavicles) in tetanurans and later groups compares well with that of archaic birds, but it became thinner and formed a deeper U shape as it became more critical in flight.

launching flight, especially through gliding. As feathers became larger over time, flapping flight evolved, and birds finally became fully airborne.

This hypothesis makes intuitive sense, but certain aspects are troubling. *Archaeopteryx* and its maniraptoran cousins have no obviously arboreal adaptations, such as feet fully adapted for perching. Perhaps some of them could climb trees, but no convincing analysis has demonstrated how *Archaeopteryx* would have climbed and flown with its forelimbs, and there were no plants taller than a few feet in the environments where *Archaeopteryx* fossils have been found. Even if the animals could climb trees, this ability is not synonymous with arboreal habits or gliding ability. Most small animals, and even some goats and kangaroos, can climb trees, but that does not make them tree dwellers. Besides, *Archaeopteryx* shows no obvious features of gliders, such as a broad membrane connecting forelimbs and hindlimbs.

The "cursorial" (running) hypothesis holds that small dinosaurs ran along the ground and stretched out their arms for balance as they leaped into the air after insect prey or, perhaps, to avoid predators. Even rudimentary feathers on forelimbs could have expanded the arm's surface area to enhance lift slightly. Larger feathers could have increased lift incrementally until sustained flight was gradually achieved. Of course, a leap into the air does not provide the acceleration produced by dropping out of a tree; an animal would have to run quite fast to take off. Still, some small terrestrial animals can achieve high speeds.

The cursorial hypothesis is strengthened by the fact that the immediate theropod ancestors of birds were terrestrial. And they had the traits needed for high liftoff speeds: they were small, active, agile, lightly built, long-legged, and good runners. And because they were bipedal, their arms were free to evolve flapping flight, which cannot be said for other reptiles of their time.

Although our limited evidence is tantalizing, probably neither the arboreal nor the cursorial model is correct in its extreme form. More likely, the ancestors of birds used a combination of taking off from the ground and taking advantage of accessible heights (such as hills, large boulders, or fallen trees). They may not have climbed trees, but they could have used every available object in their landscape to assist flight.

More central than the question of ground versus trees, however, is the evolution of a flight stroke. This stroke generates not only the lift that gliding animals obtain from moving their wings through the air (as an airfoil) but also the thrust that enables a flapping animal to move forward. (In contrast, the "organs" of lift and thrust in airplanes—the wings and jets—are separate.) In birds and bats, the hand part of the wing generates the thrust, and the rest of the wing provides the lift.

Jeremy M. V. Rayner of the University of Bristol showed in the late 1970s that the down-and-forward flight stroke of birds and bats produces a series of doughnut-shaped vortices that propel the flying animal forward. Padian and Gauthier then demonstrated in the mid-1980s that the movement generating these vortices in birds is the same action—sideways flexion of the hand—that was already present in the maniraptorans *Deinonychus* and *Velociraptor* and in *Archaeopteryx*.

As we noted earlier, the first maniraptorans must have used this movement to grab prey. By the time *Archaeopteryx* and other birds appeared, the shoulder joint had changed its angle to point more to the side than down and backward. This alteration in the angle transformed the forelimb motion from a prey-catching one to a flight stroke. New evidence from Argentina suggests that the shoulder girdle in the closest maniraptorans to birds (the new dinosaur *Unenlagia*) was already angled outward so as to permit this kind of stroke.

Recent work by Farish A. Jenkins, Jr. of Harvard University, George E. Goslow of Brown University, and their colleagues has revealed much about the role of the wishbone in flight and about how the flight stroke is achieved. The wishbone in some living birds acts as a spacer between the shoulder girdles, one that stores energy expended during the flight stroke. In the first birds, in contrast, it probably was less elastic, and its main function may have been simply to anchor the forelimb muscles. Apparently, too, the muscle most responsible for rotating and raising the wing during the recovery stroke of flight was not yet in the modern position in *Archaeopteryx* or other very early birds. Hence, those birds were probably not particularly skilled fliers; they would have been unable to flap as quickly or as precisely as today's birds can. But it was not long—perhaps just several million years—before birds acquired the apparatus they needed for more controlled flight.

Beyond Archaeopteryx

More than three times as many bird fossils from the Cretaceous period have been found since 1990 than in all the rest of recorded history. These new specimens—uncovered in such places as Spain, China, Mongolia, Madagascar, and Argentina—are helping paleontologists to flesh out the early evolution of the birds that followed *Archaeopteryx*, including their acquisition of an improved flying system. Analyses of these finds by Chiappe and others have shown that birds quickly took on many different sizes, shapes, and behaviors (ranging from diving to flightlessness) and diversified all through the Cretaceous period, which ended about 65 million years ago.

A bird-watching trek through an Early Cretaceous forest would bear little resemblance to such an outing now. These early birds might have spent much of their time in the trees

Birds living a million years ago looked a bit different from those of today. *Sinornis* **was a sparrow-sized bird that still had a mouth full of teeth and had not yet lost its wing claws.**

and were able to perch, but there is no evidence that the first birds nested in trees, had complex songs, or migrated great distances. Nor did they fledge at nearly adult size, as birds do now, or grow as rapidly as today's birds do. Scientists can only imagine what these animals looked like. Undoubtedly, however, they would have seemed very strange with their clawed fingers and, in many cases, toothed beaks.

Underneath the skin, though, some skeletal features certainly became more birdlike during the Early Cretaceous and enabled birds to fly quite well. Many bones in the hand and in the hip girdle fused, providing strength to the skeleton for flight. The breastbone became broader and developed a keel down the midline of the chest for flight muscle attachment. The forearm became much longer, and the skull bones and vertebrae became lighter and more hollowed out. The tailbones became a short series of free segments ending in a fused stump (the familiar "parson's nose" or "pope's nose" of roasted birds) that controlled the tail feathers. And the alula, or "thumb wing," a part of the bird wing essential for flight control at low speed, made its debut, as did a long first toe useful in perching.

Inasmuch as early birds could fly, they certainly had higher metabolic rates than cold-blooded reptiles; at least they were able to generate the heat and energy needed for flying without having to depend on being heated by the environment. But they might not have been as fully warm-blooded as today's birds. Their feathers, in addition to aiding flight, provided a measure of insulation—just as the precursors of feathers could have helped preserve heat and conserve energy in nonavian precursors of birds. These birds probably did not fly as far or as strongly as birds do now.

Bird-watchers traipsing through a forest roughly 50 million years later would still have found representatives of very primitive lineages of birds. Yet other birds would have been recognizable as early members of living groups. Recent research shows that at least four major lineages of living birds—including ancient relatives of shorebirds, seabirds, loons, ducks, and geese—were already thriving several million years before the end of the Cretaceous period, and new paleontological and molecular evidence suggests that forerunners of other modern birds were around as well.

Most lineages of birds that evolved during the Cretaceous died out during that period, although there is no evidence that they perished suddenly. Researchers may never know whether the birds that disappeared were outcompeted by newer forms, were killed by an environmental catastrophe, or were just unable to adapt to changes in their world. There is no reasonable doubt, however, that all groups of birds, living and extinct, are descended from small, meat-eating theropod dinosaurs, as Huxley's work intimated more than a century ago. In fact, living birds are nothing less than small, feathered, short-tailed theropod dinosaurs.

Chapter Four
The Dinosaurs' World

Introduction

Those with only a passing knowledge of dinosaurs might think that they all lived together at the same time, with *Tyrannosaurus* dining on *Brontosaurus*. Not even close. Dinosaurs were in existence for 150 million years—although this is less than one galactic year, as the solar system takes 200 million years to make one orbit—with each given dinosaur fauna in any particular place turning over every few million years at the slowest, and with different parts of the world supporting distinct faunas. Over this span of time, a super-continent split up, and the world's flora went from fairly archaic to semi-modern. David Norman gives us a survey of the changing Mesozoic scene.

The Evolution of Mesozoic Flora and Fauna

by David Norman

Appreciating Time and "Deep Time"

From the egocentric perspective of the "important" history of human civilization, one thousand years seems like a long period of time. Though we have quite reasonable historical records, it is still difficult to grasp what life was really like at the time of William the Conqueror (1066) or during the time of the Roman Empire (over 2,000 years ago). It can be argued, with some conviction, that 10 thousand years more or less encompasses the majority of human cultural evolution. But this is, by comparison, a seriously long time ago, and we have the merest glimmerings of an understanding of such early stages in our history. Yet on the scale of the history of life on Earth (measured in hundreds even thousands of millions of years), and more particularly the time of the dinosaurs, the time span of human history is utterly trivial. So different are these timescales that we actually need to think in a completely different way about history and the passage of time. A number of authors in recent years have tried to develop the notion of "deep time" to convey the immensity of time that is wrapped up in the history of the Earth, but, more often than not, our imagination simply fails against such stupendous numbers. (John McPhee appears to have coined this excellent term in 1981 as a parallel to the deep space of the astronomer/cosmologist.)

In the context of our inability to fully appreciate the depth of geological time, consider for a moment the number of years associated with the "reign of the dinosaurs." The earliest representatives of this group of animals appeared on Earth as fossils in rocks that have been dated (by measurement of radioactive decay) to about 225 million years old (during the Late Triassic period). The last of the nonavian dinosaurs existed on Earth 65 million years ago (at the close of the Cretaceous period). This gives an approximate historical range of 160 million years for the dinosaurs' time on Earth. By comparison, the entire family of human species can be traced to ancestors dating back just 4 million years. Therefore, the last dinosaurs (excluding the birds) lived about 61 million years before our human ancestors appeared on Earth, and the dinosaurs, as a group, existed for a staggering 160 million years before that. And we have genuine and understandable trouble grasping the meaning of the 10 thousand years of human civilization.

The geological timescale is simply a device (a measuring instrument of sorts) built with increasing accuracy by geologists and physicists over the past two centuries. It gives us a way of measuring the history and ages of the Earth in a meaningful way. That history stretches back about 4,500 million years to the time of the Earth's formation. This staggeringly long history has been broken down into successively smaller, and therefore more manageable, chunks of time. Dinosaurs existed on Earth during a period of time that geologists have named the Mesozoic era (literally, "middle life," comprising the Triassic, Jurassic, and Cretaceous periods). The length of the dinosaurs' time on Earth coincides with some momentous physical changes to the Earth itself, and this fact makes their time on Earth a singularly interesting one to study.

Scientists only began to grasp the physical nature and scale of changes to the structure of the Earth in the middle of the twentieth century. This resulted largely from work involving a combination of the exploration of the topography of the sea floor and the discovery of magnetic signals preserved in the rocks that form the outer layer (crust) of the Earth. These investigations showed that the crust is divided, jigsawlike, into enormous tectonic plates that are joined together along "seams" formed by either cracks (faults), deep grooves (trenches) or high mountain ranges (ridges). The main continents of the Earth (the Americas, Eurasia, Africa, etc.) float on these enormous tectonic plates, and it can be demonstrated that they have shifted over time (a process referred to as *continental drift*). Charting these movements through geological time has revealed that the geography of the Earth (the pattern and position of the continents) was very different in the distant past from what it is today. If we could travel back in time and look down on Earth as it was 220 million years ago, we would find our Earth to be unrecognizable.

Through an interesting coincidence of timing, the history of the dinosaurs covers a fascinating period in the ever changing picture of the geography of the Earth. Indeed, so strong may have been the effects of continental drift during the immense reign of the dinosaurs, that the physical changes on the Earth were likely to have affected the evolution of dinosaurs, as well as that of many other groups that existed during this time. And it is not unreasonable to suppose that some of these effects may be traced through to the present day.

Permo-Triassic Times: The World Before the Dinosaurs
Geography

Prior to the Mesozoic era the continents of the Earth formed a largely unrecognizable jumble of land and sea (compared to what we are familiar with today). But with the passage of time, the continents of this strange Palaeozoic world began to coalesce so that by the close of the Permian period (between about 250–240 million years ago) all the continents had fused to form a gigantic land area that has been named Pangea ("all Earth"). This extraordinary event—apparently unique in the history of the Earth—coincided with a period of extensive marine regression (the sea level dropped relative to the land). The regression may well have had its origin in the reduction in activity along the margins of the tectonic plates. In particular, the subsidence of the mid-oceanic mountain ranges would have greatly increased the volume of the ocean basins, causing a major sea-level drop.

Such extraordinary events, which must have taken place over a period of several million years, coincide with a major period of animal extinction. Global compilations of the diversity of animal types through the geological record of the history of the Earth agree in showing that the end Permian period marks the greatest extinction event of all time—over 90 percent of all known creatures went extinct. It even eclipses in importance the much-hyped extinction of the dinosaurs at the close of the Cretaceous period.

Biology

Such a convulsion in the biological world—caused by perfectly natural, if unusual, gradual changes in the physical world's make up—can be seen as both a devastating calamity (the loss of huge numbers of species and complete disruption of ecosystems that rely on the interaction between a range of species) or an immense opportunity. The survivors of this calamitous change inherited a world that was almost empty of species and biological competitors, meaning they had the world to themselves and, therefore, untold evolutionary opportunities to radiate and diversify.

Animals

On land, some of the most notable beneficiaries of this period of appalling mass extinction were the synapsids, a group of quite complex creatures that we might generally refer to as "reptiles." They represent a group of ancient animals that were the lineal antecedents of present-day mammals such as whales, elephants, dogs, and many others, including our very own human species. Their general anatomy, and implications that can be drawn about their physiology, tells us something about the climate and general conditions of the time.

During Permian times, the synapsids are particularly well known from sediments in North America, Russia, and South Africa. All these discoveries provide a useful insight into the life and times of the Permian period. Early in the Permian, these creatures were for the most part quite large-bodied, creatures of lizardlike proportions. Some, such as *Dimetrodon* and *Edaphosaurus* (from the Early Permian of North America and Europe) were notable for the development of unusually prominent high-spined sails on their backs. Such features have been associated with the ability of these animals to regulate their body temperature (using the sail as a combined solar panel to absorb heat or as a radiator to dump unwanted body heat). This indicates two important factors: the climate was variable; and controlling body temperature, rather than letting the body simply warming or cooling in the air, was potentially advantageous to survival.

This general observation of variable, perhaps seasonally driven, climatic conditions fits well with the observation from the southern continents that there was a major glaciation near the close of the Carboniferous period. Much of the Early Permian was affected by such variable cool-temperate conditions, and the animals that are found, such as the synapsids, show adaptations to help withstand such conditions. Many are large in size—increasing their bulk increases their thermal inertia and their ability to keep the core of the body warm.

In Mid-Late Permian times, much stouter types had replaced the large lizard-shaped synapsids. The shorter tail and more barrel-like body of these animals can again be seen as a useful adaptation to changeable conditions, because the ratio of surface area (across which heat can be lost) to body volume is considerably less than the lizard body shape.

Climatically, the Mid-Late Permian appears to have been far more equable. The southern continents (this is particularly well studied in the Karoo Basin of South Africa, which has yielded large numbers of these fossils) were drifting northward into climatic zones typical of the modern Mediterranean region: with warm, wet winters and hot summers. The animals, particularly the synapsids, are found in abundance and include a wide variety of (pig-sized or smaller) dicynodonts, as well as many wolf-sized and, in anatomical terms, rather wolflike creatures. It is a curious fact that several of these animals were burrowers—their spiral burrows have been found in the fossil record, occasionally with animals preserved inside. Burrowing was also a convenient way for an animal to avoid the extremes of heat or cold in any environment.

By Late Permian and Triassic times, the synapsids begin to show a pervasive trend toward increased mobility (longer-legged, lighter, and more cursorially adapted animals) and, linked with this, steadily smaller body size. These later synapsids, known generally as cynodonts, increasingly come to resemble later true mammals in many of their anatomical features. The very earliest mammals that have so far been recognized in the fossil record succeed these animals. Their remains are found in the very latest Karoo deposits, which are dated as Early Jurassic in age (about 200 million years old).

By the latest Triassic-Early Jurassic, the general climatic conditions, judged from the sediments in which the fossils are discovered, range from occasionally wet to extremely dry desert conditions, equivalent in most respects to Saharan and sub-Saharan Africa. This simply reflects that, geographically, the African continent has drifted northward to the equatorial regions by Late Triassic times.

Such conditions are generally extremely inhospitable to animal life. This is especially true of large-bodied creatures, unless the latter have some peculiar physiological mechanisms to aid their survival (the camel is a particular example of a large animal that is well adapted to cope with extremely hot and persistently dry conditions). The vast majority of animals that can survive in such conditions generally do so by managing to avoid the most extreme conditions (for example, by burrowing and coming out only to feed at dusk or in the early morning). Others are simply extremely well adapted to hot, dry conditions—this applies to some species of snake, lizard, and a host of beetles and other arthropods.

Animals such as the synapsids that were relatively abundant and widespread for much of this time are, consequently, largely adapted to avoidance of the extreme conditions found in the Late Triassic. As a result they tend to be small, burrowing, nocturnal animals. They had the ability to control their internal body temperature (a component of homeostasis that was steadily acquired through the Permian and Triassic times within the synapsid lineage) and refined a wide variety of senses (notably acute hearing, touch-sensitive whiskers, large eyes, and an excellent sense of smell). The intelligence to integrate these senses and react to them produced animals (genuine mammals) that were quite competent at surviving in such rigorous conditions.

The Diapsids—Another Story

While the synapsids formed an apparently dominant group of land vertebrates through much of the Permo-Triassic, there coexisted another radically different sort of animal. Collectively known as diapsids, these include animals that look like (and many of which are) lizards (technically, the group is known as lepidosauromorphs). In addition to the lepidosauromorphs, there are larger creatures that have a closer kinship with modern living crocodiles and birds (known as archosauromorphs).

During the Permian, both groups of diapsid are relatively rare as fossils. There are two reasons for this: preservational bias (lizards are generally small creatures and their bones tend to fossilize quite rarely) and genuine rarity. It may simply have been that diapsids were relatively rare in the environments that tend to preserve fossils in areas such as the Karoo Basin. What glimpses we do get of these creatures, as rare fossils, indicates generally long-bodied creatures (as is the case with living lizards and crocodiles) of various sizes. Exactly what was happening to this group as a whole is very poorly understood at present.

By the beginning of the Triassic period, the fossil record of diapsids begins to perk up a little, notably through the preservation of the larger archosauromorphs, such as the large, crocodile-shaped *Proterosuchus* and the massively built *Erythrosuchus*. Throughout the Triassic, a variety of medium-to-large-sized creatures of this type appear.

These appear to range from fast, long-legged land creatures to ones that would appear to have lived a lifestyle similar to that of modern crocodiles and alligators. The vast majority retain the primitive carnivorous habit, and very likely preyed upon the wide-ranging and abundant herbivorous synapsids of the time. Many of the latter appear, at first sight at least, to have been barrel-bodied, and comparatively slow-moving compared to the build of the predatory archosauromorphs.

The archosauromorphs, though primarily carnivorous, did produce the occasional herbivorous varieties, but these were generally not so numerous as the herbivorous synapsids of the time. Some were rather piglike and outwardly resembled the synapsid dicynodonts; known as rhynchosaurs, they flourished briefly during Middle Triassic times. However, the stagonolepids of the Late Triassic were in many respects more typical: heavily built creatures with powerful limbs and a carnivore-proof skin that was embedded with thick bony armor similar to that seen on the backs of crocodiles today.

None of these later archosaurs bears any real anatomical resemblance to the synapsids of the time. Archosaurs retain the long reptilian tail, which anchors the powerful leg-moving muscles and acts as a useful counterbalance to the front of the body. As a result, many of these animals tend to develop longer backlegs than front, and an ability to run and walk primarily bipedally (on the back legs). *Ornithosuchus*, an otherwise quite heavily built archosauromorph of the Late Triassic, is a classic example of this anatomical trend. The very large head, whose jaws are lined with long, sharp teeth, and the front part of the body are counterbalanced by the long muscular tail. The legs are also long, slender, and capable of making long strides. By contrast, the front limbs are "arms" rather than legs and are capable of holding prey instead of being used for walking.

Many archosaurs were therefore fast-moving, alert predators, but built in a classically "reptilian" manner and, for that reason, very distinct from the contemporary synapsids of the Late Triassic.

What Happened in the Late Triassic? The Rise of Dinosaurs

The first remains of true dinosaurs have been discovered in rocks dated as Late Triassic (about 225 million years ago). These include animals such as *Herrerasaurus*, *Eoraptor*, and *Pisanosaurus*. In most details, they resemble the archosauromorphs. The proportions of the body and style of locomotion are indistinguishable in nearly all respects. From this point in time onward (until the end of the Cretaceous period about 160 million years later), the archosauromorph lineage, represented mainly by the dinosaurs, became the most abundant and diverse animals on land.

This simple observation, based on the range and variety of fossils collected from rocks of Mesozoic age, is of profound importance. It flies directly in the face of everything that we might expect to see based on our experiences today. In the present day, we are very familiar with the mammalian descendants of the synapsids from the Permian and much

Artist Douglas Henderson's vision of a Late Triassic landscape, complete with a "flock" of coelophysids.

of the Triassic. Mammals today may not be the most abundant of all vertebrates, but they (along with the birds) tend to dominate most ecosystems. They are the largest and most varied animals on land (occupying the upper reaches of the ecological pyramid); and even in the seas, the whales, dolphins, and seals form much of the upper tier of the ecosystem.

We can rationalize this current situation through the observation that mammals are generally credited with being (by comparison with an "average reptile") highly intelligent, biochemically very complex, warm-blooded, and well-insulated; they also care for their young and suckle them on milk when they are most vulnerable. All these attributes allow us to single out mammals as a "successful" group, and one that we might expect to be "superior" in a general way to other types of animal, such as reptiles, amphibians, or fish. The latter animals lack many of the attributes of mammals and could support the notion that they be ranked as less successful on that account alone. For example, the reptiles (by which I mean animals such as lizards, snakes, crocodiles, and tortoises) are rather restricted on a worldwide scale. Many more are found in the warm climatic zones—rarely do they venture above about 50° north or south. By comparison, mammals are found to be globally distributed, ranging from the equatorial regions to the inhospitable polar regions.

While many of the mammalian attributes mentioned are undeniably advantageous (when looked at in general terms), in the laboratory we can test (crudely) whether mammals are really "superior" to other types of vertebrate (particularly the reptiles), based on the fossil record. The results are counterintuitive and, for that reason alone, interesting. To make the point more forcibly than is justified by the evidence, imagine the Late Triassic/Early Jurassic as a time of rather dramatic "equipoise." On the one side, the dinosaurs were just starting to appear on Earth, while on the other, the first mammals were making their appearance. These groups represent very different styles of animal. The former, in many respects, are classically reptilian: scaly, probably not insulated and of limited intelligence; the latter are, by contrast, well insulated, highly intelligent, and resourceful.

We might predict, given what we know today, that the synapsid mammals would be the group that would rise to dominate the Mesozoic world. They have all the traits that we associate with our notion of success or superiority. Perversely, the fossil record demonstrates quite clearly that the synapsids do not rise to dominance at all. For the remainder of the Mesozoic (160 million years), mammals persist as small, apparently mostly nocturnal, creatures, while dinosaurs dominated the land in both size and variety. This simple observation is a major puzzle that has caused much debate among paleontologists.

Much of the debate concerning the rise of the dinosaurs in Late Triassic times has focused on intangibles, unknowable features that have been built into unsupported stories that can be used to explain their rise to dominance. For example, it has been suggested that dinosaurs had invented a better way of running (by tucking the legs directly under the body so that they could run faster and more efficiently). On that basis, the story that was developed involved the carnivorous dinosaurs outcompeting the synapsid predators and causing their eventual extinction simply by being "better." In simple terms, the fossil record provides little evidence that would allow us to prove or disprove such a story. And even if it were true in an individual instance of one species of dinosaur compared to another synapsid, is it likely that all dinosaurs outgunned every synapsid? That is probably straining credibility.

We are left with a general pattern of succession in the fossil record. Over the space of several million years (perhaps as long as 20 million years in total), we observe the shift from synapsid dominance of vertebrate communities to dinosaur dominance. The pattern runs counter to our expectations. Why? What are the clues in the fossil record? Can we reconcile what we see with any other factors in Late Triassic life?

The Late Triassic Environment: A key?

One of the most interesting and persistent general themes that emerges when looking at a variety of sediments of Middle-Late Triassic age from around the world (Europe, North America, South America, Africa) is the presence of what are known as "red beds." These are rocks that are predominantly stained red (quite literally rust, iron oxide). The presence of iron oxide is a chemical signature associated with a persistently seasonally hot, periodically monsoonal, but arid environment. There is also widespread evidence of desert (sand dune) deposits, all of which combines to suggest that at the end, the Triassic was a particularly hot and dry place on an almost worldwide scale. Clearly, the whole world was not simply a huge desert, but desert conditions were amazingly widespread.

Because this is the time of the rise of dinosaurs and relative decline of synapsid mammals, the question that needs to be asked is "Could such widespread conditions have affected the survival potential of the two groups in different ways?" Curiously, this probes some of our deeply held prejudices. Mammals and humans are notably abundant in today's world. But would we have fared as well in the Late Triassic?

Triassic Conditions

If we consider "average" conditions in the Late Triassic, we would tend to equate them with universally hot and arid factors: very little seasonality with relatively little latitudinal climatic zonation. We know for a fact that the world was essentially a single giant continent (Pangea) and that there were no polar icecaps. Both of these factors are important. Giant supercontinental landmasses would have a predominantly dry "continental" climatic regime. And, because the ice-covered poles of today are largely responsible for the polar-equatorial climatic zones of the world, as well as the predominantly seasonal (winter-summer) conditions, the wide geographic spread of Triassic rocks indicating dry and hot conditions become very understandable. Animals living in such conditions need to be specially designed to cope. Deserts lack water, have very little vegetation, and tend to be very hot. Mammals and reptiles (generally) have very different body mechanisms that can be summarized by briefly defining ectotherms and endotherms.

Classically, reptiles are ectotherms, which is to say that they rely on external sources of heat (ultimately the sun) to keep their bodies warm. Today, their distribution around the world is strongly tied to how warm it is: they are largely restricted to tropical and subtropical zones. They keep their body temperature and chemistry relatively steady either by basking in the sun or by sheltering in the shade. This is a system that is quite simple to operate, but is not always ideal. For example, cool overcast days may mean that reptiles can never reach their preferred running temperature, hence they remain rather sluggish, even in life-threatening situations (such as the appearance of a predator). We can summarize many of the attributes of modern reptiles as follows:

- Reptiles prefer hot conditions.

- Reptiles can tolerate varying body temperatures.

- Because they do not need to supply their own body heat, reptiles do not have large appetites and eat relatively infrequently.

- Reptiles have scaly, water-impermeable skin, and a surprising array of sophisticated ways of conserving body water.

- Reptiles never urinate! Instead, they produce a urate-rich paste (the same as bird guano).

By contrast, mammals are endotherms. They generate body heat internally, using their own body chemistry as the equivalent of an internal central heating system. As a result, they are not limited by how warm it is; consequently, they are distributed across the globe, even in perishingly cold conditions. There is, however, a downside to this otherwise very useful feature: mammals can become physiologically stressed, rapidly overheat, and die in a very hot environment. Mammals monitor and control their body temperature by regulating the rate at which they lose heat from their bodies to the environment. If the outside temperature is hotter than the mammal body, heat would be absorbed, thereby adding to the heat burden of the animal. Unless cooling through evaporation (panting and sweating) is possible, the brain (which has a chemistry that is critically temperature-dependent) would rapidly overheat. Overheating of the brain would upset its biochemical processes, and the mammal would die through loss of brain function. (This is precisely why medical practitioners are always anxious to reduce the body temperature of patients with serious infections; just a few degrees of temperature change can induce fever, coma, or death.)

Generating body heat through chemical reactions inside the body is also an expensive business, and mammals, pound for pound, have much larger appetites than reptiles (food is the fuel for this process). Generated heat is also a precious commodity, and mammals use a combination of subcutaneous fat and fur as insulation to reduce the rate of heat-loss from the body to the environment. Related to, and in addition, mammals have large brains to control their lives and activities; these also demand a constant supply of oxygen, water, minerals, and food. Mammals are, in truth, very sophisticated creatures but, compared with the reptile model, rather "expensive" (in terms of needed elements) to run satisfactorily. To sum up, mammals, in contrast to reptiles

- prefer variable, warm-cool environments;

- have very little tolerance of body temperature changes—their thermostatic control has quite a narrow setting;

- have large appetites to fuel their endothermic physiology;

- have generally soft porous skins that allow them to sweat (lose water evaporatively) as one part of their cooling system;

- excrete their waste as urea, which has to be dissolved in water to form urine; this is another inevitable drain on their body water supply.

Using these simple checklists, it becomes immediately apparent that the best choice of lifestyle for an animal living in consistently hot, dry habitats with very little plant food available (the Late Triassic) would be that exhibited by modern reptiles. They are clearly well-adapted for survival in hot conditions because not only do they work better at high ambient temperature, but (even more significantly) they are able to cope with temperature change within their bodies without physiological stress; they are also well able to survive on minimal quantities of food (because they are ectotherms), and are very well designed to conserve water (they have a waterproof scaly skin, and therefore no sweat glands, and virtually no urinary water losses).

Perhaps the rise of dinosaurs (or more accurately a "reptilian style of life") was favored by the environmental conditions that prevailed at the close the Triassic period. Such conditions would seem to have conspired against what we would intuitively expect to be "superior" mammals. It would also adequately explain why the early mammals predominated at the beginning of the reign of dinosaurs as small, nocturnal animals, such as *Megazostrodon*. Such small creatures could live in cool, damp burrows during the heat of the day, and, by coming out only at night, would be well adapted (keen sight, sense of smell and touch) to catch food, such as the abundant insect life that is able to thrive in desert conditions. If this scenario (and, sadly, it can be no more than a scenario, though one using moderately sophisticated levels of interpretation) is broadly true, it suggests that the world and its environment (which can be shaped by remarkable physical processes—such as continental drift—working on an immense timescale) has the potential to exert a major influence over the evolution and succession of organisms throughout the history of the Earth.

Whatever the true pattern of events that took place across the few million years that mark the close of the Triassic period, the product of this process was a radically different picture of vertebrate diversity compared to a snapshot taken at the beginning of the Late Triassic. The synapsids, large archosaurs, strange piglike rhynchosaurs, and many other groups had become extinct. In their place, a range of dinosaurs and tiny mammals came to dominate the land, by day and night, respectively.

The dinosaurs of the Late Triassic are an interesting assortment of creatures. To date, predatory forms such as *Eoraptor* and *Herrerasaurus* are known from relatively rare discoveries only from Argentina. However, far more numerous and more widespread in the latest Triassic are the remains of a delicately built theropod carnivore known as *Coelophysis*. Numerous skeletons of this animal have been discovered in America (Arizona), and a very similar creature has been found in South Africa; there are also very *Coelophysis*-like footprints known from Britain (South Wales). These discoveries are important because they confirm what the geologists tell us, that animals could walk (or run) on land from areas like the equivalent of America in Late Triassic times all the way to Europe and on to South Africa. These little dinosaurs may truly have had a worldwide distribution.

Another Late Triassic type of dinosaur belonged to the sauropodomorph group. Many (such as *Massospondylus* or *Plateosaurus*) were probably omnivorous (able to eat plant or animal remains) and therefore more versatile than the exclusively meat-eating forms. Their geographic distribution is also interesting, with closely related forms having been discovered widely across North and South America, Europe, and into Asia. Quite often, these dinosaurs have been found in considerable abundance.

From these discoveries, it therefore seems that that world was inhabited by many dinosaurs (judged from the large numbers of skeletons that have been found), that they were exceptionally widespread, but that they were not particularly diverse (they did not exist in great variety). In part, this might be accounted for by their ability to migrate so widely across the world, leading to rather uniform communities of animals everywhere; it may also be an artifact of fossil preservation and collecting (something that we must always be wary of, though in this instance the large numbers of skeletons do seem to offer a fairly clear message); finally, it may tell us something about the dynamics of the evolutionary rise of the dinosaurs (that early pioneering forms were actually rather limited in variety).

The Jurassic Period
Climate and Plants

The onset of the Jurassic period (about 200 million years ago) is marked by a progressive shift from the widespread arid conditions of the Triassic to relatively more pleasant (equable) conditions. One of the most consistent features, geologically, can be traced in the style and types of sediment that were laid down to form Jurassic rocks. The red beds largely disappear except at equatorial latitudes and are replaced elsewhere by muds, lime-stones, and sandstones that were deposited in shallow, warm-water marine environments. This is a clear indication that sea levels began steadily to rise from the unusually low lev-els that had persisted during much of the Permo-Triassic. The sea-level change may well be a reflection of increased tectonic activity. In simple terms, the seams between the great tec-tonic plates, which had been passive following the formation of the supercontinent of Pangea, began to buckle and strain. This was in response to slow but inexorable forces exerted on the undersurface of the crust of the Earth by friction caused by movements of the fluid rock in the Earth's mantle. Such activity was a prelude to the gradual breakup of the supercontinent of Pangea and the beginning of changes that would lead in the end to the formation of the continents that we recognize on the Earth today.

Sea-level rise and the consequent flooding of the edges of the continents and low-lying lands across continents would have had a gradual but important influence on the cli-mate of the Jurassic. The more extensive shallow seas would increase rainfall levels across the land and encourage the development of more extensively vegetated (forested) areas. Certainly this occurred in mid-high latitudes worldwide. The fact that there are significant Jurassic coal deposits in various parts of the world links neatly with this gen-eral climatic trend. Coal, rather than oil, is formed in areas of the world that are low-lying and reasonably heavily forested for long periods of time. Such areas, if subjected to stable environmental conditions, build huge thicknesses of dead and decaying vegetation (peat) that are occasionally inundated by floodwaters and thereby covered by sandy sed-iments. Continued burial of these organically rich layers traps, compresses, and heats the plant material. If the conditions are right, the plant remains decompose slowly in what is almost a slow-burn process (a little like making charcoal from wood in an earth kiln today) and can be converted to coal over very long periods of time.

As was the case in the Triassic period, there is no evidence of polar ice in the Jurassic. The world was considerably warmer than it is today and would have had equatorial deserts, but wetter subtropical and warmer temperate zones. Conditions in these latter areas appear to have been perfect for the growth and development of plants of far greater range and variety than in the Triassic. The high latitudes also had very significant plant cover, but this was warm/cool-adapted, and subject to winter low-light regimes, rather that to freezing cold temperatures.

Dominant components of the plant community during much of the Jurassic differ little from what are known from the Triassic. There were a variety of gymnosperm trees (of which recognizable varieties today include various conifers such as pines, cypress, and cedars), as well as the rather exotic-looking araucarias (monkey puzzle trees), ginkgos, cycads, cycadeoids, and a wide variety of ferns (ranging from tree-sized individuals to the small low-growing varieties that are more generally familiar today).

Preceding page: Charles Knight's study of sauropods, done around the turn of the 20th century. The concept of the giant plant-eaters rearing up to feed high on trees is once again being seriously considered by modern paleontologists. Right: North American dinosaurs from the Late Cretaceous in Knight's typically misty scene, including a variety of plant-eaters. Note that, aside from *Struthiomimus* in the center, all of the dinosaurs' tails and feet are on the ground.

Left: Arguably the best known of the classic-era dinosaur images, Knight's *Triceratops* and towering *Tyrannosaurus* meet face to face. The horizontal, tail-off-the-ground pose of the foreground tyrannosaur is reasonably modern, but the great beasts are slow-moving and the tyrannosaur in the background seems unconcerned about the action.

This page. Top left: Plowing through the marsh with widespread forelegs, this tail-dragging *Apatosaurus*, although beautifully execute[d] by Bill Berry, is of the old school. Top right: Berry's massive-legged *Brachiosaurus* seems barely able to lumber out of the water. Bottom right: Executed along with the rest of these works in the 1960s, Berry's snorkeling *Diplodicus* are among the most evocative illustrations of a concept that was on the verge of becoming obsolete: aquatic sauropods. Bottom left: In contrast, this restoration o[f] *Camarasaurus* head about to chomp down on some conifers is quite up to date. Opposite page. Top left: While Berry's camptosaur[s] graze in the sunny field, a pair of little *Dryosaurus* browse in the shady thicket. Top right: A herd of *Camptosaurus* graze peacefully i[n] a glade. Note the lack of cheeks covering the tooth rows, a subject that still arouses controversy. Bottom: Berry's *Allosaurus* attacki[ng] an *Iguanodon* has a modern look and feel to it, with tails held straight back and centers of balance over rear hips.

Above: One of the early large theropods was double-head-crested *Dilophosaurus* from the southwest of the United States, shown here bouncing on its tail kangaroo-style while lashing out at an unseen enemy. Opposite page. Top: the small theropod *Coelophysis* travels in a pack through a Late Triassic forest in North America. Below: The primitive ornithopods *Heterodontosaurus* walk across a mud flat in North America, leaving trackway imprints.

Top: As the Jurassic progressed, dinosaurs became more numerous and larger. A newly discovered Chinese fauna includes the elephant-sized, tail-clubbed sauropod *Shunosaurus*, the theropod *Gasosaurus*, and the early stegosaur *Huayangosaurus*. Left: A pair of giant Chinese *Yangchuanosaurus* take a needed after-meal break from feasting on their latest kill. A rising full moon completes this twilight tableau. Opposite page: By the Late Jurassic, dinosaurs were becoming really large, as exemplified by Chinese *Omeisaurus* shown here browsing in a forest of tall monkey puzzle conifers. Just how high such sauropods could raise their necks is still a matter of debate.

Top: The famed Morrison Formation remains the premier site for Late Jurassic dinosaur fossils. Massively built *Apatosaurus* seems to have been constructed to allow it to rear up and feed high in the conifers, such as this monkey puzzle. Bottom: Two Morrison classics, *Allosaurus* and *Stegosaurus*, which is spinning to keep its spiked tail in the theropod's face.

Top: Showing small dinosaurs covered in feathers was a hotly controversial subject until the spectacular Yixian dinosaurs showed that it was real. Here is *Sinosauropteryx* in the deep woods, as envisioned by Michael Skrepnick. Bottom: Skrepnick's restoration of the bird-like *Caudipteryx*, a stubby-tailed Yixian theropod with well developed contour feathers on its arms and tails.

Above: Newly discovered in the Sahara Desert, *Jobaria* was well suited for standing on its hind legs in order to better interact with its fellow sauropods.

Top: *Suchomimus* is another new, big theropod from the Sahara. Its long snout has prompted some to argue that it was a fisher, but this remains somewhat speculative. Bottom: Another mud-flat trackway, this one being left by a herd of *Brachiosaurus*, the giant Jurassic sauropod.

Top: A *Maiasaura* colony of nest mounds, located in Montana. While a pair of parents tend to their hatch-lings (left), an adult adjusts the temperature of its fermenting nest by removing some soil. Bottom: Also in Montana, two *Krtiosaurus* watch over and tend their nest mounds. The nests contained fermenting vege-tation that heated and incubated the eggs within the mound.

Top: Argentina recently produced one of the oddest theropod dinosaurs, the short-snouted *Carnotaurus*. Here artist Luis Rey portrays them using their massive brow horns to engage in shoving matches.
Bottom: Mass bonebeds in Alberta show that the ceratopsian *Centrosaurus* moved in great herds. This scene suggests the way the bonebeds may have been formed: mass drownings at river crossings.

Above: Artist Peter Trusler's vision of southern dinosaurs. A group of herbivores gather beneath the Southern Lights in a south-polar latitude, where long nights and cooler climate would have made endothermy a distinct advantage for the dinosaurs.

Age of True Herbivores and "Super-Herbivores"

The diversification and greater productivity of plants following the onset of the less extreme Jurassic climatic regime provided a potential spur to the evolution of specialist herbivores. Late Triassic sauropodomorph dinosaurs probably were not specialist herbivores. Their size and extended reach of their long necks did mean that they could browse on foliage that few animals had been able to reach (apart from a few nimble leaf-eating climbers) before them in the history of life. It is not particularly surprising then to discover that the Jurassic is notable for a rise in both abundance and diversity of specialist herbivorous dinosaurs.

Super-Herbivores

Sauropodomorphs had already shown the way in the Triassic as large-bodied (up to 26 feet long), long-necked animals with relatively small heads. They were equipped with sets of rather simple, coarse-edged teeth (such as in *Plateosaurus*) that could be used equally well for stripping foliage (combined with stone-laden gizzards for pulping the plant remains) or for scavenging meat from carcasses. During the Jurassic, these types were superseded by descendants (generally called sauropods) that were some of the most amazing-looking herbivores that have ever lived on Earth. Growing to lengths of 100 feet or more and weighing anywhere between 15 and 50 tons, these were simply extraordinary plant-processing factories. They include such well-known animals as *Diplodocus, Apatosaurus, Brachiosaurus, Barosaurus,* and *Mamenchisaurus, Seismosaurus*. Enormous skeletons of these dinosaurs grace some of the most famous natural history museums in the world.

The sauropods of the Jurassic were clearly at the apex of the food web among terrestrial herbivores, and they were the approximate ecological equivalent of the large mammals of the present day (elephants, rhinos and ungulates). It does appear that dinosaurian herbivores were much larger than their mammalian equivalents today, but that may be more apparent than real for two reasons: anatomy and humans.

Anatomically, sauropod dinosaurs were built on the classic "reptile" blueprint. As a result, they had large, muscular tails that provided the power for walking and running (pulling backward on the hind legs and thereby driving the body forward). The very large tail was anchored to the sturdy body of the animal, which was massive to support a huge gut for digesting plants. In front, the long neck was supported through the weight of the tail—connected by long spinal ligaments. In a sense, the long neck "came free" with the long tail because the one supported the other, like a gigantic seesaw. The head, small though it might seem for such a mighty creature, was adequate for constantly pruning vegetation that it supplied to its huge stomach, and the long neck simply provided a wide and deep sweep of the head through the foliage.

The Late Jurassic was the age of sauropods, which lived in forests dominated by conifers. In this Chinese scene the

The long neck and tail provided these animals with their visually staggering proportions of length and overall size, but this is deceptive. Look at any large sauropod in a museum, remove the neck and tail (in your imagination!) and what are you left with? The body of a pretty large elephant or one of the huge extinct rhinoceroses of the past, such as *Baluchitherium*. Mammals, however, are built differently from dinosaurs; they do not need the huge tail or tail muscles, so their body proportions differ accordingly. Nevertheless, the sauropods were impressive animals in size, number, and variety in Jurassic times, and adults must have been invulnerable to attack by predators, much as elephants are today.

Range and Variety of Ornithischian Herbivores

In contrast to the sauropods, the remainder of the herbivores belong to the ornithischian group; they comprised ankylosaurs, stegosaurs, and ornithopods, all of which made up

featured sauropod is *Mamenchisaurus*, a 75-foot-long giant.

the bulk of the small to medium-sized herbivores of the Jurassic. Ankylosaurs were comparatively rare as fossils, but they were distinctively heavily armored, with a skin studded by conical bony lumps and spikes to deter the most persistent of carnivorous theropod. Stegosaurs had a very distinctive set of bony plates running in rows down their spine and fearsome sharp spines on their shoulders and, more particularly, on their tails; these latter could be flailed at would-be assailants. Ornithopods were the least conspicuous group of herbivores, having neither armor nor extravagant spines or spikes for protection. Most were small and bipedal, relying on speed and size to avoid predators—perhaps the equivalent of gazelles and antelopes of the time.

The Carnivores (Theropods)

The predators of the Jurassic were far less varied. The large forms, though they varied from place to place in subtle details, were essentially 20- to 33-foot-long, allosaurlike large predators; at the smaller end of the scale, they were 6½ feet long, lightly built, fast-pursuit predators such as *Coelurus*. The carnivores (rather like modern carnivores) tended to represent variations on a pretty constant theme of body design. Size varied depending on prey preference, and presumably in part on whether they acted as packs of hunters or individuals. Unfortunately, the evidence from the fossil record does not allow us to discriminate on such fine bahavioral points.

Birds Emerge

Of lesser significance in world terms, but of major significance in the larger picture of the history of life on Earth, a group of theropod dinosaurs (similar to *Coelurus*) became small, highly active predators of small animals (primarily preying on highly agile lizards, equally active small mammals, and flying or fast-moving insects). These theropods lived very active lives, sustained themselves as highly intelligent, well-coordinated "terrorists," and, because of their small size (ranging from approximately 8–20 inches in body length), maintained their activity levels by being endothermic (warm-blooded), and were insulated with a shaggy filament-like covering. A few of these—one well-known example of which is *Archaeopteryx*—developed such filaments into feathers and became the animals we would later recognize as birds. The early origins of birds are currently shielded from scientific study largely by bias in the fossil record. The bias preserves small, lightly boned creatures much less well and less often than those heavily built dinosaurs whose bones are so much more robust, and hence survive scavenging and the vagaries of fossilization.

Mammals in Nooks and Crannies

Throughout the Jurassic, we find evidence of mammals. They are always small creatures, presumably nocturnal and probably burrow-dwelling, pursuing a style of life similar to that of living shrews and rodents. Though rather rare in the fossil record, there is a clear bias against them; they were probably very abundant as small creatures (in the way that mice and rats are today). Their persistence in the fossil record has to be marveled at and admired simply because they maintained the thread of continuity of mammals, which would eventually produce animals such as ourselves long after the bulk of the dinosaurs had become extinct.

The Jurassic Summed Up

The climate for most of the Jurassic (from the late Early Jurassic onward) was less harsh (less widespread hot, dry conditions) than that which seems to have applied at the very close of the Triassic. Milder conditions supported greater vegetational cover and greater productivity. Such conditions also provided an ecosystem that would support a far greater variety of herbivores; these ranged from normal low-level browsers (stegosaurs and ankylosaurs) to low-/medium-level browsers (ornithopods), and those with extraordinary ranges from low to very high, represented by a variety of sauropods. The huge sweep of the neck of sauropods enabled these forms to reach broadly and to great heights among the trees of the time (the first genuine high arboreal herbivores in the history of life on Earth).

The sharp rise in diversity among herbivores in Jurassic times provided opportunities for the carnivores, which increased in size range and somewhat in variety, but their adaptations for attack were not nearly as varied as the defensive adaptations shown by their prey. This lag on the part of the predators was compensated for during the Cretaceous period, which showed diversity in both types of dinosaur.

The Jurassic also saw a physically smaller "play" being acted out. Though smaller in a literal sense, this was, in the longer term, much more significant for the long-term history of life. The mammals persisted as the equivalent of rats and mice throughout Jurassic times, while the dinosaurs produced a line of predatory forms (theropods) that combined high activity levels with very small size. This combination of attributes demanded not only behavioral intelligence (a relatively large brain), but high energy levels and insulation to keep their bodies warm, and led to the appearance of the first birds.

The Cretaceous Period

The Jurassic world was dominated by the persistent grouping of the continents in a sprawling, but more or less connected, Pangea-like arrangement. Although continental rifting was starting to occur, the initial phases that took place during Jurassic times did not produce any major regional geographic changes. Notably, a moderately large oceanic basin was beginning to form in the North Atlantic (as North America and Europe/Africa started to separate), and the joined southern continents of Africa and South America were separating from the group comprising India, Antarctica, and Australia.

Geography and Climate

The Early Cretaceous period saw geographic changes that were subtle but, in their cumulative effect, far more profoundly important; these altered not only continental positioning but ocean distributions and their circulation patterns. For the first time since the Late Devonian, there developed an east-west oceanic link at equatorial level that separated the northern (Laurasia) and southern (Gondwana) continents. Somewhat later and further south, the essentially polar continents of Antarctica, India, and Australia were separated from Africa-South America by a circumpolar ocean. Increasing fragmentation of the continents throughout the Cretaceous was further exaggerated by continued sea-level rise (no doubt partly caused by the widespread tectonic (ridge) activ-

The Early Cretaceous was a time when iguanodonts were common. In this scene, three medium-sized theropods cross the river while hunting the herbivores, while the latter maneuver in unison to avoid them.

ity associated with such global rifting activity); this resulted in extensive flooding of low-lying areas by shallow epicontinental seas (such as that which separated eastern from western North America, regionalized Africa, subdivided Europe from Asia, etc). The detailed climatic implications derived from modeling of all these geographic changes are far from being fully understood at present. The fossil flora and fauna collected from Cretaceous rocks reflect some of these changes and hold up a mirror (albeit one that is rather cracked and distorted).

The absence of polar ice sheets throughout the Cretaceous period reflects persistent higher average global temperatures than in the present day; we would find the world of the dinosaurs uncomfortably and persistently hot. Whereas the absence of an uninterrupted east-west equatorial seaway in Jurassic times ensured persistent, very arid equatorial conditions across the continents, the Cretaceous showed a disruption of this pattern. The northern continents in particular showed evidence of wetter and more seasonal conditions. The northern fringes of the southern continents (Africa in particular) shows faunal and floral evidence of persistently hot arid conditions. However, middle latitudes globally show persistently wetter seasonal conditions that were capable of supporting significant plant biomass (including Cretaceous coal fields). Whether this included the more classic canopied rainforest conditions is less clear. However, since rainforests typically do not support large diversity of large-bodied herbivores, it seems likely that the vegetational regimes were predominantly those associated with a range from savannah scrub through mixed woodland to warm temperate forest, all of which were capable of supporting a wide diversity of large herbivores.

While the geography of the Cretaceous was generating changing environments and climates, other changes have also been noticed in the fossil record of plants. While the Early Cretaceous seems to show little if any change in the dominant plant types (conifers, cycads, ginkgos, extinct gymnosperm trees, and considerable varieties of ferns) and, therefore, represent a continuation of the Jurassic regime, the Mid-Cretaceous marks an important event. Most dramatically, in the floral world, the flowering plants (angiosperms) that are so widespread and abundant today make their first appearance in the form of fossil pollen and preserved flowers. Evidence from fossil pollen shows a very rapid and abrupt diversification of flowering plants species at this time. While varied, the Cretaceous flowering plants probably did not rise to dominate plant communities instantly but formed a range of weedy and scrubby vegetation. The huge ecological advantage that angiosperms have over the gymnosperms is to be found in their more rapid reproduction and growth. The capacity to grow weedlike in disturbed areas of ground may have been of profound importance. It is reasonably certain that forest fires were somewhat frequent in Cretaceous times, and that these were capable of clearing large areas of land, making them ready for recolonization. Of perhaps similar significance, the activities of the larger dinosaurs may well have had a similar effect on areas of land. It is possible that these very large herbivores behaved in part like modern elephants, using their sheer bulk to push over and then strip the foliage from large trees, as well as trampling vegetation underfoot. This is more likely if the multi-ton herbivores fed as herds. Footprint evidence and mass-death assemblages in the fossil record seem to suggest that some groups of herbivores, notably ornithopods, ceratopians, and sauropods, moved, as appears to have been the case in at least some instances, as large herdlike aggregations. In this way, angiosperms (flowering plants) can be seen to be opportunistic (genuinely weedlike), capable of invading and rapidly colonizing areas of scorched or disturbed habitat. They were perfectly designed to take full advantage of a range of conditions that resulted from a combination of climatic conditions (favoring periodic brush fires similar to those that even today regularly affect southern Africa, Australia, and parts of North America) and the regular habitat disturbance or destruction that may have resulted from the activities of dinosaur-sized herbivores.

Coevolution?

The appearance of a major new type of food source (the flowering plants) during the reign of dinosaurs probably offered new opportunities for some groups of herbivores. Although there is only very limited information about the diets of particular herbivores in the fossil record (the rare discovery of preserved stomach contents, for example), it is interesting to note that, following the first fossil remains of flower plants, there is a very marked increase in the range and variety of dinosaurian herbivores in several parts of the world.

Looked at from a larger evolutionary perspective, a number of possibilities emerge. On the one hand, some Cretaceous dinosaurs may indeed have been responding to the appearance of the varieties of new plants. But, in addition, perhaps the activities of dinosaurs during Jurassic and Early Cretaceous times may have been at least one of the factors that created the opportunities or circumstances (for example, the ecological space in the form of disturbed habitats) that favored the evolution of the angiosperms in the first place. Various aspects of these questions are currently being investigated in an attempt to disentangle some of these signals from data in the fossil record. A number of ecological and evolutionary associations can be demonstrated in relationships among animals and plants in the present day, and it seems not unreasonable to expect that broad-scale relationships might also be demonstrated in the past such that the evolutionary histories of plants and major groups of herbivorous dinosaurs might be linked in a coevolutionary way.

The Range and Variety of Cretaceous Herbivores

The Early Cretaceous is notable for the abundance of ornithopod dinosaurs of a range of sizes (from 6½–36 feet in length). It is clear, from the large numbers of fossils of these creatures that are found that they simply must have been abundant in many parts of the world. One of the best known of the Early Cretaceous ornithopods is *Iguanodon*. Best known from Britain and Europe, where its remains are particularly abundant, the animal has recently been reported in parts of North America and Asia (Mongolia). So similar are these remains that it seems very likely that the dinosaur *Iguanodon* could range across the northern continents, all of which were in pretty close connection during Early Cretaceous times. The southern continents appear to contain a fauna that was isolated from the north and progressively diverges in variety and appearance over time, though the localities for such dinosaurs is currently relatively rare. For example, in the late Early Cretaceous of North Africa, there are ornithopods (*Ouranosaurus, Lurdusaurus*) that are large-bodied and closely related to but characteristically distinct from the Northern Hemisphere *Iguanodon*. This suggests a short period of evolution in isolation following dispersal or isolation of populations of ornithopods in North Africa in Early Cretaceous times.

In addition to the ornithopods (some of which were very fast-running creatures similar to *Dryosaurus*), there were also relatively rare heavily armored ankylosaurs; stegosaurs were either extinct or very rare at this time. Sauropods are represented in Early Jurassic times, although they are less frequently found than in Jurassic times, with the notable exception of South America. Through much of the Cretaceous, this part of the world, although it was physically connected to Africa, seems to maintain its own regionally distinct flora and associated fauna. In some respects, the South American region seems to represent a "relict" area of the world—an area that seems almost anachronistic, as though it had become stuck in the Jurassic period. The flora was persistently dominated by gymnosperms (a hangover from Jurassic times), and the fauna was dominated (as in Jurassic times) by an abundance of sauropods (many of which bore bony body armor reminiscent of that seen in the skin of ankylosaurs), a variety of small ornithopods, and some bizarre-looking large theropod carnivores known as abelisaurids. It is notable that stegosaurs and ankylosaurs seem to be absent from South America in Early Cretaceous times.

The Late Cretaceous marks a striking period in dinosaur evolution. It coincides with a period of progressive sea-level rise, which resulted in the flooding of low-lying continental areas to the extent that there was considerable fragmentation and isolation of the continental areas upon which the dinosaurs lived. The Late Cretaceous was a remarkable provincial world by any standards, and some of the high-standing diversity of dinosaurs found at this time undoubtedly reflects the fact that so many populations were isolated in their own geographic areas, and began to evolve in their own specific ways.

At the beginning of the Late Cretaceous period, two new groups of ornithischian dinosaur appear for the first time, but only in the Northern Hemisphere: the pachycephalosaurs and the ceratopians, both of which have clear origins from within the ornithopod lineage of the Jurassic and Early Cretaceous.

Pachycephalosaurs are notable for the development of a greatly strengthened and frequently thickened and domed skull roof. Frequently, the skull roof is additionally ornamented with a fringe of knobs and spikes. These herbivores are relatively rare as fossils but may well have used their thickened skulls as a means of behavioral posturing and butting. More striking is that these dinosaurs were the ceratopians. Early members of the group, known as psittacosaurs, are known in the late Early Cretaceous. They strongly resemble ornithopods and are about 6½ feet long, but are distinctive because they have a peculiarly narrow and high snout, and a somewhat parrotlike beak. Later forms grew to considerable size (26–30 feet in length), developed an even more exaggerated hooked beak, and had comparatively huge heads, with elongate horns on the nose and above the eyes, and an enlarged frill of bone at the back of the head that covered the neck as far back as the shoulder. The gigantic skulls and massive jaws, equipped with dense blades comprising as many as 1,000 teeth, were clearly designed to chop plant food. The group as a whole was evidently highly successful in Asia, but more particularly in western North America, where there is a great deal of evidence of large herds of these animals having existed at times.

Associated with the very abundant ceratopians was a very significant range and variety of ornithopods commonly called hadrosaurian or duck-billed dinosaurs. Very few of these had anything remotely resembling ducklike bills, but they did have large horny beaks at the front of the snout and impressive arrays of grinding teeth on the sides of their jaws for chewing plant food. A considerable range and variety of such duck-bills were also found in America and Asia at this time, alongside the ceratopians. There is also strong evidence of vast herds of these creatures.

Finally, among the ornithischians, the ankylosaurs, a group that can be traced back to the Early Jurassic, began to make an impression on dinosaur communities. In the Late Cretaceous of America and Asia, a considerable number of these large, tanklike creatures were to be found. Not only did these animals have heavily armored skins, they also developed an interesting defensive weapon in the form of a large bony club that could be swung from the end of the tail at the legs of their attackers.

In sharp contrast to these abundant dinosaurs in the Late Cretaceous, the sauropods are comparatively rare members of any fauna (apart from South America). They are still found as occasional skeletons worldwide, but they are far rarer than either the ceratopians or duck-bills.

Trends

Among herbivores, there is a clear shift away from the very large, multi-ton herbivores (sauropods) of the Jurassic in favor of the generally smaller, more mobile, low-browsing forms of dinosaur (ornithopods, ceratopians, ankylosaurs) that were capable of cutting or grinding up their fodder in the mouth far more completely than any sauropod could. Whether this reflects a shift to more varied sorts of plants, and perhaps reflects the rise of the angiosperms especially in the very latest Cretaceous, is uncertain, but the coincidence is certainly there to be observed. It is also true that the geographic areas in which such animals "played" were considerably reduced in Late Cretaceous times, and this may well have affected the maximum size and sustainability of large vertebrate herbivores in smaller ecosystems (a broad equivalence of the miniaturization of large mammals (dwarf elephants, hippos, and rhinos), factors occasionally observed among mammals in small island communities in the present day or relatively recent past). The fact that sauropods persisted in abundance and variety in South America during the Cretaceous, where there is some evidence for the retention of a more typically "Jurassic style" floral regime is perhaps instructive. It certainly suggests that some component of the floral changes that affected the rest of the world may be linked to the strikingly different variety of herbivores that are seen.

Range and Variety of Carnivores

Just as the herbivores are seen to be more more varied and interesting in the Cretaceous, it is no surprise to discover that the predators that fed on at least some of these dinosaurs also became significantly more varied—some became positively bizarre! They also show a similarly strong north-south divide.

Northern Predators

A number of different predator families (some of which appear to have been rare elements of Jurassic communities) became established in Cretaceous times across the Northern Hemisphere. Notable among these were the large and highly specialized tyrannosaurs: *T. rex*, *Albertosaurus*, and others; and the much smaller but equally highly specialized dromaeosaurs, the curiously birdlike, sickle-clawed predators such as *Deinonychus* and *Velociraptor*.

One of the more bizarre themes pursued by some of the theropods of the Cretaceous was the loss of teeth and the development of very birdlike beaks. Ornithomimosaurs are well known for their close resemblance to modern ostriches in both size and shape (except for the dinosaurian tail and three-clawed arms); even more bizarre were the oviraptorosaurs with their very peculiar high-crested short-snouted skulls and curiously birdlike stumpy tails. But perhaps the weirdest of all theropods known to date are the therizinosaurs of the Late Cretaceous of Asia. These animals were large, and had rather barrel-like bodies, short tails, powerful back legs, long necks, and small heads with rather tiny teeth; but their arms were large and powerful and ended in the most enormous flattened, scythe-shaped claws. To add further puzzlement to their nature, it has been reported recently that some therizinosaurs (perhaps all) had a shaggy, hairlike covering.

Southern Hemisphere Predators

Southern predators are far less well known, but were probably no less varied or outlandish in their form than their northern equivalents. Abelisaurids from South America are well known, and include forms such as the bull-horned *Carnotaurus*. Not only did this dinosaur have bull-like horns on its head, but it had the most extraordinarily stunted arms (similar, but built completely differently from the famously small arms of tyrannosaurs). *Giganotosaurus* is another South American form that equaled the largest

In the Late Cretaceous of North American tyrannosaurs, duck-billed hadrosaurs and horned ceratopsids were the dominant dinosaurs. Here they are represented by *Gorgosaurus*, *Kritosaurus* and *Centrosaurus* respectively. Most

tyrannosaurs in size. Unfortunately, the fossil record of other Southern Hemisphere forms is not well understood.

North-South Migrators?

Occasionally, evidence appears that dinosaurs apparently crossed the north-south divide during Cretaceous times. Among the herbivores, some duck-billed ornithopods managed to spread into South America near the end of the Cretaceous. A bizarre group of theropods known as spinosaurs are also known from Europe, North Africa, and South America, based upon *Spinosaurus* (first discovered in Egypt), *Irritator* (from Brasil), and *Baryonyx* (which is found as species in both Niger and Britain). They had peculiarly elongate and narrow-snouted skulls that were vaguely reminiscent of those of crocodiles; in addition, they had extraordinarily powerful shoulders and arms. Currently, it is thought that these were specialist fish-eating dinosaurs.

large trees continued to be conifers, but flowering plants were beginning to become common.

If this proved to be correct, then it would be an extremely unusual way of life for such large creatures, though an affinity with water might help to explain their rather unusual and unexpected geographic distribution.

World of the Dinosaurs

There is no single descriptor for the world of the dinosaurs, and any attempt to characterize their time on Earth is hopelessly flawed. The dinosaurs (excluding the birds) lived on Earth and dominated all terrestrial ecosystems for 160 million years. This is such an enormous span of time that imperceptible factors, such as the incredibly slow movement of the continents (Europe is currently moving away from the eastern seaboard of North America at a rate of less than an inch per year) assume considerable importance. The interplay between continental movement, the relative positions of the sea and land, ocean circulation, the patterns of winds, and the topography of the land all affect the

229

relative success/failure, spread, isolation, evolution, and interactions of the flora and fauna across the entire globe.

The dinosaurs happen to have lived during an immensely important and interesting phase in the history of the position of the continents—or perhaps they lived *because* of the position of the continents at that time. They arose on the supercontinent of Pangea and the relatively harsh and biologically challenging continental climatic regime (seasonally very hot monsoonal weather with predominantly desert conditions). Such conditions (in the Late Triassic) may well have favored the biological attributes of conventional reptiles and given them the edge in survival terms, culminating in their rise and dominance.

The subsequent Jurassic and Cretaceous periods demonstrate rather milder climatic conditions, though climatically the world was persistently much warmer, more arid, and less seasonal than the present day. The Jurassic demonstrates the rise of a range of sophisticated large-bodied herbivores (notably the sauropods), all of which suggest that plant productivity was sufficiently high to have sustained such ecological monsters.

By contrast, the Cretaceous world first seemed little different from Jurassic times. But as the Cretaceous unfolded, a fascinating diversity of new types of herbivorous (notably the pachycephalosaurs, ceratopians, and ornithopods) and carnivorous (oviraptorosaurs, therizinosaurs, spinosaurs) forms appeared, some of which are just as puzzling as the abundant multi-ton herbivores of the Jurassic, but clearly reflected a changed ecological regime. The latter is marked by the rise of a new plant type (the flowering plants) and the increasing provinciality of the Late Cretaceous world. Not only was the ecology new and changing, but the areas where these animals lived was far more restricted than previously. This latter may well have fueled the evolutionary process that led to the extraordinary diversity of dinosaurs that appeared at this time.

The disappearance of the nonavian dinosaurs 65 million years ago comes as a shock to the system in many ways. Dinosaurs were rapidly diversifying and adapting to a range of new and challenging conditions. They were, however, decimated by a profound ecological perturbation; the only survivors were the specialized feathered dinosaurs that we call birds today. Perhaps the range and diversity of bird anatomies and behavior offers a distant echo of at least some of the dinosaurs (the small theropods) and their world.

Chapter Five
Dinosaur Behavior

Introduction

Lacking time travel, how can we determine what dinosaurs did on a daily basis, as they lived their long-ago lives? Fortunately, as Jim Farlow explains, dinosaurs left numerous footprints and trackways, which, when preserved in the sediments, serve as the closest thing we have to recordings of their movements. Is there evidence that herbivorous dinosaurs moved in great herds and that theropods attacked these groups in organized packs? Farlow recently worked with the late David Thomas to examine one particularly interesting set of dinosaur footprints that appears to preserve the moment a giant predatory dinosaur attacked an even more gigantic sauropod, circa 100 million years ago.

In order to power all that activity, dinosaurs had to eat. David Norman has been working to understand how dinosaurs dined. He explores the great diversity of feeding types found among the group, from knife-toothed predators to cowlike herbivores. As usual with dinosaurs, there are areas of mystery and controversy when it comes to what some of them ate. Were the strange oviraptors and the ostrichlike ornithomimids chowing down on flesh as most other theropods, or did they eat plants? Less ambiguous is how *Tyrannosaurus* made its living. Gregory Erickson presents the latest research on what—despite the discovery of larger theropods—continues to be the most potent killing machine yet known to have dwelled on land.

Dinosaurs ate plants and flesh for the same reason all animals do: to acquire the energy needed to reproduce. Our knowledge of how dinosaurs bred and spawned has exploded in the last few decades, as a result of finding vast numbers of eggs, some in nests and some even containing the skeletons of the embryos that are so crucial to identifying the dinosaur that laid the eggs. David Varricchio covers the reproduction of dinosaurs from around the world, whose modes of reproducing may have been more diverse than that of any other comparable vertebrate group. Most interesting are those theropods that appear to have practiced a preavian form of nesting, during which they incubated their eggs—still more evidence for the dinosaurian heritage of birds.

Tracking Dinosaur Society

by James O. Farlow

Nonavian dinosaurs are completely extinct, so reconstructing their social behavior requires a significant stretch of the scientific imagination. Were they solitary animals, or did they move around and interact in groups? Did different species of dinosaurs live in different kinds of social groups? Answering such questions is difficult; we cannot go into the field and observe the behavior of dinosaurs the way that specialists on living animals can. But this does not mean that we are reduced to pure speculation about dinosaur lives. We can begin to think about dinosaur behavior by considering the behavior of their living relatives, crocodilians and birds.

Adult alligators, crocodilians, and gharials are generally solitary animals, and can be downright unfriendly during territorial disputes or quarrels over mating privileges. On the other hand, adult crocodilians are often very protective of their young, ferociously guarding them from a variety of predators (including larger individuals of their own species). Crocodilian behavior therefore suggests that interactions among members of the same dinosaur species could similarly have included both nasty and tender components.

Modern bird species show a considerably greater range of body form and lifestyles than do modern crocodilians, and it is therefore not surprising that bird species show a great variety of social behaviors. Some birds are solitary and provide the bare minimum of care for their offspring. Other species are gregarious and even colonial. Adults of many bird species house, protect, and feed their chicks from the time they hatch as helpless animated appetites until the time they are ready to leave the nest.

Judging from the range of behaviors seen in extant crocodilians and birds, we have considerable latitude in imagining what dinosaur societies may have been like. Some dinosaur species may have consisted of grumpy individualists, while members of other species could have been quite social animals. Individuals of some species might have been gregarious and social at some stages or times in their lives, but were otherwise loners.

Skeletal anatomy, too, hints at features of dinosaur behavior. Many dinosaur species sport a variety of bumps, crests, knobs, horns, or frills on their heads. In many living animals—deer and antelope, for example—such features enable individuals to distinguish members of their own species from other species, and to identify the sexes, which is highly useful. Long-limbed sauropods, supporting a greater percentage of mass on their forelimbs, such as *Brachiosaurus*, had weaker forelimbs; the diplodocids had stronger forelimbs, owing to their short arms and low percentage of mass supported by the forelimbs. These structures become operational only as sexual maturity was being approached; perhaps adolescence was as hormonally stormy a time in dinosaurs as it is in humans.

If many dinosaur species went to so much trouble to sprout horns, bumps, frills, crests, or whatever to make themselves appealing to other members of their own kind, this suggests that they spent enough time together to make these visual signals useful. Of course, they might have aggregated only when spring was in the air and a young dinosaur's fancy turned to . . . well . . . you know what. As we shall see, however, there are other reasons for thinking that, in many species, dinosaur interactions involved more than one-night stands.

But not all interactions among individuals of dinosaur species were so friendly. The bones of therapods not infrequently show fractures and other injuries. While some of these were undoubtedly incurred during struggles with prey, many of them were probably incurred during battles with other carnivorous dinosaurs including members of their own species. Some theropod skulls and jaws have impressive bite marks, indicating that fights among predatory dinosaurs involved nipping and chomping at opponents' faces. Meat-eating dinosaurs may often have been as ill-tempered as modern crocodiles.

Interactions among individuals of the same theropod species may at times have been fatal. Skeletons of the small theropod *Coelophysis* have been found that contain the bones of smaller individuals of the same species in the gut region. These smaller dinosaurs are too big to be unborn embryos, which means that they were instead dinner. It is possible that the eaten individuals were scavenged, but we cannot overlook the more grisly possibility that the smaller *Coelophysis* were the victims of cannibalism. This suggests an interesting parallel with modern crocodilian behavior: hatchlings are often protected by adults, but once the young reptiles grow to a certain size, the protective instinct switches off, and adolescents become fair game for larger individuals.

Fights among dinosaurs may not have been restricted to carnivores. Up to about 1 percent of the bones in bonebeds (more about these in a moment) dominated by the ceratopsians *Centrosaurus* and *Pachyrhinosaurus* show fractures, and ribs seem to have been most vulnerable to such damage. Perhaps, as in modern bison, these injuries were received during fights when one dinosaur rammed another.

Some of the most compelling evidence for dinosaur sociality comes from monospecific bonebeds. Bonebeds are deposits where, in contrast to the usual isolated or scattered bones or skeletons of individual animals, one instead finds the jumbled remains of many

How times change. Hadrosaurs such as *Anatotitan* used to be imagined as sluggish dwellers of swamps, but they are rarely found in such deposits. The new view based on analysis of their locomotion, feeding adaptations,

location of preservation, and vast hadrosaur bonebeds indicate that they were ungulate-like, herd-forming, land herbivores capable of moving great distances at high speeds in search of food.

individual animals. Such assemblages are often dominated by the remains of a single species, and so are said to be monospecific. Bonebeds of this kind are usually the result of some horrific disaster, the nature of which is not always apparent—catastrophic floods, droughts, and volcanic eruptions are among the popular suspects. However they formed, bonebeds provide a snapshot of a dinosaur population from an instant of time, and so provide important clues to dinosaur ecology and behavior.

One would expect catastrophes to be equal-opportunity assassins, nuking all of the animals in a particular area. It is hard to see how a catastrophe could kill numerous animals of one species but few of any other, and then concentrate all of the victims in one spot, if the stricken animals had been widely dispersed as individuals across the landscape when the lights went out. It is much more plausible to think that the victims died together, all at the same place. Now, if a drought were responsible for mass mortality, it is possible that normally solitary animals might be concentrated in a small area, desperately seeking the final scraps of food. But if such were the case, why would the concentrated animals be more tolerant of members of their own, as opposed to other, species? It seems more reasonable to think that the animals that died together to make a monospecific bonebed also lived together, as members of a herd or other social group.

Bonebeds of this kind are known for prosauropods, ceratopsians, hadrosaurs, and theropods. Some dinosaur species are represented by more than one such bonebed, strengthening the inference that social groups were not freakish occurrences for these kinds of animals. In some monospecific bonebeds, there are hundreds, maybe even thousands, of individual animals represented, which raises some interesting ecological questions. If these dinosaurs were indeed members of a herd, how could there be so many of them? How could the landscape have supported so many hungry mouths on big bodies? The question is particularly puzzling when one considers that there would probably have been numerous other species of equally hungry dinosaurs living in the area at the same time.

Dinosaur skeletons fall into the general category of body fossils, the actual remains of once-living organisms. Every body fossil represents a tragedy for its erstwhile owner— the beast had to die to donate its skeleton to the fossil record—and bonebeds indicate tragedy on a grand scale. In contrast, trace fossils are sedimentary records of ancient creatures going about their daily business, generally without the threat of imminent death. Fossilized trackways of dinosaurs are a type of trace fossil that provides yet more evidence about their social lives.

Interpreting dinosaur trails involves a degree of uncertainty that doesn't arise with reasonably complete, well-preserved skeletons. With sufficiently diagnostic bony elements, it is possible to determine exactly what kind of dinosaur once occupied the skeleton. In contrast, even the best-preserved footprints generally provide only vague information about the animals that made them. Because the foot structures of the various dinosaur groups are usually fairly conservative within those groups, footprint shapes are also pretty generic within a group. One may be able to say that a given footprint was made by a large theropod, or a big ornithopod, or a sauropod, but it is seldom possible to say exactly which species of theropod, ornithopod, or sauropod we're dealing with.

We can narrow the range of possibilities by considering the age and geographic location of the geological formation in which the footprints occur. Thus, large theropod footprints in a Late Cretaceous rock unit in western North America are unlikely to have been made by a theropod species whose bones are known from the Late Triassic of Australia. On the other hand, if the skeleton of a particular kind of theropod is known from rocks from the same region and geologic age as our theropod footprints, and if the shape of the dinosaur's foot skeleton is consistent with the shape of the footprints, we can argue that it is very likely that this theropod species was our trackmaker—but we can't be completely sure.

This consideration is important because our interpretations of what dinosaurs were doing at a given footprint site will depend on whether the trackmakers were members of the same or of different species. Suppose that we saw two trackways whose footprints were similar in shape but different in size, and the two trackways were side by side and heading in the same direction. If we thought that the two trackmakers were members of the same species—perhaps an adult and a juvenile—we might think that the site recorded a dinosaurian parent and child moving across the site together. On the other hand, if we thought the two trackmakers were members of different species, then it would seem considerably less likely that the two animals were walking side by side.

One of the world's most spectacular dinosaur footprint sites is way out in the middle of nowhere, in the outback of Queensland, Australia. This dramatic locality tells the story of more than 100 small bipedal dinosaurs from the Early Cretaceous that were seemingly stampeded into mass flight by the approach of a much larger bipedal dinosaur, probably a carnivore. The paleontologists who originally described the site thought that the small bipedal dinosaurs represented two different species, one a theropod and the other an ornithopod. If true, the co-occurrence of the two kinds of trackway at the same spot suggests that members of each species tolerated the presence of the other, at least at times. However, some paleontologists have suggested that the two trackway types are more likely to have been made by members of the same species, in which case, our behavioral interpretation need not invoke mixed-species flocks. The point, then, is that with trackways we can't tell for sure whether we are dealing with members of the same species, and so our behavioral interpretations must be rather tentative.

Caveats duly noted, we can now consider what dinosaur trackways suggest about their makers' social behavior. At the famous Early Jurassic locality at Dinosaur State Park

At the top, the large number of theropod trackways at this Early Jurassic New England site are heading on the same course. This may record the passage of a pack of the predatory dinosaurs. The bottom shows the Early Cretaceous British Columbia locale, where four iguanodonts appear to have moved as a unit, changing directions at the same time.

One of the world's best known set of dinosaur footprints is this concentration of Early Cretaceous sauropod trackways from Texas. Since every individual trackway is in the same direction and in close association with others the probability that the makers were moving together is high. It is not certain whether there were one or two groups involved, and past claims that the young were in a protective center have not borne out. You can see that just two sauropods were large adults of about 18 tons, the rest ranged in size down to over a ton.

near Hartford, Connecticut, numerous theropod trackmakers traveled every which way, with no obvious pattern to their courses. Quite possibly these dinosaurs were loners. At other sites, the pattern is very different. At an Early Jurassic dinosaur tracksite near Holyoke, Massachusetts, at least 20 bipedal dinosaurs, all theropods, traversed the site all heading the same way. They were not constrained to move in the same direction by any obvious physical barrier, because a few other trackways (which appear to have been made by different kinds of dinosaurs, as best can be told from print size and shape) at the same site head in different directions. The easiest explanation for this pattern is that the 20 dinosaurs were moving together as a group.

A common pattern displayed by trackways at dinosaur footprint sites is for about half of the trails to be heading in one direction and the other half in the opposite direction. If we think about the conditions under which footprints are likely to be formed and preserved, the reason for this pattern will be apparent. Footprints require soft substrates in order to be formed, but some such soft-substrate situations are more likely to preserve prints than others. Footprints can easily be made in dry sands well away from watercourses, but the likely fate of such prints is that they will eventually be gone with the wind. Footprints have a much better chance of survival if they are made in wet sediments, along the margins of streams, lakes, or seas, where they can eventually be buried beneath other sediments.

Although animals will sometimes wander into or out of the water, and so move across the shoreline, more often than not, when an animal finds itself at the water's edge, it will move parallel to the shoreline. So even though the water's edge isn't a physical barrier constraining the movement of animals, the way the walls of a canyon are, animals will still tend to drift along the shore. Over a period of time, we would expect about as many animals to move in one direction along the shoreline as the other, and so trackways are likely to be arranged in just the mirror-image pattern that is so often recorded at dinosaur tracksites.

Now, the mirror-image pattern could readily be generated if groups of animals were to move in either direction along the shore over time. However, the same pattern would be created if solitary animals individually tended to turn in one direction or the opposite when they came to the water's edge. Thus, the mirror-image pattern does not provide compelling evidence that dinosaurs were moving in groups. The famous Early Cretaceous dinosaur footprint sites of the Paluxy River, in what is now Dinosaur Valley State Park near Glen Rose, Texas, provide a good example of this for one kind of trackmaker. The great majority of footprints displayed in the limestone bed of the river are big three-toed jobs likely made by large theropods. The trackways of the big theropods nicely show the mirror-image pattern. We cannot reject the possibility that the theropods were walking along the shore of the ancient Gulf of Mexico in groups, but neither can we eliminate the possibility that the theropod trails were made one at a time, with individual carnivores moving one way or the other.

As impressive as the theropod prints are, they are puny alongside the tracks that made the Paluxy River famous: prints made by enormous quadrupeds, with the hindfoot impressions a yard or more in length. These prints were made by sauropods, and at least a dozen of them moved across the main tracksites in the park. Unlike the theropod trails, which go in both directions along the inferred ancient shoreline, nearly all of the sauropod trails head in only one of those directions. If the sauropods had been moving alone, traveling through the area at different times, we would expect about half of them to go in one direction and half in the other. The fact that they don't suggests that the reason nearly all of the sauropods headed in the same direction is that they were wandering in a herd.

We can, of course, think of other reasons why the sauropods might have moved in only one direction. Maybe the Paluxy River site was made during a season in which sauropods, but not theropods, were migrating, and so moving in the same direction, one animal at a time, independently of one another. However, unidirectional sauropod tracksites are common enough in Texas and elsewhere to suggest that these dinosaurs really did move about in groups; it doesn't seem likely that every sauropod tracksite would happen to have formed during a hypothetical migratory season.

On the other hand, if individuals of sauropod species did commonly travel in groups, we might often expect to find monospecific bonebeds dominated by a single kind of sauropod. A few such occurences have been reported, but most such monospecific sauropod

Before the revolution in dinosaur science, *Triceratops* and *Tyrannosaurus* were usually portrayed as solitary creatures that fought one-on-one combat. Evidence from single species bonebeds and trackways suggest that horned dinosaurs moved in herds, and that at least some giant predatory dinosaurs hunted in packs.

bonebeds have only a few individual animals in them, nothing like the enormous monospecific death assemblages known for some hadrosaurs and ceratopsians. Most bonebeds that have sauropods generally contain more than one sauropod species. Could it be that sauropod herds sometimes were multi-species aggregations?

At some dinosaur tracksites, the arrangement of trackways, and not just the direction of travel of the trackmakers, suggests that the animals were moving as a group. A splendid example comes from the Early Cretaceous of British Columbia, Canada, at a site with several trails made by large ornithopods. At one spot are four trackways that are fairly close together, all of which follow a sinuous path, bending first to the right and then to the left, but not crossing each other. It is hard to resist the idea that these trackways were made by dinosaurs walking side by side, and that when one of them changed direction, the others had to make corresponding changes to avoid unneighborly collisions. This impression is reinforced if we picture the trackmakers in their trails: the overall body size of the dinosaurs relative to their footprints would be such that the trackmakers would have been very close to each other had they been walking together.

To summarize, the social behavior of extant archosaurs, crocodilians, and birds gives us ample justification for thinking that some dinosaur species could have been gregarious animals. This conclusion is reinforced by the fossil record. Dinosaur skeletal anatomy includes cranial features that, by analogy with living animals, are likely to have been employed during interactions among members of the same species. Most compelling of all, monospecific bonebeds and the arrangement of trackways at dinosaur footprint sites provide evidence that is difficult to explain if certain dinosaurs did not live, at least at times, in groups. Of course, this does not mean that all dinosaurs were social animals. Some species may well have comprised solitary individuals that interacted only during the mating season.

If some dinosaurs did live in societies, what were those social aggregations like? Extrapolating from what we see in living crocodilians and birds, it is plausible that some dinosaur groups were family structures, consisting of a parent and its young, or a group of juveniles that had become large enough to get by without their parents, but that stayed together, at least for a time, for mutual protection and foraging. But what about the enormous herds suggested by some ceratopsian and hadrosaur bonebeds? Did their members live together for extended periods of time, or were these immense aggregations temporary affairs? At present we just don't know. There is much that remains to be learned about dinosaur society.

Tracking a Dinosaur Attack

by David A. Thomas and James O. Farlow

American paleontologists seeking dinosaur fossils made some notable finds in the years just before World War II. One of the excavations undertaken during this heyday produced no bones at all, but it nonetheless proved to be rich in information about dinosaurs. That dig, along the banks of the Paluxy River in central Texas, unearthed a dinosaur trackway, a collection of footprints left on an ancient mud flat. Those fossilized impressions owe their preservation to the sediments that encased them, hardening and then eroding 100 million years later to reveal the wanderings of dinosaurs in now solid rock.

Dinosaur trackways of this kind are invaluable to paleontologists. Although there has been a great deal of speculation about dinosaur behavior, fossilized footprints provide the only direct evidence of how dinosaurs actually moved. By studying excavated trackways, paleontologists have been able to glean information about dinosaur gaits and postures. They have used such fossil footprints to determine how quickly different species walked and to deduce that many kinds of dinosaurs traveled in herds.

One particularly intriguing observation from the Paluxy River trackway was made by Roland T. Bird, who discovered this curious set of dinosaur prints in 1938 and partially excavated them in 1940. Before cutting huge slabs of this footprint-studded rock to ship back to his employer, Barnum Brown of the American Museum of Natural History in New York City, Bird mapped and photographed the parallel tracks most thoroughly. He saw immediately that one set of footprints, from a two-footed carnivorous dinosaur (probably an *Acrocanthosaurus*, weighing perhaps 2 or 3 tons), ran parallel to the trail left by an even larger, four-footed herbivore (most likely a giant *Pleurocoelus*), which was apparently traveling in a herd. And he later noticed that the carnivorous dinosaur seemed at one point to have taken a strange skipping stride, leaving two consecutive right footprints in the mud.

Bird believed these two sets of footprints with a peculiar hop in the middle represented the moment the smaller carnivorous dinosaur struck at its larger herbivorous cousin. Most paleontologists with an interest in dinosaur tracks initially scoffed at that rather dramatic interpretation. But some now think Bird may have been correct. That revision in thinking came about because of an unlikely string of events that prompted us to reevaluate this decades-old find.

Following in the tracks of its gigantic herbivorous prey as it crosses a carbonate mudflat, the great *Acrocanthosaurus* bites the leg of the huge *Pleurocoelus* while the latter's tail swings to dislodge its tormentor.

Digging Up Papers

Six years after Bird's death, in 1984, Texas Christian University undertook to publish Bird's autobiography, *Bones for Barnum Brown*. They contacted one of us (Farlow) to read the manuscript and to act as its scientific editor. And it came as a surprise that Bird's writing referred to various charts and a film of the excavation that paleontologists had not known existed.

Interviews with Bird's wife and sister revealed that he had stashed away quite a bit of unpublished information about the Paluxy River trackway. Bird's nephew soon discovered a canister with the lost film of the excavation; it was neatly stored in a basement refrigerator. A box in Bird's attic provided countless notes, along with some large charts of the footprints in question. These diagrams were key finds, because the trackway had deteriorated considerably since 1940: not only were large pieces no longer in place (Bird sent slabs both to New York and to the Texas Memorial Museum in Austin), but seasonal flooding of the Paluxy River had eroded the rock surface and covered it with a mantle of sediment.

Fortunately, Bird's charts and photographs showed the placement of each of the prints in fine detail, enabling Farlow to study the site anew. That work resulted in an extensive report, published in 1987, which stated that the two parallel tracks represented one dinosaur following the other, just as Bird had originally surmised. But the interpretation of the strange hop remained open to debate within the community of professional dinosaur paleontologists—prompting a working artist to get involved.

Sculpting Science

In 1983, the city of Albuquerque commissioned one of us (Thomas) to produce a life-size dinosaur statue for the New Mexico Museum of Natural History and Science, a job that led to many further opportunities to cast dinosaurs in bronze and fiberglass for other museums around the world. To give those creations realistic stances, it was necessary to investigate how animals moved in general and how dinosaurs walked in particular.

A quick study shows that a four-legged mammal, such as an elephant, routinely steps on its own tracks as it walks, with a rear foot often landing in the spot that the corresponding front foot has just vacated. It can do so because both feet on one side of the animal can be off the ground at once. But a modern reptile never places a rear foot exactly where its front foot has trod. It walks with diagonal legs moving together, so the rear foot on one side lands before the front foot on that side leaves the ground.

Interestingly, the tracks of four-legged dinosaurs indicate that these creatures must have moved with gaits similar to those of living mammals and distinct from those used by most contemporary reptiles. This association is evident from the record unearthed from the banks of the Paluxy River: the four-footed herbivorous dinosaurs left imprints from their rear feet that commonly overlap their front footprints.

Might modern mammals have more to teach about the extinct reptiles that once roamed this ancient strand? In an effort to assess Bird's hypothesis, Thomas studied the way mammalian predators hunt. Attackers typically match the speed and direction of their game before they can strike. But often they do more. A carnivore on the attack will usually adjust its stride until it comes into exact rhythm with the running animal that it hopes to bring down.

For example, a lion, cheetah, or other swift cat will accelerate until it has caught up with its target. The predator then changes the length of its stride to match that of its prey. Only by keeping pace in this manner can the attacker reduce its motion relative to its quarry, which frequently is a much bulkier animal galloping furiously in an effort to escape. Eliminating relative motion is critical. Otherwise the predator would have diffi-

Fossil imprints of the Paluxy River trackway were unearthed by Roland T. Bird and his co-workers in 1940. A large slab taken from the excavation was sent to the American Museum of Natural History in New York City for exhibit, where it can still be seen by visitors. Bird also provided the Texas Memorial Museum in Austin with a segment of the trackway, one that contains a print showing a conspicuous drag mark made by the four-legged herbivore, perhaps just after being struck.

culty striking safely and effectively when, finally, it lunges.

In an informal study of recorded attacks of various African predators (lions, leopards, cheetahs, and hyenas), there proved to be only a few instances in which these animals clearly did not come into rhythm with their prey. Some of these exceptional occurrences involved an attacker executing a sudden ambush. In such cases, the need for surprise outweighs the desire to strike more carefully. Other examples in which the attacker failed to come into matching cadence were generally restricted to pairings of small prey and larger attackers, a combination for which it is neither practical nor necessary for the predator to harmonize its stride.

Some caution is clearly warranted in using these observations to help understand the fossilized Paluxy River tracks. After all, modern four-legged mammals are quite different from two- and four-legged dinosaurs. For instance, all the modern predators examined were in a gallop during the hunt, whereas the two-legged carnivorous dinosaur in question was probably in a fast walk or slow run, and the four-legged herbivorous dinosaur was in a mammal-like walk. Still, it would seem that in the early Cretaceous period, just as today, predators must have learned the advantages of matching rhythm with their prey.

100-Million-Year-Old Attack

One might imagine that the two dinosaurs under consideration had taken similar routes along what is now the Paluxy River simply because both were following the ancient shoreline, with their passages separated by many hours or even days. But detailed analysis of Bird's chart of the trackway shows that the proximity of the two sets of tracks could not have resulted from such happenstance.

Not only do the two trails run parallel, detailed examination reveals that the paths of the two animals wavered slightly and in the same fashion. So the subtle movements of one animal must have influenced the other. And something that is not seen at all provides additional evidence that the two sets of tracks were made at the same time. Near the end of the excavated lane, the tracks of the carnivore and herbivore both head toward the left. Had either animal continued in that direction, its footprints would have crossed into the adjacent sets of tracks. But they do not. Both animals must have turned right after leaving the area that Bird ultimately excavated. So, in all, the two trajectories make several jags and two broad turns together. These animals were undoubtedly interacting.

It indeed seems most likely that the carnivorous dinosaur was following the herbivore. The predator probably approached its prey from directly behind, lingering, at first, some steps back as it measured its quarry. The carnivore could then have come into rhythm by adjusting the length of its stride, just as mammals do today. Evidence for this behavior comes from a stretch of the trackway where the footprints for the two dinosaurs form an amazingly symmetrical array. For about a dozen steps, the carnivore placed its right foot near or into the print made by the left hindfoot of the herbivore. This pattern is just what one would expect if the carnivore were trailing the herbivore as closely as possible without colliding—just a few steps back and slightly off to the left.

The rest of the scenario suggested by Bird—that the carnivore actually struck at its prey midway along the trackway—is more speculative, but there is good reason to believe it to be true. Hypothesizing an attack of this kind not only explains the uncanny similarity in the spacing of footprints between the two sets of tracks, it also accounts for the missing left footprint as a hop made by the carnivore, and it elucidates one additional piece of the puzzle unearthed more than half a century ago.

Bird's original observations show that the herbivore's right rear foot dragged at one point. This drag mark can be clearly seen in the slab housed at the Texas Memorial Museum. It occurs a few steps ahead of the spot where two consecutive right carnivore tracks occur. The mark suggests that the carnivore indeed hopped as it set upon the larger beast walking slightly ahead, because it makes sense that the animal struggling to escape would have faltered just as it was hit.

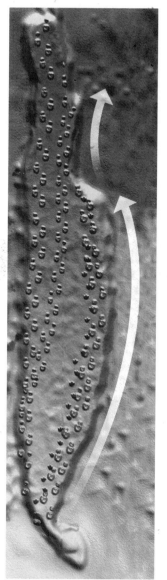

Bird's chart of the Paluxy River Trackway shows how the tracks of the two-legged carnivore (smaller size prints curving around left turn arrow) closely parallel one of several sets left by four-legged herbivores through a broad turn to the left as illustrated by the bend in the lower arrow. Both dinosaurs must also have veered right (upper arrow on diagram) where the excavated tracks end, because their imprints do not cross the other two sets of tracks. The footprints of the predator and the prey also show remarkable symmetry for about a dozen paces where it appears they moved in synchrony, except for one point where the carnivore apparently hopped, leaving no left footprint. This gap can be seen roughly midway through the track sequence.

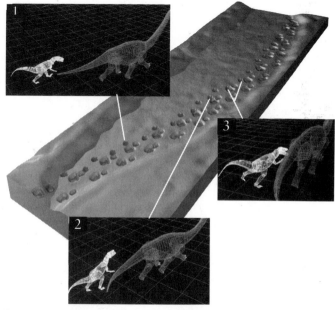

1. Reconstructed attack sequence suggests that the carnivore approached in step with its prey.
2. Closer positioning probably required the carnivore to shift strides so that it moved its right leg forward just as the herbivore advanced its left legs.
3. Apparent hop (two consecutive right footprint) may mark the point on the trackway where the carnivore first struck at its prey.

The drag mark and consecutive right footprints support the notion that the carnivore and herbivore moved over this patch of ground together, separated by only a few steps when the carnivore struck. And the location of these features points to a possibility that the herbivore attempted to carry out a defensive maneuver. It might have tried to throw its weight into the attacker just before being hit, exaggerating the defensive motion by turning left.

We do not know with certainty that such an attack or defense actually happened, how many carnivores joined the chase, or why this particular herbivore was chosen to be culled from the herd. Too much of the record of this intriguing episode remains buried underground. But it now appears perfectly clear that about 100 million years ago, on a limey mudflat in what is now Texas, at least one swift carnivore singled out and possibly attacked a huge, lumbering herbivore. It seems that Bird was not only lucky enough to find remarkable evidence of this incident of natural history but that he was also wise enough to recognize, document, and excavate part of the record of this ancient hunt left on a sodden plain, now turned to stone.

Feeding Adaptations in the Dinosauria

by David Norman

The three distinctive anatomical groupings of dinosaurs—theropods, sauropodomorphs, and ornithischians—are also the basic divisions in feeding types, with some exceptions. The last of all these dinosaurs lived on Earth 65 million years ago. Such long-dead creatures cannot speak to us directly about what they did or did not like to eat. To discover their feeding preferences, paleontologists seek a variety of clues. The most direct would be evidence from their preserved droppings (called coprolites); however, the question that we are often left with is "What left this here?" Unfortunately, dinosaurs did not leave personal identifiers in their droppings! Equally rarely, stomach contents may be preserved within the skeleton, but even this type of evidence raises questions: Did its last meal kill the animal? Were these bits and pieces later washed into the abdominal area of the carcass of the dead animal before it was buried?

More often than not, paleontologists have to follow a more circuitous path of deduction, to take into account what we know of the structure and habits of living animals—the shape of skulls, jaws and teeth, the size of the animal relative to its head, the size of the abdomen, and all manner of other clues. This really is a whodunit on a grand scale, and even today we only know a few of the answers.

Putting It into Context

Humans exhibit a strong tendency to categorize and compartmentalize things in the world around them. This is entirely understandable. The world is a very large place full of a bewildering variety of objects, be they animals, plants, rocks and minerals, books, postage stamps, or cars. The choice and variety is practically endless. What we are often looking for is some way of organizing or making sense of what we see in a more systematic way, to discover whether there are underlying patterns—some might call it looking for a rhyme or reason—to what appears at first glance to be a bewildering diversity of life.

Taking, for example, the feeding habits in animals, we might readily categorize animals as, on the one hand, those that eat meat (the carnivores) and on the other, those that are vegetarian (the herbivores). At first, this seems a nice and neat division, and coincides with what we know of some of the more obvious animals that we know of: cats and crocodiles eat meat, whereas rabbits, elephants, and horses eat plants. Unfortunately, life is a little more complex than that. A moment's reflection and we realize that there are animals that don't fit neatly into our two categories—where, for example, would we put humans? Indeed, there are countless other examples of animals that eat a variety of foodstuffs, so we really need at least one other category, for those that eat more or less anything (we call these omnivores). Further pondering reveals that there is also a variety of organisms that feed in very specialized ways: some eat only ants, others eat a variety of insects; some eat only fish, others filter plankton in the oceans. Perhaps we should recognize these specialists as well.

I have stressed the difficulties of such an approach (which at first sight seemed obvious) simply to make a general point. The world is such a varied place, and it is truly amazing the range and diversity of things that are eaten, and the curious adaptations (ways and means) by which even the most unpalatable of things are devoured (imagine, if you can, preferring to chew pieces of stony coral, as some fish do). So trying to stuff animals into imaginary boxes on the basis of what they prefer to eat is both understandable and, at the same time, futile; it also risks missing the point about the variety and richness of life and nature, which often stubbornly refuses to bend to our will to systematize it.

In the living world that we inhabit, it is possible to wonder about such issues almost at leisure. As scientists, we can—through observation of the habits and dietary preferences of animals—at least test our theories or simple ideas about the range and variety of feeding habits. We can watch what animals eat, check our observations against the shape of their teeth and jaws, and, if we wish, probe the intricacies of their digestive system for tell-tale signs associated with particular diets.

Dinosaur Fossils: Clues and Deductions

Fossil animals, such as dinosaurs, however, present us with a huge number of problems when it comes to trying to establish their diet and food preferences. We have to resort to techniques that would probably have gained the grudging approval of the archetypal fictional detective Sherlock Holmes: we must eliminate what is clearly impossible, and whatever we are left with should, not matter how improbable, be close to the truth. We are helped in our investigations by some of the results of several branches of science, and through the way that these can be interwoven with clues from the fossil record and the intellectual spark ignited by some apparently simple, and yet deeply complex, questions.

How did the earliest dinosaurs feed? What was their diet?

We are confident that dinosaurs first evolved on Earth during Middle Triassic times. The animals most closely related to these early dinosaurs were small to medium-sized (2–10 feet long) long-legged creatures with sharp claws on their hands and feet, large forward-

Medium-sized prosauropods such as South American *Riojasaurus* set the pattern for sauropodomorphs in using small heads at the end of long necks to reach their food, in this case the then-common *Dicroidium*. Unlike the later sauropods, prosauropods retained supple-fingered hands with which to grasp branches.

looking eyes, and jaws with narrow, curved teeth with serrated edges, rather like the cutting edge of a modern steak knife.

From this evidence we can deduce that these predinosaurs were likely to have been swift, beady-eyed predators. Long, strong legs imply that they could run fast; but that does not indicate diet, even though it is useful as circumstantial evidence. Large eyes that look ahead are characteristic of predators today: think of a cat or an eagle; they need to get a good look at their prey. In this context, prey animals might have large eyes; but rather like rabbits or horses, these are positioned on the sides of the head so that they have a very wide range of vision to the side as well as behind (often the direction of the predator's ambush). Most convincing, however, is the shape of the teeth; these are curved, narrow, pointed blades with serrated edges that are only good for one thing: slashing and slicing flesh.

Modern observation of large predatory lizards such as goannas, monitors, and the Ora or "Komodo dragon" supports this view: they have just these sorts of teeth and an unquestionable slash-and-slice method of killing and eating their prey.

Theropods and the Very Earliest Dinosaurs

On that basis we can be reasonably sure that the earliest dinosaur-like animals in the Triassic were predatory carnivores. *Eoraptor* and *Herrerasaurus*, both of which have been discovered in rocks of Late Triassic age in Argentina, are among the earliest, well-preserved dinosaurs, and conform exactly to this style of animal. They ranged from 3–10 feet in body length and probably fed on a range of small to medium-sized prey items. We have no idea whether they were pack-hunting animals and capable of bringing down even larger prey (such as the 15–25-foot-long sauropodomorphs of the time)—such evidence is simply not available in the fossil record. The earliest theropod dinosaurs (such as *Coelophysis*) are even more slender and lithe animals than either *Eoraptor* or *Herrerasaurus*.

Were there any other types of early dinosaur? Were they all predators?

During the Late Triassic, other radically different sorts of dinosaur do appear in the fossil record. These include small and frustratingly poorly known forms such as *Pisanosaurus* from Argentina, whose fossil fragments comprise a piece of jaw, with teeth, and broken parts of the backbone and hip region and legs. The clues from this specimen are tantalizing. The body was lightly built like the predinosaur type: this was clearly a runner. The jaw is quite solid and short; and, instead of having sharp, curved teeth, they are rather blunt and chiselled off at the top. The solidity of the jaw and the shape of the teeth are typical of a plant-eater, and this may well represent (though the evidence is very slender at the moment) one of the earliest of the group known as ornithischian dinosaurs. As we shall see, ornithischians appear to have been exclusively plant-eating in habit.

Early Sauropodomorphs

The Late Triassic also reveals a number of larger, and surprisingly widespread, dinosaurs such as *Plateosaurus*. These animals range in size up to about 25 feet in length. They have a posture and general shape that is different from, or at least an obvious modification from, that seen in the earliest dinosaurs. These are, by virtue of their size, quite heavily built, and therefore more lumbering than scampering in their style of movement. The backlegs are longer than their front and used for walking; the front legs are short and strong and have large sharp claws, but may also have been used for walking at times. The neck of these dinosaurs is unusually long, and linked with this, the head is quite small, surprisingly so for the size of the complete animal. The jaws are slender and lined with tall teeth with serrated edges. But there are important differences. The teeth are tall, narrow, and not curved backward for gripping prey; in fact, the teeth are quite coarsely serrated, and not as well suited for slashing and slicing as in a creature such as *Coelophysis*.

What was the diet of an early sauropodomorph such as *Plateosaurus*? The evidence is far from clear. These types of dinosaur are clearly not slim and lithe—as might perhaps be expected of a pursuit predator. They also had a large abdomen, rather than the slender "wasp-waisted" proportions seen in most meat-eaters. A large abdomen is frequently typical of plant-eaters because they eat bulkier food that takes longer to digest. However, the claws, particularly those on the hands, are large, curved talons similar to those seen in *Herrerasaurus*. The teeth, though they are serrated, are not curved; similarly serrated teeth have been seen in living plant-eating lizards such as iguanas. The head is also unusually small for the size of the animal—clearly this animal could not take huge bites out of its prey, as is often the case with predators. And, finally, its eyes were not forwardly positioned, but were capable of scanning to the side and backward while the head was facing forward.

Typically, these dinosaurs are thought to be plant-eaters—not that that view has never been challenged. A plant diet agrees with the bulk of the animal, the small size of its head, the positioning of its eyes, and the relative simplicity of its teeth. This view has been further bolstered by the observation that some dinosaurs of this type appear to have swallowed pebbles and stored them in an area of the gut (gizzard) where they could be used to grind up food. Nevertheless, we should be a little wary of these observations. For example, crocodiles have a stone-laden gizzard and they are certainly not herbivores. And iguanas (with their similarly shaped, but smaller, teeth), though they do habitually nibble plants, are perfectly content eating bird eggs, young chicks, and mice and will scavenge almost any dead animals when they can find them.

On this basis, the evidence is not entirely compelling either way. Perhaps plateosaurs and their relatives were omnivore, perfectly capable of feeding on plants, but equally willing to eat meat or scavenge if the opportunity arose. Such wide-ranging feeding habits can be decidedly advantageous when food of one sort or another is in short supply. Perhaps these were genuine Late Triassic opportunists; their general abundance at this time in Earth history may be a reflection of this versatility.

So, by the close of the Triassic, it would appear that dinosaurs had reinvented some of the major feeding strategies recognized today: *carnivore* (theropod), *herbivore* (ornithischian), and *omnivore* (sauropodomorph), and that these types clearly had evolved from a small, fast-moving, essentially carnivorous ancestor.

Once dinosaurs appeared, did they conform to modern concepts of feeding types for the rest of their 180-million-year reign on Earth?

The answer to the question is the typically enigmatic, "Well, yes and no!" Subsequent Jurassic (210–145 million years ago) and Cretaceous times (145–65 million years ago) saw a far greater variety of dinosaurs and, by logical implication, a greater variety of feeding techniques and preferences. And this is the point at which our observations and deductions become decidedly more interesting. Throughout the time of the dinosaurs, there were groups of dinosaurs that fit the categories of carnivore and herbivore perfectly, while others are not nearly so easy to categorize.

Theropods are found throughout the remainder of the Mesozoic era. Many are rather conservative in their body structure and proportions—they differ relatively little from the Late Triassic examples we have seen already. If we had to find a comparison today, we might equate them with the cat family (ranging from feral cats of the backyard through a range of body sizes culminating in the African lion). All these animals are unmistakably catlike; they just differ in relatively subtle ways (coloring, ear shape, tail shape) or more obviously in body size.

Carnivores

The range of theropods includes many small forms, such as the wonderfully delicate 2-foot-long *Compsognathus* of the Late Jurassic, which might be considered the predatory analogue of the modern feral (or domestic) cat. Small theropods tend to have, as one might expect, light, agile bodies; nimbleness and speed of reaction would have been a premium in the survival stakes. The long, whippy tail counterbalances the body, and not only aids it in balancing, but provides a means of changing direction while in hot pursuit of prey. The neck tends to be long, slender, and very flexible, rather like that of a modern swan or goose and ends in a light, and again rather slender and pointed skull. These animals appear to have been resourceful predators feeding on a very wide range of small creatures. This is confirmed in the case of *Compsognathus* through the discovery of one specimen with a lizard (*Bavarisaurus*) folded neatly in its gut (quite literally evidence of its last meal). The lightness of build and relatively small size of the skull, teeth, and jaws suggest that these dinosaurs did not necessarily immediately subdue their prey with powerful, paralyzing bites. They have quite long arms and powerful hands, tipped with sharply taloned claws, which, undoubtedly, they used to catch and immobilize prey before it was killed. The lizard, for example, may well have been swept into the mouth of *Compsognathus* in one very rapid movement of the hands. If you have ever tried to catch a small modern lizard, then you will appreciate how fast and skillful a hunter this dinosaur must have been.

In contrast, theropods such as *Allosaurus* (ranging up to 25 feet in body length) might be regarded (in some sense) as Late Jurassic lions. They are, apart from their size, very similar in general body plan to the much smaller, but roughly contemporary, *Compsognathus*. Differences are largely mechanical—the product of building a much larger creature and scaling its skeletal support appropriately. Allosaurs have much stouter legs to support the heavier body; they also differ markedly in the shape of the head and neck, undoubtedly a consequence of differences in diet and predatory habits. An animal with a head that is nearly 3 feet long would not be expected to poke among the stones for lizard brunch—or have much success if it tried (the inertia of its head and the relative slowness of its reactions would preclude that). These predators undoubtedly preyed upon other dinosaurs, and their heads reflect that: they are large, heavily muscled, and extremely strongly built. Their jaws are long and armed with extremely sharp, viciously curved, serrated teeth, which were undoubtedly used to slash at their prey before bolting down large pieces of meat. To aid their gluttony, the lower jaws of these dinosaurs could bow outward thanks to a special hinge found in the middle on either side. The strong, sharply clawed hands and feet almost certainly helped in subduing and tearing apart their victims with speed and efficiency.

The enormous forces generated in the skull during the kill and dismemberment of the prey had to be transmitted back to the body through the neck. A long, snaky neck would not have been very suitable. So, allied to the large, strong head, these dinosaurs had short, thick, heavily muscled necks. The principal differences in body shape seen when comparing the skeletons of small and large theropods are explained largely in terms of simple mechanics associated with body size and different styles of feeding (a similar situation to the cat family of today). We may note subtle differences in the shape of the head, in odd bumps, crests, or ridges, or in the proportions of the arms or hands, but these are not so important in the scheme of feeding adaptations. I have little doubt that such features were very important to the animals themselves because they probably helped in recognition (how else would they know others of their species to mate with?) but that is another story.

Hyper-Carnivores

Among the large theropods, tyrannosaurs of the Late Cretaceous fit, unquestionably, into their own "hyper-carnivore" category. We have never, before or since, seen anything quite like these creatures. Ranging up to 45 feet in length, they equal the largest of all carnivores ever known—and they may well represent the largest bipedal carnivore that nature will allow. The huge reinforced head 4–5 feet in length has equally long jaws lined with gigantic banana-like teeth that were used for piercing and crushing their prey with terrifying strength. Excellent binocular vision (seen in the skull), high intelligence (judged from the size of the brain cavity), long, powerful back legs for rapid movement, and very powerful talons all bear testament to its predatory abilities. The curious and noteworthy feature of all tyrannosaurs is the apparently absurd shortness of the arms: each is equipped with two rather puny clawed fingers.

Some have speculated that these arms were used as gaffs to hold their prey close to the head, which seems manifestly absurd. Alternatively, it has been suggested that they were used as grapples to grip the ground so that these animals could stand up, after resting, without the indignity of pitching forward on to their snouts (also an absurd, if rather imaginative, idea). Modern biology may come to our aid in interpreting these structures. Many ground-dwelling birds have much reduced wings (i.e., arms) and use their heads and feet quite effectively for foraging and feeding. In the past, some giant birds, such as phororhacids, were very large-headed predators, not unlike the greater theropods in some respects. Such creatures proved perfectly capable predators, despite having puny "arms." Tyrannosaurs used their well-equipped heads and legs for foraging and feeding in the same way, and had little use for arms at all; their reduction probably reflects this fact—they may well have had some incidental role in the life of tyrannosaurs, but what that was we can only guess.

Tyrannosaurs have, fortunately, left some tell-tale evidence of their feeding habits which falls into what might euphemistically be called the "smoking gun" category. Greg Erickson and Karen Chin have found compelling evidence of the feeding habits of tyrannosaurs. Erickson reported a partially chewed pelvis of a large ceratopian dinosaur with a range of bite marks, some of which could be molded to reveal the unmistakable shape of tyrannosaur teeth. Chin reported an extremely large coprolite of Late Cretaceous age, full of broken bone fragments and tissues. Judged by the size of the specimen, its geological age, and its content, this seems likely to have been left by a large tyrannosaur.

Puzzles

As noted, not all theropods fit neatly into the categories described above. I will only hint at a few types, but they do unquestionably blur the boundaries when trying to establish such simple facts as the feeding habits of theropods.

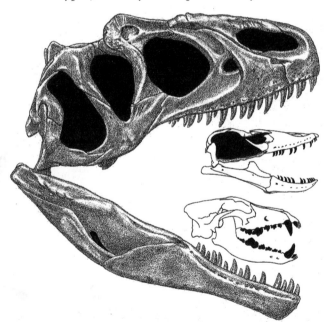

With long rows of big serrated teeth, there is no doubt that theropods such as *Allosaurus* were fleash-eaters. The configuration of theropod skulls was more similar to that of lizards like the Komodo monitor than it was to carnivorous mammals such as wolves, which have fewer, more complex teeth set in more solidly built skulls. Neither living reptiles or mammals can compare in size to the great dinosaurian killers.

Ornithomimosaurs are one rather odd-looking group of theropods of the Cretaceous period. Ranging between about 7–14 feet in length, these animals have the proportions of generalized small theropods, such as *Compsognathus*. They have the long legs, long tail, long arms, and long neck typical of small theropods. But, unexpectedly, they lack teeth, and their jaws appear to have been lined by a birdlike beak. Their general resemblance to modern emus or ostriches is very striking despite the long bony tail and long, clawed arms.

Speculations concerning the feeding habits of these creatures are

256

many and varied, and entirely unsupported by facts. The evidence is absolutely minimal; for example, we have no evidence from stomach contents or coprolites. We can point to the feeding habits of modern, large, flightless birds for comparison and suggest that the ornithominids were omnivores—feeding upon everything and anything (ostriches are legendary in that regard). All that we can do is rule out their being able to kill or bring down large dinosaurian prey in the manner of large theropods, since they lacked the necessary strength and weaponry.

Oviraptorosaurs are a bizarre-looking group from the Late Cretaceous. Like ornithomimosaurs, these theropods are medium-sized and have long legs, arms, and necks, and a rather unusually abbreviated tail (incipiently birdlike). The reason for the "bizarre" epithet relates to two observations, one historically misleading, the other just plain strange. Oviraptorosaurs were first recognized among the remains of dinosaur eggs and nest recovered from Mongolia in the 1920s by crews from the American Museum of Natural History in New York. A complete skull was found, slightly crushed, atop a nest of eggs laid by the dinosaur *Protoceratops*. The name *Oviraptor* (i.e., egg thief) indicates the original belief that this dinosaur had been caught in the act of robbing a nest of eggs. This seemed to make sense in a general way not only because of its incriminating fossilized position, but because its jaws lacked teeth and were very short and stout—ideal, or so it seemed, for cracking open thick dinosaur eggshells. These dinosaurs also look genuinely bizarre because they have an extraordinary gargoylelike crested head.

Return trips to the Mongolian desert in the 1990s by other teams from the American Museum of Natural History (Mike Novacek, Mark Norell, and colleagues) have made some startling discoveries. Not only were complete *Oviraptor* skeletons found, but also clear evidence of *Oviraptor* brooding clutches of its eggs. Never, or so it seems, was a more inappropriate name given to a dinosaur. So what were the feeding habits of *Oviraptor*? We have absolutely no idea! Its tortoiselike beak was very strong and would have given it a powerful bite. Modern tortoises are plant-eaters, but are hardly a good comparator for the lithe *Oviraptor*. Perhaps, like ornithomimosaurs, they were opportunistic feeders, capable of tackling most items as long as they were not too big. Incidentally, we still cannot entirely rule out the possibility that they might have eaten the eggs of other dinosaurs.

Therizinosaurs are another uncomfortably puzzling group of theropods. Rather like the oviraptorosaurs, the remains of these dinosaurs have been gradually put together from chance discoveries made over a number of years. It now appears that these are quite large, very oddly proportioned creatures. The head is relatively quite small, and the teeth are small and leaf-shaped, rather than daggerlike, which is suggestive of a diet of plants. But the arms are powerful and terminate in extraordinarily long sickle-

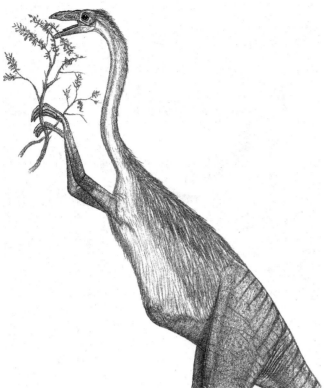

shaped claws—good weapons for a carnivore. The somewhat rotund appearance of the abdomen is again indicative of a preference for plants, which would be decidedly unusual for any group of theropods. But it has to be said that these are highly unusual theropods. Again, our evidential framework is creaky; but though the evidence is rather weak, this may have been the only group of herbivorous theropods.

Sauropodomorphs Ate Plants

During the Jurassic and Cretaceous periods, the sauropodomorphs produced some of the most spectacular animals that have ever lived on Earth. They are generally referred to as sauropods, and include such familiar animals as *Diplodocus*, "Brontosaurus" (i.e., *Apatosaurus*), *Seismosaurus*, *Mamenchisaurus*, and *Brachiosaurus*. At more than 60 feet in length and weighing several tens of tons, these were the largest land animals ever known. They are particularly notable for the extraordinary length of their necks and the absurdly small size of their heads in comparison with the vast bulk of the body. The enormous size of the chest and abdomen of these dinosaurs indicates very plainly that these animals had very large guts, a feature invariably associated with a diet of plants. In addition, the head and jaws strongly support this type of diet.

Most theropods were flesh-eaters par excellence, but not ornithomimids such as *Struthiomimus*. These dinosaurs were so similar to ostriches in size and form that they made have had similar, plant-oriented feeding habits. These theropods differed from the ostriches in having long arms and clawed fingers with which to help manipulate food items.

Taking *Diplodocus* as a well-known example, the head of this dinosaur is remarkable for a number of very obvious reasons. The skull has unusual proportions: the muzzle is very long, the nostrils are found on top of the head just in front of the eye sockets, and the braincase and rear of the skull are twisted downward. The lower jaw is remarkably slender, and all the teeth (peculiarly pencil-like and have chiselled tips) are clustered along the front edge of the mouth and meet a similar dense cluster of teeth in the front part of the upper jaw.

Because of the position of the rear of the skull the muscles that closed the jaws would have exerted a backward, rather simple upward, pull. This would have helped the teeth, which form a rakelike array at the front of the mouth, to strip vegetation from stems and branches, in combination with a backward pull provided by the neck and its associated muscles. These dinosaurs seem to have used their heads as no more than simple food-gathering devices. Clearly, they had no teeth capable of chewing the food to extract nourishment. This had to occur in the capacious stomach and intestine, which combined to form a grinding mill (using swallowed pebbles as in the case of *Plateosaurus*), and large fermentation chambers (roughly equivalent to the complex stomachs of cattle).

In the past, some paleontologists have speculated that these dinosaurs may have fed exclusively on soft water weeds (because their teeth seemed weak) or that they lived on shellfish, which they plucked from the mud. Neither of these ideas is sustainable from what we know of the ecology of sauropods—and do more to demonstrate the inventiveness of the human mind than any genuinely scientific insight.

Other sauropods, notably brachiosaurs and camarasaurids, have less specialized heads than *Diplodocus* and teeth that are more spoon-shaped and robust. Such sauropods clearly had a more powerful bite, as reflected in the patterns of wear preserved on the edges of the teeth. These sauropods did not chew their food in any sense; they merely had a more effective cropping technique, which may reflect differences in the types of plants upon which they preferred to feed. Such differences, however, cannot be precisely identified as a particular menu at present, though attempts are being made to try to narrow down the number of possibilities within the known range of fossil plant types found associated with sauropods.

Ornithischians Ate Plants

The Early Jurassic shows the first really convincing evidence of the presence of an extremely important group of plant-eating dinosaurs—the ornithischians. They were important because they were probably major prey items for the many theropod species and because, from an evolutionary point of view, they really seem to have experimented with techniques for coping with a diet of plants to a far greater extent than either the sauropodomorphs or a few of the more aberrant theropods.

Clearly, from the previous examples chosen from the sauropodomorphs (*Plateosaurus* and *Diplodocus*), plants present a number of problems for the plant-eaters themselves. We have seen that plant-eaters tend to be bulky—they need a large gut divided into chambers to crush, ferment, and digest plant material. This is all necessary because plants are singularly well protected. They are built of cellulose (the tough, fibrous carbohydrate material of plant cell walls), which animals have never developed the chemical machinery to digest, as well as even more resistant lignins and other reinforcing materials, some mineral grains, and occasionally metabolic poisons. All these are a part of the defensive armory that they have evolved to protect themselves from being preyed upon by herbivores—and that does not begin to cover the external defenses (spikes and thorns and friendly ants, wasps, etc.) they are capable of erecting.

To get the nourishment out of plant material animals have a range of options, but, for the sake of this essay, we can limit them to those that have been employed by sauropodomorphs and ornithischians, respectively:

- *The sauropodomorph option.* This involved crushing up the tough, resistant plant tissues using swallowed stones lodged in a muscular gizzard, and fermenting the soupy pulverized remains using micro-organisms that are stored in its gut (which had the appropriate chemical machinery to break down cellulose and detoxify plant chemicals). This combination obviously required a very bulky gut and, consequently, a large abdomen, plus suitable equipment (the head and neck) for capturing as much unprocessed vegetation as possible. This clearly worked as a strategy—sauropodomorphs lived continuously through the entire era of the dinosaurs, though they may have waxed and waned from time to time.

- *The ornithischian options.* These dinosaurs seem to have explored a wide range of strategies for exploiting plants, from the sauropodomorph solution at one extreme to systems reminiscent of those seen in modern mammals (such as elephants, horses, cows, and sheep) and some entirely of their own making. To explore these options, let's first look at the key anatomical innovations that came with the evolution of ornithischian dinosaurs more generally.

Key Innovations in Ornithischians

One of the constant features of ornithischian anatomy is to be seen in the structure of the pelvis (hip bones). Notably, the pubis, the bone that normally points forward and downward from the hip joint in most tetrapods (land-walking vertebrates) has a radical new position. The key aspect of this anatomical change is that the pubis, in shifting its orientation, greatly enlarged the area available for the gut of ornithischians, and positioned it between the legs so that the heavy gut could be neatly balanced beneath the animal's center of balance, positioned just below the hip joint. This important change meant that early ornithischians not only could have a large gut, but could continue to run bipedally, and, therefore, were able to avoid the attention of at least some of the contemporary dinosaurian predators.

Beaks

Another key feature associated with all known ornithischians, is a horn-covered beak or bill. This structure, at the tip of the upper and lower jaws, tends to vary in shape between a broad almost ducklike bill and the much stronger and sharp-edged beak seen in modern tortoises. The beak was limited to the toothless parts of the upper jaw, near

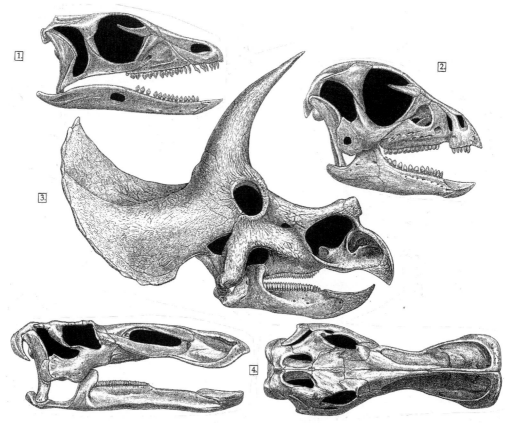

Ornithischian skulls show improvements and specializations in feeding capabilities. All had beaks for cropping plant parts. The Early Jurassic little ornithischian *Lesothosaurus* (1) presents a rather unsophisticated food processing apparatus, because the teeth were widely spaced and did not form a well developed dental battery. It is not certain whether cheeks were strongly developed. In the small Early Cretaceous ornithopod *Hypsilophodon* (2) the teeth were more tightly packed. The enormous skulls of ceratopsids featured parrot-like beaks; long rows of vertically slicing teeth indicate that ceratopsids (3) were able to bite off and dice up tough plant materials. The big hadrosaur *Anatotitan* (4) had the complex dental battery that characterized other hadrosaurs, but it took to an extreme the elongation and broadening of the bill. This duck-like feature may have been used to crop ground cover in the manner of grazing geese, which was then processed in the well-developed grinding battery.

the front of the mouth, and a unique toothless bone at the tip of the lower jaw, known as the predentary. This type of equipment is a clear adaptation for a diet of plant. Plants are typically very abrasive and can wear teeth down at a high rate. This can be biologically costly: building teeth is an expensive business, and one way to avoid wearing out teeth is to have a constantly renewed beak that grows continuously throughout life (as nails, claws, and hair do) and is self-sharpening.

Teeth

To cope with a diet of plants, teeth need to be modified in special ways. Often, the teeth are broadly expanded and shaped a little like the outline of a beech leaf. The rough edges are ideal for tearing vegetation prior to swallowing, but they do not form proper tooth-to-tooth cutting surfaces. Some ornithischians (cf. *Heterodontosaurus*/*Hypsilophodon*) develop teeth that are chisel-edged; this forms a much more effective cutting surface. Others, like the hadrosaurs, went so far as to develop large arrays of interlocking teeth that were used to grind plant fibers as part of a chewing action. The range and variety of tooth types is quite large in ornithischians, and probably reflects the range of plants that were eaten by different species.

Cheeks

The majority of ornithischian dinosaurs have a very distinctive depression or recess running along the side of the face adjacent and parallel to the teeth along the jaws. It is now generally agreed that these recesses correspond to the position of fleshy cheeks, broadly analogous to the cheeks of living mammals and that these permitted the repetitive chewing of plant food and avoided the risk of continually losing food from the sides of the mouth with each bite. The presence of laterally placed cheek pouches also makes better use of the mouth cavity, which would otherwise be extremely narrow if confined to the area between the jaws.

In recent years, it has been suggested that fleshy cheeks did not really exist in ornithischians. This is based largely on the fact that cheeks are unknown in present-day reptiles. It has been proposed that the edges of the ornithischian mouth were covered by elevated rims of horn (as a simple extension of the anterior beak). Though interesting, the logic supporting this alternative is not very robust, for these reasons:

- It has never been suggested that the cheeks of ornithischians were the same as those of mammals; they are considered to have been structures that are unique to ornithischians.

- The supposed outer rims of keratin would have functioned only if able to abrade (the upper cutting against the lower); otherwise, continuous growth of these structures would have welded up the side of the face with keratinous sheets.

- If the keratinous cheeks did abrade, they could have acted as cutting surfaces, and this may have superseded the teeth, yet these latter are never lost.

Chewing

Finally, from the point of view of ornithischian feeding, there is good structural evidence that some ornithischian dinosaurs developed the ability to chew plant food repetitively prior to swallowing it, and allowing the main gut digestive processes to take over. Detailed analysis of the structure of the skull of some ornithopods—such as *Hypsilophodon*, *Iguanodon,* and various hadrosaurs—has revealed a mechanical arrangement that allows the upper jaws to hinge outward against the skull roof in a mechanism named pleurokinesis. The details do not need to be explained, but the result of this unique system is that these dinosaurs were capable of developing a power stroke. As a result, the teeth of upper and lower jaws were able to slide past one another as the jaws closed, to achieve the cutting and simultaneous crushing of plant fiber. This is analogous to the chewing system that humans use today, and it is far more familiar in the chewing action of cattle, sheep, and many other mammalian herbivores.

Repetitive chewing clearly assumes that food could be recycled in the mouth using a combination of a muscular tongue and cheeks of one sort or another to prevent food from spilling out of the sides of the mouth (a compelling reason for believing that cheeks were present in these animals).

The Evolution of Diet in Ornithischians

The combination of many of these key innovations in many of the currently recognized groups of ornithischians undoubtedly played a part in the evolutionary history of this important group of herbivores. It is possible to chart the history of the Ornithischia in very general terms in a simple phylogenetic chart.

The armored dinosaurs (ankylosaurs and stegosaurs) retained an essentially primitive style of herbivory, with relatively simple leaf-shaped teeth (in some instances, arranged in mini-batteries), but with relatively little jaw mobility and little, if any, power stroke. It can be surmised that these animals were bulk feeders (a little like sauropodomorphs) browsing almost continuously and using their bulky frame to support a large and complex gut where the majority of the food processing occurred prior to digestion. For relatively slow-moving dinosaurs, this strategy of herbivory (where there is no great demand for high quantities of energy on tap) may have been perfectly adequate for their day-to-day needs.

The other major radiation of ornithischians consists of pachycephalosaurs (clearly a group of active ornithischians, judged by their anatomy) that seemed to employ a feeding mechanism little different from that seen in the armored forms. It is not clear whether, or how, such dinosaurs were able to supply their metabolic needs, especially when there was a rapid metabolic need. One possibility is that they were specialist feeders, seeking high-energy sources of food that needed little food processing before digestion could take place. At present, this idea cannot be tested adequately. The ornithopods were another clearly active group of ornithischians, notable for the development of sophisticated methods for orally processing plant food (chewing in various ways) prior to swallowing; this method predigests the food and makes it far easier for the gut to extract nutrients rapidly to supply metabolic demands. The ceratopians (the horned and beaked dinosaurs) are notable for their massive heads, powerful jaws, and huge curved beaks. In many respects, these were also highly efficient feeders upon plants, employing the massive beak, extensive cheeks, and batteries of sharp-edged teeth to process their food. Consistent differences in the form of the beak and teeth in the ceratopians and ornithopods suggest that their diets were significantly different, and they represent both ends of an ecological spectrum of herbivores, especially toward the close of the reign of the dinosaurs, a time when both groups were particularly abundant and diverse.

Conclusions

A diet of meat is, energetically, one of the easiest to sustain. Meat naturally corresponds biochemically to the constituents of an animal's body, and digestion and assimilation are maximally matched—provided prey can be caught. Predators tend, as a consequence, to be built along a range of generally similar lines. We see this among the theropod dinosaurs, where, with relatively few exceptions, their skulls, teeth, and bodies are surprisingly alike even if they vary somewhat in size. The exceptions to this general pattern prove to be very interesting simply because they test out reasoning ability more than most.

By contrast, feeding on plants proves to be far more challenging biochemically and physiologically. Plants are made of tough materials that are largely immune to normal digestion in vertebrates. Extraction of nutrients for assimilation can be made possible mechanically by using stones in a gastric mill or by slicing or crushing food in the mouth (which itself requires a range of complicated changes in skull anatomy). The guts of herbivores also show major modification compared to carnivores: most obviously guts are capacious (herbivores have large bellies!) to accommodate the chemical factories (formed of symbiotic microbes) necessary to digest plant materials before absorption and assimilation can take place. Such challenges are reflected in the diversity of plant-feeding dinosaurs, ranging from the giant multi-ton sauropods to the tiny, fleet-footed ornithopods, with almost every imaginable type in between these two extremes.

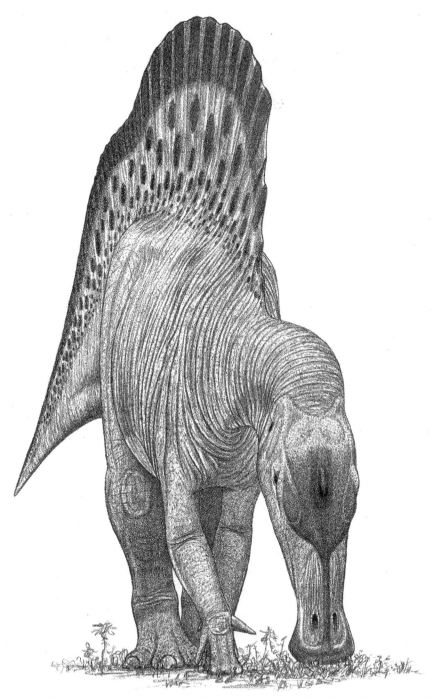

Ouranosaurus was intermediate in form to iguanodonts and hadrosaurs, and like its relatives had well-developed dental batteries that were probably covered by cheeks. It may have used its broad, square-tipped bill to graze like a dinosaurian wildebeest. Here the spectacular dorsal spines are shown as a thin fin, rather than carrying thick fat deposits as has been suggested.

Examining dietary preferences is a surprisingly instructive way of learning to appreciate the staggering diversity and range of adaptations exhibited by those slow, sluggish old monsters that were doomed to extinction. We now appreciate dinosaurs to have been marvelously designed and wonderfully diverse creatures, apparently beautifully in tune with the Mesozoic world they inhabited.

Breathing Life into Tyrannosaurus Rex

by Gregory M. Erickson

Dinosaurs ceased to walk the Earth 65 million years ago, yet they still live among us. *Velociraptors* star in movies, and *Triceratops* clutter toddlers' bedrooms. Of these charismatic animals, however, one species has always ruled our fantasies. Children, Steven Spielberg, and professional paleontologists agree that the superstar of the dinosaurs was and is *Tyrannosaurus rex*.

Harvard University paleontologist Stephen Jay Gould has said that every species designation represents a theory about that animal. The very name *Tyrannosaurus rex*—"tyrant lizard king"—evokes a powerful image of this species. John R. Horner of Montana State University and science writer Don Lessem wrote in their book *The Complete T. Rex*, "We're lucky to have the opportunity to know *T. rex*, study it, imagine it, and let it scare us. Most of all, we're lucky *T. rex* is dead." And paleontologist Robert T. Bakker of the Glenrock Paleontological Museum in Wyoming described *T. rex* as a "10,000-pound roadrunner from hell," a tribute to its obvious size and power.

In Spielberg's movie *Jurassic Park*, which boasted the most accurate popular depiction of dinosaurs ever, *T. rex* was, as usual, presented as a killing machine whose sole purpose was aggressive, bloodthirsty attacks on helpless prey. *T. rex*'s popular persona, however, is as much a function of artistic license as of concrete scientific evidence. A century of study and the existence of 22 fairly complete *T. rex* specimens have generated substantial information about its anatomy. But inferring behavior from anatomy alone is perilous, and the true nature of *T. rex* continues to be largely shrouded in mystery. Whether it was even primarily a predator or a scavenger is still the subject of debate.

Over the past decade, a new breed of scientists has begun to unravel some of *T. rex*'s better-kept secrets. These paleobiologists try to put a creature's remains in a living context—they attempt to animate the silent and still skeleton of the museum display. *T. rex* is thus changing before our eyes, as paleobiologists use fossil clues, some new and some previously overlooked, to develop fresh ideas about the nature of these magnificent animals.

Rather than draw conclusions about behavior based solely on anatomy, paleobiologists demand proof of actual activities. Skeletal assemblages of multiple individuals shine a light on the interactions among *T. rex*, and between them and other species. In addition, so-called trace fossils reveal activities through physical evidence, such as bite marks in bones and wear patterns in teeth. Also of great value as trace fossils are coprolites, fossilized feces. (Remains of a herbivore, such as *Triceratops* or *Edmontosaurus*, in *T. rex* coprolites certainly provide "smoking gun" proof of species interactions!)

One assumption that paleobiologists are willing to make is that closely related species may have behaved in similar ways. *T. rex* data are therefore being corroborated by comparisons with those of earlier members of the family Tyrannosauridae, including their cousins *Albertosaurus*, *Gorgosaurus*, and *Daspletosaurus*, collectively known as albertosaurs.

Solo or Social?

Tyrannosaurs are usually depicted as solitary, as was certainly the case in *Jurassic Park*. (An alternative excuse for that film's loner is that the movie's genetic wizards wisely created only one.) Mounting evidence, however, points to gregarious *T. rex* behavior, at least for part of the animals' lives. Two *T. rex* excavations in the Hell Creek Formation of eastern Montana are most compelling.

In 1966, Los Angeles County Museum researchers attempting to exhume a Hell Creek adult were elated to find another, smaller individual resting atop the *T. rex* they had originally sought. This second fossil was identified at first as a more petite species of tyrannosaur. My examination of the histological evidence—the microstructure of the bones—now suggests that the second animal was actually a subadult *T. rex*. A similar discovery was made during the excavation of "Sue," the largest and most complete fossil *T. rex* ever found. Sue is perhaps as famous for her $8.36-million auction price following ownership haggling as for her paleontological status (see "No Bones about It," News and Analysis, *Scientific American*, December 1997). Remains of a second adult, a juvenile, and an infant *T. rex* were later found in Sue's quarry. Researchers who have worked the Hell Creek Formation, myself included, generally agree that long odds argue against multiple, loner *T. rex* finding their way to the same burial place. The more parsimonious explanation is that the animals were part of a group.

An even more spectacular find from 1910 further suggests gregarious behavior among the Tyrannosauridae. Researchers from the American Museum of Natural History in New York City working in Alberta, Canada, found a bonebed—a deposit with fossils of many individuals—holding at least nine of *T. rex*'s close relatives, albertosaurs.

Philip J. Currie and his team from the Royal Tyrrell Museum of Paleontology in Alberta recently relocated the 1910 find, and are conducting the first detailed study of the assemblage. Such aggregations of carnivorous animals can occur when one after another gets caught in a trap, such as a mud hole or soft sediment at a river's edge, in which a prey animal that has attracted them is already ensnared. Under those circumstances, however, the collection of fossils should also contain those of the hunted herbivore. The lack of such herbivore remains among the albertosaurs (and among the four-*T. rex* assemblage that included Sue) indicates that the herd most likely associated with one another naturally and perished together from drought, disease, or drowning.

From examination of the remains collected so far, Currie estimates that the animals ranged from 13 to 29 feet in length. This variation in size hints at a group composed of juveniles and adults. One individual is considerably larger and more robust than the others. Although it might have been a different species of albertosaur, a mixed bunch seems unlikely. I believe that if *T. rex* relatives did indeed have a social structure, this largest individual may have been the patriarch or matriarch of the herd.

Tyrannosaurs in herds, with complex interrelationships, are in many ways an entirely new species to contemplate. But science has not morphed them into a benign and tender collection of Cretaceous Care Bears: some of the very testimony for *T. rex* group interaction is partially healed bite marks that reveal nasty interpersonal skills. A paper just published by Currie and Darren Tanke, also at the Royal Tyrrell Museum, highlights this evidence. Tanke is a leading authority on paleopathology—the study of ancient injuries and disease. He has detected a unique pattern of bite marks among theropods, the group of carnivorous dinosaurs that encompasses *T. rex* and other tyrannosaurs. These bite marks consist of gouges and punctures on the sides of the snout, on the sides and bottom of the jaws, and occasionally on the top and back of the skull.

Interpreting these wounds, Tanke and Currie reconstructed how these dinosaurs fought. They believe that the animals faced off, but primarily gnawed at one another with one side of their complement of massive teeth rather than snapping from the front. The workers also surmised that the jaw-gripping behavior accounts for peculiar bite marks found on the sides of tyrannosaur teeth. The bite patterns imply that the combatants maintained their heads at the same level throughout a confrontation. Based on the magnitude of some of the fossil wounds, *T. rex* clearly showed little reserve, and sometimes inflicted severe damage to its conspecific foe. One tyrannosaur studied by Tanke and Currie sports a souvenir tooth, embedded in its own jaw, perhaps left by a fellow combatant.

The usual subjects—food, mates, and territory—may have prompted the vigorous disagreements among tyrannosaurs. Whatever the motivation behind the fighting, the fossil record demonstrates that the behavior was repeated throughout a tyrannosaur's life. Injuries among younger individuals seem to have been more common, possibly because a juvenile was subject to attack by members of his or her own age group, as well as by large adults. (Nevertheless, the fossil record may also be slightly misleading, and simply contains more evidence of injuries in young *T. rex*. Nonlethal injuries to adults would

The skull of the *T. Rex*, the undisputed champion of "bite force." Gregory Erickson and bioengineer Dennis Carter of Stamford University simulated the production of feeding bite marks using a cast of a *T. Rex* tooth on cow pelvises. They estimated 2,900 pounds of force for one side of the mouth, greater than the biting potential of a wolf, shark, lion or alligator.

have eventually healed, destroying the evidence. Juveniles were more likely to die from adult-inflicted injuries, and they carried those wounds to the grave.)

Bites and Bits

Imagine the large canine teeth of a baboon or lion. Now imagine a mouthful of much larger canine-type teeth, the size of railroad spikes and with serrated edges. Kevin Padian of the University of California at Berkeley has summed up the appearance of the huge daggers that were *T. rex* teeth: "lethal bananas."

Despite the obvious potential of such weapons, the general opinion among paleontologists had been that dinosaur bite marks were rare. The few published reports before 1990 consisted of brief comments buried in articles describing more sweeping new finds, and the clues in the marred remains concerning behavior escaped contemplation.

Nevertheless, some researchers speculated about the teeth. As early as 1973, Ralph E. Molnar of the Queensland Museum in Australia began musing about the strength of the teeth, based on their shape. Later, James O. Farlow of Indiana University–Purdue University Fort Wayne and Daniel L. Brinkman of Yale University performed elaborate morphological studies of tyrannosaur dentition, which made them confident that the "lethal bananas" were robust, thanks to their rounded cross-sectional configuration, and would endure bone-shattering impacts during feeding.

In 1992, I was able to provide material support for such speculation. Kenneth H. Olson, a Lutheran pastor and superb amateur fossil collector for the Museum of the Rockies in Bozeman, Montana, came to me with several specimens. One was a 3¼-foot-wide, almost 5-foot-long partial pelvis from an adult *Triceratops*. The other was a toe bone from an adult *Edmontosaurus* (duck-billed dinosaur). I examined Olson's specimens and found that both bones were riddled with gouges and punctures up to about 5 inches long and several inches deep. The *Triceratops* pelvis had nearly 80 such indentations. I documented the size and shape of the marks and used orthodontic dental putty to make casts of some of the deeper holes. The teeth that had made the holes were spaced some 4 inches apart. They left punctures with eye-shaped cross-sections. They clearly included carinas, elevated cutting edges, on their anterior and posterior faces. And those edges were serrated. The totality of the evidence pointed to these indentations being the first definitive bite marks from a *T. rex*.

This finding had considerable behavioral implications. It confirmed for the first time the assumption that *T. rex* fed on its two most common contemporaries, *Triceratops* and *Edmontosaurus*. Furthermore, the bite patterns opened a window into *T. rex*'s actual feeding techniques, which apparently involved two distinct biting behaviors. *T. rex* usually used the "puncture and pull" strategy, in which biting deeply with enormous force was followed by drawing the teeth through the penetrated flesh and bone, which typically produced long gashes. In this way, a *T. rex* appears to have detached the pelvis found by Olson from the rest of the *Triceratops* torso. *T. rex* also employed a nipping approach in which the front (incisiform) teeth grasped and stripped the flesh in tight spots between vertebrae, where only the muzzle of the beast could fit. This method left vertically aligned, parallel furrows in the bone.

Many of the bites on the *Triceratops* pelvis were spaced only a few inches apart, as if the *T. rex* had methodically worked his way across the hunk of meat, as we would nibble an ear of corn. With each bite, *T. rex* appears also to have removed a small section of bone. We presumed that the missing bone had been consumed, confirmation for which shortly came, and from an unusual source.

In 1997, Karen Chin of the U.S. Geological Survey received a peculiar, tapered mass that had been unearthed by a crew from the Royal Saskatchewan Museum. The object, which weighed approximately 15½ pounds and measured approximately 7¼ by 6¼ by 5 inches, proved to be a *T. rex* coprolite. The specimen, the first ever confirmed from a theropod, and more than twice as large as any previously reported meat-eater's copro-

King-size coprolite, 17 inches long, is the largest of its kind from a carnivorous animal, more than twice the size of any previously reported. Its size, age, contents and geographic context rule out anything other than a tyrannosaur, and most likely a *T. rex,* as its producer.

lite, was chock-full of pulverized bone. Once again making use of histological methods, Chin and I determined that the shattered bone came from a young herbivorous dinosaur. *T. rex* did indeed ingest parts of the bones of its food sources and, furthermore, partially digested these items with strong enzymes or stomach acids.

Following the lead of Farlow and Molnar, Olson and I have argued vehemently that *T. rex* probably left multitudinous bite marks, despite the paucity of known specimens. Absence of evidence is not evidence of absence, and we believe two factors account for this toothy gap in the fossil record. First, researchers have never systematically searched for bite marks. Even more important, collectors have had a natural bias against finds that might display bite marks. Historically, museums desire complete skeletons rather than single, isolated parts. But whole skeletons tend to be the remains of animals that died from causes other than predation and were rapidly buried before being dismembered by scavengers. The shredded bits of bodies eschewed by museums, such as the *Triceratops* pelvis, are precisely those specimens most likely to carry the evidence of feeding.

Indeed, Aase Roland Jacobsen of the Royal Tyrrell Museum recently surveyed isolated partial skeletal remains and compared them with nearly complete skeletons in Alberta. She found that 3.5 times as many of the individual bones (14 percent) bore theropod bite marks as did the less disrupted remains (4 percent). Paleobiologists therefore view the majority of the world's natural history museums as deserts of behavioral evidence when compared with fossils still lying in the field waiting to be discovered and interpreted.

Hawk or Vulture?

Some features of tyrannosaur biology, such as coloration, vocalizations, or mating displays, may remain mysteries. But their feeding behavior is accessible through the fossil record. The collection of more trace fossils may finally settle a great debate in paleontology—the 80-year controversy over whether *T. rex* was a predator or a scavenger.

When *T. rex* was first found a century ago, scientists immediately labeled it a predator. But sharp claws and powerful jaws do not necessarily a predator make. For example, most bears are omnivorous and kill only a small proportion of their food. In 1917, Canadian paleontologist Lawrence Lambe examined a partial albertosaur skull and ascertained that tyrannosaurs fed on soft, rotting carrion. He came to this conclusion after noticing that the teeth were relatively free of wear. (Future research would show that 40 percent of shed tyrannosaur teeth are severely worn and broken, damage that occurs in a mere two to three years, based on my estimates of their rates of tooth replacement.) Lambe thus established the minority view that the beasts were in fact giant terrestrial "vultures." The ensuing arguments in the predator-versus-scavenger dispute have centered on the anatomy and physical capabilities of *T. rex*, leading to a tiresome game of point-counterpoint.

Scavenger advocates adopted the "weak tooth theory," which maintained that *T. rex*'s elongate teeth would have failed in predatory struggles or in bone impacts. They also contended that its diminutive arms precluded lethal attacks and that *T. rex* would have been too slow to run down prey.

Predator supporters answered with biomechanical data. They cited my bite-force studies that demonstrate that *T. rex* teeth were actually quite robust. (I personally will remain uncommitted in this argument until the discovery of direct physical proof.) They also note that Kenneth Carpenter of the Denver Museum of Natural History and Matthew Smith, then at the Museum of the Rockies, estimate that the "puny" arms of a *T. rex* could curl nearly 103-plus pounds. And they point

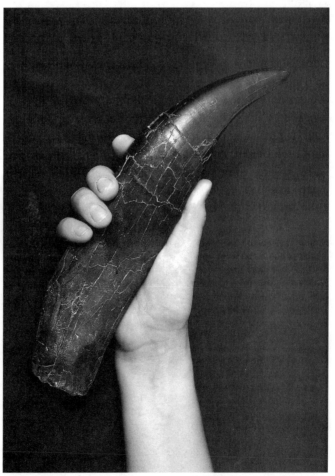

Massive tooth of a tyrannosaur: only about 25 percent of the tooth (smooth section at right) would have been visible above the gum line.

to the work of Per Christiansen of the University of Copenhagen, who believes, based on limb proportion, that *T. rex* may have been able to sprint at almost 30 miles per hour. Such speed would be faster than that of any of *T. rex*'s contemporaries, although endurance and agility, which are difficult to quantify, are equally important in such considerations.

Even these biomechanical studies fail to resolve the predator-scavenger debate—and they never will. The critical determinant of *T. rex*'s ecological niche is discovering how and to what degree it utilized the animals living and dying in its environment, rather than establishing its presumed adeptness for killing. Both sides concede that predaceous animals, such as lions and spotted hyenas, will scavenge, and that classic scavengers, such as vultures, will sometimes kill. And mounting physical evidence leads to the conclusion that tyrannosaurs both hunted and scavenged.

Within *T. rex*'s former range exist bonebeds consisting of hundreds and sometimes thousands of edmontosaurs that died from floods, droughts, and causes other than predation. Bite marks and shed tooth crowns in these edmontosaur assemblages attest to scavenging behavior by *T. rex*. Jacobsen has found comparable evidence for albertosaur scavenging. Carpenter, on the other hand, has provided solid proof of predaceous behavior, in the form of an unsuccessful attack by a *T. rex* on an adult *Edmontosaurus*. The intended prey escaped with several broken tailbones that later healed. The only animal with the stature, proper dentition, and biting force to account for this injury is *T. rex*.

Quantification of such discoveries can help determine the degree to which *T. rex* undertook each method of obtaining food, and paleontologists can avoid future arguments by adopting standard definitions of predator and scavenger. Such a convention is necessary, as a wide range of views pervades vertebrate paleontology as to what exactly makes for each kind of feeder. For example, some extremists contend that if a carnivorous animal consumes any carrion at all, it should be called a scavenger. But such a constrained definition negates a meaningful ecological distinction, as it would include nearly all the world's carnivorous birds and mammals.

In a definition more consistent with most paleontologists' common-sense categorization, a predatory species would be one in which most individuals acquired most of their meals from animals they or their peers killed. Most individuals in a scavenging species, on the other hand, would not be responsible for the deaths of most of their food.

Trace fossils could open the door to a systematic approach to the predator-scavenger controversy, and the resolution could come from testing hypotheses about entire patterns of tyrannosaur feeding preferences. For instance, Jacobsen has pointed out that evidence of a preference for less dangerous or easily caught animals supports a predator niche. Conversely, scavengers would be expected to consume all species equally.

Within this logical framework, Jacobsen has compelling data supporting predation. She surveyed thousands of dinosaur bones from Alberta, and learned that unarmored hadrosaurs are twice as likely to bear tyrannosaur bite marks as are the more dangerous horned ceratopsians. Tanke, who participated in the collection of these bones, relates that no bite marks have been found on the heavily armored, tanklike ankylosaurs.

Jacobsen cautions, though, that other factors confuse this set of findings. Most of the hadrosaur bones are from isolated individuals, but most ceratopsians in her study are from bonebeds. Again, these beds contain more whole animals that have been fossilized unscathed, creating the kind of tooth-mark bias discussed earlier. A survey of isolated ceratopsians would be enlightening. And analysis of more bite marks that reveal failed predatory attempts, such as those reported by Carpenter, could also reveal preferences, or the lack thereof, for less dangerous prey.

Jacobsen's finding that cannibalism among tyrannosaurs was rare—only 2 percent of albertosaur bones had albertosaur bite marks, whereas 14 percent of herbivore bones did—might also support predatory preferences instead of a scavenging niche for *T. rex*, particularly if these animals were in fact gregarious. Assuming that they had no aversion to consuming flesh of their own kind, it would be expected that at least as many *T. rex* bones would exhibit signs of *T. rex* dining as do herbivore bones. A scavenging *T. rex* would have had to stumble on herbivore remains, but if *T. rex* traveled in herds, freshly dead conspecifics would seem to have been a guaranteed meal.

Coprolites may also provide valuable evidence about whether *T. rex* had any finicky eating habits. Because histological examination of bone found in coprolites can give the approximate stage of life of the consumed animal, Chin and I have suggested that coprolites may reveal a *T. rex* preference for feeding on vulnerable members of herds, such as the very young. Such a bias would point to predation, whereas a more impartial feeding pattern, matching the normal patterns of attrition, would indicate scavenging. Meaningful questions may lead to meaningful answers.

Over this century, paleontologists have recovered enough physical remains of *Tyrannosaurus rex* to give the world an excellent idea of what these monsters looked like. The attempt to discover what *T. rex* actually was like relies on those fossils that carry precious clues about the daily activities of dinosaurs. Paleontologists now appreciate the need for reanalysis of finds that were formerly ignored, and they recognize the biases in collection practices, which have clouded perceptions of dinosaurs. The intentional pursuit of behavioral data should accelerate discoveries of dinosaur paleobiology. And new technologies may tease information out of fossils that we currently deem of little value. The *T. rex*, still alive in the imagination, continues to evolve.

The Teeth of the Tyrannosaurs

by William L. Abler

Understanding the teeth is essential for reconstructing the hunting and feeding habits of the tyrannosaurs. The tyrannosaur tooth is more or less a cone, slightly curved and slightly flattened, so that the cross-section is an ellipse. Both the narrow anterior and posterior surfaces bear rows of serrations. Their presence has led many observers to assume that the teeth cut meat the way a serrated steak knife does. My colleagues and I, however, were unable to find any definitive study of the mechanisms by which knives, smooth or serrated, actually cut. Thus, the comparison between tyrannosaur teeth and knives had meaning only as an impetus for research, which I decided to undertake.

Trusting in the logic of evolution, I began with the assumption that tyrannosaur teeth were well adapted for their biological functions. Although investigation of the teeth themselves might appear to be the best way of uncovering their characteristics, such direct study is limited; the teeth cannot really be used for controlled experiments. For example, doubling the height of a fossil tooth's serrations to monitor changes in cutting properties is impossible. So I decided to study steel blades whose serrations or sharpness I could alter, and then compare these findings with the cutting action of actual tyrannosaur teeth.

To measure the cutting properties of the blades, I mounted them on a butcher's saw operated by cords and pulleys, which moved the blades across a series of similarly sized pieces of meat that had been placed on a cutting board. Using weights stacked in baskets at the ends of the cords, I measured the downward force and drawing force required to cut each piece of meat to the same depth. My simple approach gave consistent and provocative results, including this important and perhaps unsurprising one: smooth and serrated blades cut in two entirely different fashions.

The serrated blade appears to cut meat by a "grip and rip" mechanism. Each serration penetrates to a distance equal to its own length, isolating a small section of meat between itself and the adjacent serration. As the blade moves, each serration rips that isolated section. The blade then falls a distance equal to the height of the serration, and the process repeats. The blade thus converts a pulling force into a cutting force.

A smooth blade, in contrast, concentrates downward force at the tiny cutting edge. The smaller this edge, the greater the force. In effect, the edge crushes the meat until it splits, and pulling or pushing the blade reduces friction between the blade surface and the meat.

After these discoveries, I mounted actual serrated teeth in the experimental apparatus, with some unexpected results. The serrated tooth of a fossil shark (*Carcharodon* megalodon) indeed works exactly like a serrated knife blade does. Yet the serrated edge of even the sharpest tyrannosaur tooth cuts meat more like a smooth knife blade, and a dull one at that. Clearly, all serrations are not alike. Nevertheless, serrations are a major and dramatic feature of tyrannosaur teeth. I therefore began to wonder whether these serrations served a function other than cutting.

The serrations on a shark tooth have a pyramidal shape. Tyrannosaur serrations are more cubelike. Two features of great interest are the gap between serrations, called a cella, and the thin slot to which the cella narrows, called a diaphysis. Seeking possible functions of the cellae and diaphyses, I put tyrannosaur teeth directly to the test and used them to cut fresh meat. To my knowledge, this was the first time tyrannosaur teeth have ripped flesh in some 65 million years.

I then examined the teeth under the microscope, which revealed striking characteristics. (Although I was able to inspect a few *Tyrannosaurus rex* teeth, my cutting experiments were done with teeth of fossil albertosaurs, which are true tyrannosaurs and close relatives of *T. rex*.) The cellae appear to make excellent traps for grease and other food debris. They also provide access to the deeper diaphyses, which grip and hold filaments of the victim's tendon. Tyrannosaur teeth thus would have harbored bits of meat and grease for extended periods. Such food particles are receptacles for septic bacteria—even a nip from a tyrannosaur, therefore, might have been a source of a fatal infection.

Another aspect of tyrannosaur teeth encourages contemplation. Neighboring serrations do not meet at the exterior of the tooth. They remain separate inside it down to a depth nearly equal to the exterior height of the serration. Where they finally do meet, the junction, called the ampulla, is flask-shaped rather than V-shaped. This ampulla seems to have protected the tooth from cracking when force was applied. Whereas the narrow opening of the diaphysis indeed put high pressure on trapped filaments of tendon, the rounded ampulla distributed pressure uniformly around its surface. The ampulla thus eliminated any point of concentrated force where a crack might begin.

Apparently, enormously strong tyrannosaurs did not require razorlike teeth, but instead made other demands on their dentition. The teeth functioned less like knives than like pegs, which gripped the food while the *T. rex* pulled it to pieces. (This so-called puncture-and-pull technique is also discussed in the previous essay by Gregory Erickson, "Breathing Life into *Tyrannosaurus rex*".) And the ampullae protected the teeth during this process.

An additional feature of its dental anatomy leads to the conclusion that *T. rex* did not chew its food. The teeth have no occlusal, or articulating, surfaces and rarely touched one another. After it removed a large chunk of carcass, the tyrannosaur probably swallowed that piece whole.

Work from an unexpected quarter also provides potential help in reconstructing the hunting and feeding habits of tyrannosaurs. Herpetologist Walter Auffenberg of the University of Florida spent more than 15 months in Indonesia studying the largest lizard in the world, the Komodo dragon. (Paleontologist James O. Farlow has suggested that the Komodo dragon may serve as a living model for the behavior of the tyrannosaurs.) The dragon's teeth are remarkably similar in structure to those of tyrannosaurs, and the creature is well known to inflict a dangerously septic bite—an animal that escapes an attack with just a flesh wound is often living on borrowed time. An infectious bite for tyrannosaurs would lend credence to the argument that the beasts were predators rather than scavengers. As with Komodo dragons, the victim of what appeared to be an unsuccessful attack might have received a fatal infection. The dead or dying prey would then be easy pickings to a tyrannosaur, whether the original attacker or merely a fortunate conspecific.

If the armamentarium of tyrannosaurs did include septic oral flora, we can postulate other characteristics of its anatomy. To help maintain a moist environment for its single-celled guests, tyrannosaurs probably had lips that closed tightly, as well as thick, spongy gums that covered the teeth. When tyrannosaurs ate, pressure between teeth and gums might have cut the latter, causing them to bleed. The blood in turn may have been a source of nourishment for the septic dental bacteria. In this scenario, the horrific appearance of the feeding tyrannosaur is further exaggerated—their mouths would have run red with their own bloodstained saliva while they dined.

Reproduction and Parenting

by David J. Varricchio

Early-morning sunlight slants through an open woodland of southern Wyoming some 145 million years ago. Among a clatter of high-pitched insect, pterosaur, and bird calls, a low, barely perceptible drum rumbles. Two giant *Brachiosaurus* emerge from the trees and lumber toward an encounter. The heads of these massive sauropods rise some 40 feet above the ground. As the dinosaurs approach each other, their long necks begin to sway rhythmically. Synchronous, languid head-bobbing follows and the courtship ritual begins. Crashing classical movements or, for those tuned to a more modern imagination, a slow hip-hop groove might be a fitting accompaniment for the subsequent fury and sound.

Reproduction includes a wide variety of behaviors and physiological changes. A description of reproduction in a living animal might focus on courtship displays, territoriality, mating, mating systems, nests and nest construction, male and female reproductive anatomy, egg production, hormonal controls, timing of reproductive events, care of eggs or young, embryonic development, growth of young, or any number of other topics. For an extinct animal such as a dinosaur, most of these aspects of reproduction will leave no significant mark on a fossil record largely consisting of bones and sedimentary rocks. Any portrayal of dinosaur reproduction, such as the fanciful *Brachiosaurus* example above, are simply imaginative speculation.

Nevertheless, the fossil record does include several types of evidence that provide a limited but significant view of reproduction: adult skeletal morphology, bone histology (the microscopic aspects of bone tissue), eggs, eggshell, clutch arrangements, and nests. These represent to varying degrees the direct products of either reproductive physiology or behavior. For example, eggs reflect ovary and oviduct function, and bone histology reflects the growth of an individual. Trace fossils such as clutch arrangement and nest construction represent the activities of adults and provide the best means for "observing" the reproductive behavior of dinosaurs.

Living animals provide important additional aids in understanding dinosaur reproduction. Most significant are the two groups sharing the closest ancestry with dinosaurs, namely crocodilians (crocodiles, alligators, gharials) and birds. Crocodilians, birds, dinosaurs, and other extinct groups comprise the clade Archosauria. With some caution,

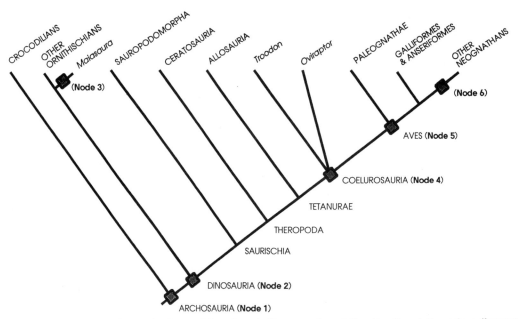

A cladogram of the evolutionary development from the primitive ancestors of dinosaurs though the various dinosaur families and superfamilies, with the logical inclusion of Aves (birds) as their direct descendants.

scientists can use these modern groups to interpret those aspects of dinosaurs that do not typically fossilize, such as soft tissue and behavior. Dr. Larry Witmer of Ohio Univesity has formalized this methodology as the Extant Phylogenetic Bracket. "Extant," the biological opposite to extinct, refers to still living groups. The "phylogenetic bracket" consists of those groups that, on an evolutionary tree, fit to either side of our extinct group of interest. Crocodilians and birds form the living bracket for dinosaurs. On an evolutionary tree, crocodilians sit on one side as dinosaurian "cousins," and birds occupy the other side as the descendants of theropod dinosaurs. Technically, birds belong to the clade Dinosauria, and both birds and dinosaurs are dinosaurians.

With the Extant Phylogenetic Bracket, one can hypothesize the presence of a normally unpreservable feature in dinosaurs if that same feature occurs universally in both crocodilians and birds. This argument gains strength if the feature of interest has a tight link with skeletal structures present in both the bracket (crocodilians and birds) and the group of interest (dinosaurs). As a reproductive example, crocodilians and birds possess an "assembly-line" oviduct, where functional regionalization occurs within the reproductive tract. Among living vertebrates this feature occurs uniquely in archosaurs. As an egg passes down the oviduct, it gains parts in an assembly-line fashion, with each region adding a specific layer to the egg. Since the assembly-line oviduct exists in both birds and crocodilians, the simplest interpretation predicts it evolved within the common ancestor of Archosaurs. Consequently, dinosaurs should have inherited this feature. Support for this argument comes from dinosaur eggshell, which shares features with crocodilian and bird eggs—such as a hard clacite construction and other microscopic features. Simply put, the end products of the dinosaur reproductive tract, eggs, closely resemble those of crocodilians and birds. So the oviducts likely did not differ significantly.

Although crocodilians and birds aid in the interpretation of dinosaurs, their behaviors and physiologies cannot simply be substituted for any information lacking from their extinct relatives. Significant amounts of evolutionary history and major ecological differences separate the groups. The line leading to modern crocodilians diverged from that giving rise to the dinosaurian lineage some 240 million years ago. Over 140 million years separate modern birds from their theropod ancestors. Ecologically, crocodilians remain semi-aquatic, while all modern birds descended from a flying ancestor. Further, the vast majority of living birds have a body mass smaller than the smallest dinosaur. Undoubtedly, flight, aquatic habitats, and small body size affected the reproductive habits of living archosaurs. Given the amazing diversity of dinosaurs in terms of shape, size, habitat, and geologic age, they likely exhibited a wide range of reproductive behaviors and physiology. One should not expect any dinosaur to have behaved precisely like any living archosaur. For example, no evidence suggests that any dinosaur nested like a bird, with exposed eggs free of sediments or vegetation in an open cuplike construction. Dinosaur eggs exhibit a greater range of sizes, shapes, and micro-structure than found among crocodilians and birds, which suggests a wider range of reproductive behaviors as well. Interestingly, the study of dinosaur reproduction has important implications for extant animals, particularly birds, as we will see later on.

Here then, is a summary of the current knowledge of dinosaur reproduction, focusing on those aspects that have their basis in body or trace fossils, and ignoring those topics (e.g., courtship rituals, mating systems, hormonal controls) that presently lie beyond the physical fossil evidence and could only be discussed conjecturally. We'll also look at those few species for which good reproductive information has been gathered.

Sexual Display and Dimorphism

Two anatomical conditions, sexual display features and sexual dimorphism, provide some insight on dinosaur mating and mating systems. Display features include those anatomical features used to attract the opposite sex and to dismay rivals. Examples from living vertebrates include lizard dewlaps, mountain goat beards, and the brightly colored feathers of a peacock. Sexual dimorphism simply refers to differences between males and females of a species. For any given dinosaur species, recognition of sexual display features and dimorphism depends upon a full understanding of variation. Individuals may differ because of age, sex, genetics, environmental conditions, or because they represent separate species. Fossil specimens may also accumulate differences due to burial-caused plastic deformation. Determining the cause of variation remains a difficult task for dinosaur paleontologists, where specimens are few and often poorly preserved.

Despite typically poor sample sizes, many dinosaurs appear to possess sexual display features. These structures give dinosaurs their characteristic flare: horns and frills of ceratopsians; inflated nostrils and crests of hadrosaurs; armor plates, spikes, and clubs of stegosaurs and ankylosaurs; thick-domed and spiked heads of pachycephalosaurs; and cranial bosses and crests of theropods. Paleontologists have often interpreted these features as antipredator devices, warning signals and weapons for countering attack. Another alternative hypothesis views these features as aids in thermoregulation. Although these structures likely functioned in both roles, two arguments strongly support them as significant sexual display features. First, these structures exhibit delayed growth, showing full development only late in maturity. Display features in living vertebrates show a similar pattern. For example, gulls typically achieve adult plumage only on reaching sexual maturity, long after (two to three years) becoming full-sized. Further, display features often serve as the only means of distinguishing closely related species.

Ceratopsians and hadrosaurs probably represent the best examples among dinosaurs. These two diverse groups largely consist of species that, from neck to tail, look quite similar. The unique aspects of each species include primarily the horns, frills, and crests, features only fully developed in adult individuals.

Display features serve several roles. They likely served as recognition clues, labeling an individual as belonging to a specific species. These same aspects help paleontologists distinguish varieties of dinosaurs as well. As living antelopes or beetles use their horns, some dinosaurs may have used these structures to fend off rivals. Pathologies present on the skulls of some ceratopsians may represent battle scars from such conflicts. The final role of horns, frills, and crests would be in advertising sexual maturity and health. Biologists generally consider sexual display features as the result of sexual selection—an evolutionary mechanism separate from natural selection. Mate choice and same-sex competition characterize this mechanism. Given the elaborate and widespread occurrence of display features in dinosaurs, sexual selection apparently played a significant role in their evolution.

The simple question of whether a specific dinosaur skeleton represents a male or female remains largely unanswerable. Although sexual dimorphism can express itself as size and shape variation, it is often limited to less preservable features like hair, feathers, scales, coloration, vocalizations, and physiology. Further, the sexes of many living vertebrates show no variation other than reproductive organs. Paleontologists have eagerly sought sex-specific skeletal characters for dinosaurs. One feature, much discussed in recent years, consists of the position of the first chevron at the base of the tail. Throughout the tail, Y-shaped chevron bones extend from the bottom of the vertebral column.

These bones support muscles of the tails and protect the region's main blood vessels within the arms of the Y. The first chevron's position may correspond to the presence or absence of copulatory structures, such as the penis of male crocodilians. Some paleontologists, using chevron locations and the degree of skeletal robustness have argued that larger, stouter *Tyrannosaurus rex* skeletons represent females. This data remains inconclusive. Information supporting the correlation between chevron position and sex in living crocodilians awaits publication. Additionally, although *T. rex* represents one of the better known theropods, only a handful of articulated skeletons exist. Demonstrating dimorphism minimally requires statistical separation of two morphologies within a population; this cannot yet be done for *T. rex*.

A few dinosaur species, known from larger sample sizes, exhibit clearer cases of dimorphism. For example, the early Jurassic theropod, *Syntarsus rhodesiensis* displays gracile and robust forms. The occurrence of greater numbers of robust forms suggests they are females. Populations of *Protoceratops mongoliensis* exhibit two clear morphologies. The sexes appear to differ in the degree of nasal horn and frill development. Other dinosaurs—large ceratopsians, pachycephalosaurs, and hadrosaurs—also exhibit dimorphism in their sexual display features.

Sexual dimorphism can reflect a number of ecological and behavioral factors such as niche partitioning. Females often outweigh males where selection favors high egg output or where males challenge the safety of the young. Monogamous pairs, where sexes often share equally in parental care, typically display no significant size dimorphism. In contrast, males increasingly outsize females in species with multi-male social units and particularly where males maintain harems. Consequently, comprehension of sex identification and sexual dimorphism may eventually yield insight on dinosaur mating systems.

Eggs

The best and most abundant information on dinosaur reproduction comes from eggs, their microscopic structure, and their overall arrangement relative to other eggs and sediments. In 1859, a French priest, Father Pouech, made what is considered the first discovery of dinosaur eggshell. Pouech found and described eggshell fragments from the Pyrenees of southern France. He thought these belonged to a giant bird, a reasonable conclusion given the newness of dinosaurs, which had only been named in 1842.

More misidentified eggshell turned up in the United States 50 years later. In 1913, Charles Gilmore, while working for the United States National Museum, found an unusual black clamshell bed. These clams later proved to be eggshells. The American Museum of Natural History expeditions to Mongolia led by Roy Chapman Andrews discovered the first whole eggs and egg clutches in 1923. These come from the Cretaceous dinosaur-rich deposits of the Flaming Cliffs.

Throughout the last half of the twentieth century, paleontologists continued to make significant egg and eggshell discoveries. In 1978, John Horner made an important find in the Late Cretaceous of Montana: three new egg varieties, each with embryonic remains, the first dinosaur egg-embryo associations. Many of the dinosaur-bearing areas of the world have now produced significant eggshell or eggs, including North America from Mexico to Canada; major areas of Mongolia, China, and Central Asia; and numerous countries in Europe and South America. Australia and Africa have yielded the least material, with only a few eggs known from Africa.

A problem that continues to trouble work on eggs and eggshell is their assignment to a specific dinosaur species. Positive identification requires eggs preserved with late-term embryos. Taxonomic assignment of eggs based on other criteria, such as a predominance of a particular dinosaur species in the general area, have proved unreliable. *Oviraptor* and *Troodon* eggs represent noteworthy cases of misidentification.

Given the inherent problems in identifying eggs to species, the Chinese paleontologist Zhao Zi-Kui established a parataxonomic system for naming eggs. This system parallels normal taxonomic schemes. Paleontologists can assign a parataxonomic species name to an egg variety much as they can name a new dinosaur skeleton. The species can then be grouped with similar species into a parataxonomic genus and then family. Paleontologists define egg species on a variety of characters: egg size and shape; eggshell surface texture (ornamentation), thickness and a number of microscopic features; and clutch arrangement.

The parataxonomic system allows for the definition of specific egg types, aids scientific discussion of otherwise unidentifiable eggs, and facilitates comparison of egg types from distant regions. For example, separate researchers have assigned eggs from the Cretaceous of southern France and Argentina to different species within the parataxonomic genus, Megaloolithus. French paleontologists had long considered their eggs as those of sauropods. The occurrence of sauropod embryos in the Argentine eggs and the recognized egg similarity supports the French hypothesis. Likely, in the coming years, the parataxonomic system will change to better incorporate evolutionary thinking.

Surprisingly, several references designate eggs and eggshell as a type of trace fossil, along with such things as burrows, footprints, and scat. This represents an egregious mistake. Eggshell forms an integral part of a young dinosaur's "body." It provides protection from crushing, prevents dessication, allows gas exchange with the external environment, and supplies necessary calcium for bone growth. Technically, since eggs and eggshells represent body fossils, names established on their basis would hold as much validity as names based on skeletons or teeth. Because matching adult skeletons with eggs typically proves difficult (again, this can only be done if mature embryos occur), the use of the parataxonomic system allows definition of parallel but somewhat unofficial names that do not affect standard dinosaur names.

Like living archosaurs, dinosaurs produced hard, somewhat brittle and largely calcitic eggshell. In contrast, lizards and snakes lay "leathery" shell eggs. Microscopic thin sections show the shell to consist of vertically oriented shell units of calcite crystals with a network of organic matrix. Pore canals pass through the shell to permit embryonic gas exchange. Although the eggshell of birds, crocodilians, and dinosaurs shares a basic similarity of construction, dinosaur eggshell exhibits a much greater variation in microscopic morphology than either of its living relatives possesses.

To date, paleontologists have named nearly 80 parataxonomic egg species in over 30 genera. To facilitate their recognition as egg taxa, these names usually end in "oolithes" or "oolithus," meaning "egg-stone." These many species fall primarily into three broad categories: spherulitic, prismatic, and ornithoid. Spherulitic eggs have microscopic shell units with a spherocrsytalline form where crystal growth radiates from a central point. A variety of shell unit and pore morphologies occur within this diverse group. Generally, an irregular ornamentation marks the shell surface. Most spherulitic eggs are spherical to slightly oval with diameters between 5–8 inches. Slightly more oval forms occur within the parataxonomic genus Ovaloolithus. Importantly, embryos occur in a few varieties. Sauropods, hadrosaurs, and possibly the unusual therizinosaurs possess spherulitic eggs. Clutches consist of irregularly arranged eggs in single to multiple layers.

Prismatic eggs have a two-part shell unit, an elongate to oval shape and a typically smooth surface potentially marked by fine lineations. In the Jurassic of Portugal, unidentified embryonic theropod remains occur with a clutch of 34 small (less than 6 inches long) eggs. In contrast, the prismatic asymmetric eggs of Troodon stand nearly vertical within sediments, pointy-end down. The dinosaur egg varieties of the Orntihoid type exhibit strong similarities to bird eggs. Their shell units, separate at the base, coalesce toward the exterior to form a continuous layer. These elongate eggs come in a wide range of sizes from 3 to over 18 inches in length and show a surface texture marked by distinct linear to branched tubercles. Eggs of the theropod Oviraptor belong to this category. Some ornithoid clutches consist of egg pairs arranged in a ring or in a straight line. The rings of Oviraptor may have as many as three layers.

The observable variation in dinosaur egg size, shape, surface texture, crystal morphology, and pore form and density likely has implications for the method of incubation and the development of young. By the 1870s, paleontologists had begun the microscopic examination of dinosaur eggshell. Despite this and many recent studies, the functional aspects of dinosaur eggshell remain largely unexplored.

Dinosaur egg-to-adult-body-size ratios typically exceed those of living reptiles but fall short of avian values. Those theropods with a close ancestry with birds have the largest egg/adult size ratios, but even these are only half those of modern birds. Interestingly, the lambeosaur *Hypacrosaurus* has an egg over four times larger than that of the closely related hadrosaur *Maiaisaura*. Asymmetrical eggs occur rarely in reptiles, where eggs form en masse, but typify birds where eggs develop singularly. The occurrence of asymmetric eggs, for example, those of *Troodon*, may indicate the presence of a more birdlike reproductive tract.

The high pore density of some dinosaur eggshell types more closely matches that of extant reptiles. This may imply that these dinosaurs used vegetation mounds or sediments for incubation. In living animals, high pore densities facilitate gas exchange in these oxygen-poor, highly humid nest environments. Many surface textures and microscopic crystal structures found in dinosaurs are unique and lack modern analogs. These features may also play a role in gas exchange and reflect the incubation mode employed (vegetation, sediments, direct body contact), but this awaits further research. Another area that may provide insight into the reproductive systems of dinosaurs concerns pathologic eggshell. These shells, with one to several extra layers of partial to complete shell, result from abnormal egg retention in the oviducts. Their form and frequency may reflect overall reproductive tract anatomy and function.

Knowledge of dinosaur nests largely comes from the arrangement of egg clutches, not from independently preserved nesting structures. The latter preserve only rarely. Most clutches consist of irregular masses in one to several layers, possibly representing incubation by burial. More orderly arrangements, for example, the ring of eggs for *Oviraptor* and the tight oval in *Troodon*, appear to reflect brooding behavior. Other arrangements, such as the potential arcs in some spherulitic *Megaloolithus* eggs and paired rows of ornithoid type, defy definitive explanation.

Growth and Care of Young

Hatchlings of living animals emerge from the eggs in a wide range of developmental states. At one extreme, all living crocodilians and the more primitive groups of extant birds produce precocial young. These offspring hatch fully capable of locomotion and feeding themselves. In contrast, advanced birds generally produce altricial young. These hatchlings are incapable of leaving the nest and rely completely on their parents for food. Although exceptions exists, other aspects of reproduction strongly correlate with the developmental state of hatchlings. For example, species with altricial young tend to have smaller eggs, grow faster, and require more parental care than precocial species.

Despite nearly 80 described varieties of eggs, few dinosaur embryos exist. To date, those associated with eggs includes only the hadrosaurs *Maiasaura* and *Hypacrosaurus*, an unidentified Argentine sauropod, an unidentified Portuguese theropod, a possible therizinosaur, *Oviraptor*, and *Troodon*. Examination of embryonic bones reveals that only one of these dinosaurs, *Maiasaura*, produced truly altricial young.

Dinosaurs' closest living relatives, crocodilians and birds, differ from other extant egg-layers in showing a greater amount of parental care. Both groups tend their eggs during incubation. Crocodilians guard and adjust their nests, while nearly all birds brood their eggs. Both groups also exhibit a significant amount of post-hatching care of young. Using the extant phylogenetic bracket, one would expect a similar amount of care in dinosaurs, but demonstrating such behavior in fossils remains fairly difficult. Nevertheless, some data, such as bone histology, egg size, and the occurrence of adult and juvenile material within fossil assemblages, suggests the presence of some parental care among dinosaurs. Excellent specimens from Mongolia and China preserve adult *Oviraptors* perched atop egg clutches in a birdlike brooding pose. Similar but less well-preserved specimens also exist for two more theropods, *Troodon* and *Deinonychus*.

Horner has argued on the basis of small eggs, poorly ossified juvenile bones, and an assemblage of juveniles, that *Maiasaura* young were altricial and nest-bound. This interpretation implies that adult *Maiasaura* tended and fed their altricial young for some time after hatching. Fossil assemblages from the Jurassic of Wyoming contain an abundance of tooth-marked bone and shed *Allosaurus* teeth, both juvenile and adult. Located within very fine-grained sediments, these accumulations may represent lairs, sites where adult *Allosaurus* brought food back to feed their young. A few assemblages contain remains of adults and juvenile skeletons, and may imply some type of family group. Species represented include: the ceratopsians *Centrosaurus* and *Pachyrhinosaurus*, the ornithopod *Tenontosaurus*, the sauropods *Apatosaurus* and *Camarasaurus*, and the theropod *Troodon*. More ambiguous for parental care are all-juvenile assemblages known for some hadrosaurs, *Psittacosaurus*, *Protoceratops*, and the ankylosaur *Pinacosaurus*. Although some of these contain young individuals, the possibility exists that juveniles formed adult-free associations upon leaving the nest.

Although dinosaurs occupied feeding niches most similar to modern mammals, their reproductive behavior differed drastically. Larger mammals produce a few large young that receive significant amounts of parental care, including lactation. In contrast, egg-laying in dinosaurs produced many small young; these received less parental attention. Whereas mammalian populations generally include a majority of adults, the higher reproductive output of dinosaurs would have created a population dominated by juveniles. The dinosaur reproductive style would permit a more rapid rate of population increase but may have also made them more susceptible to over-population and population crashes.

Examples

The species that provide the best information on dinosaur reproduction include: *Maiasaura*, Hypacrosaurus, an unidentified Argentine sauropod, *Oviraptor*, and *Troodon*. Complete eggs, clutches, and embryos exists for each of these species.

Maiasaura

Horner's discoveries of the hadrosaur *Maiasaura* in 1978 and subsequent years represented a significant advancement in the study of dinosaur reproduction. These finds included the first embryonic remains and the first efforts to interpret reproductive behavior based on the associations of bones, eggshell, and sediments.

Maiasaura eggs come from the Upper Cretaceous of Montana and belong to the Spheroolithus egg variety. The spherical eggs have a diameter of 5 inches and a total volume of only 55 cubic inches. Within clutches, eggs occur in at least two layers and may number 16 or so. The most interesting *Maiasaura* specimens are two concentrations of juveniles. In the first, a mass of broken eggshell contains parts of seven individuals only 1½ feet long. The second specimen consists of 3-foot-long individuals found within a bowllike depression. From these two assemblages, Horner hypothesized that *Maiasaura* young hatched out at a small size (a foot long) then remained in the nest until they reached a length of 4 feet. The small *Maiasaura* eggs, smaller than an ostrich egg, and the histology of juvenile bones support the altricial interpretation of *Maiasaura* young. Nestling bones lack well-ossified ends. Instead they terminate with massive calcified cartilage pads. This construction matches that of young birds and suggests high growth rates typical of altricial young.

Hypacrosaurus

Although the lambeosaur *Hypacrosaurus* shares a fairly close relationship with *Maiasaura*, it exhibits some significant differences reproductively. Like those of *Maiasaura*, *Hypacrosaurus* eggs have spheroidal shape and a similar microscopic structure. Both belong to the Spheroolithus egg variety. *Hypacrosaurus* eggs differ in being over four times bigger with a diameter of roughly 8 inches and a volume of 244 cubic inches. Clutches consist only of one layer of eggs, and one clutch contains 22 eggs. These larger eggs minimally held individuals 2 feet long. Nothing more is currently known about *Hypacrosaurus* nests. A fluvial deposit contains the disarticulated skeletons of more than 6-foot-long juveniles. This assemblage may represent a nesting ground or possibly an adult-free juvenile group.

Sauropod

Recently discovered eggs from the Late Cretaceous of Argentina have produced the first unequivocal embryonic remains of sauropods. These support the sauropod interpreta-

When male dinosaurs fought one another for breeding rights, they may have used horns as safer alternatives to teeth and beaks to settle disputes. Horns would have been useful both for display and for actual combat. A couple of *Chasmosaurus* lock horns and engage in a shoving match to test their strength.

In the new view of dinosaurs it is widely accepted that at least some species may have taken care of their young, most especially the larger brained, more bird-like theropods. This vignette, extrapolated from the behavior of ostriches, of a pair of Late Cretaceous *Gallimimus* tending to their crèche of little ones is therefore speculative but plausible.

Even sauropods were cute when they were little babies fresh out of the egg. This is a titanosaur hatchling; the skin is based upon traces found within the eggs. Whether new sauropods received intensive parental care is doubtful.

tion of other eggs from the Jurassic and Cretaceous of North America, Europe, and India. The Argentine eggs remarkably preserve not only excellent skeletons but also fossil skin casts. Skeletal features indicate that the embryos probably represent a titanosaur.

These megaloolithid eggs have a maximum diameter of 5–6 inches and a volume of 49 cubic inches. Considering the massive size of sauropods, these eggs seem surprisingly small. Eggs occur massed in clutches. Thousands of eggs cover an area greater than half a square mile, suggesting colonial nesting.

Oviraptor

The history of *Oviraptor* represents a case of mistaken identity and a good example of why only embryos can be used in the identification of dinosaur eggs. In the 1920s, the American Museum expeditions to Mongolia discovered the first whole dinosaur eggs at the Flaming Cliffs. The expedition also found a number of skeletons at this site; 96 percent (101 out of 105) were the ornithischian *Protoceratops*. It seemed a reasonable conclusion that the eggs belonged to *Protoceratops*. In 1923, the expedition discovered an unusual theropod lying atop an egg clutch. Presumably, this carnivore died while raiding the nest and was given the name *Oviraptor*, the "egg seizer." This story changed when the American Museum returned to Mongolia in the 1990s and unearthed well-preserved oviraptorid (likely *Oviraptor*) embryos inside the "protoceratops" eggs. These embryos have well-ossified bones and would have hatched as precocial young. The new expeditions also produced an exquisitely preserved adult skeleton perched atop a nearly complete egg clutch. Two additional adult-with-eggs specimens have recently turned up, providing a fairly accurate reconstruction of *Oviraptor* reproduction.

The nest consists of elongatoolithid eggs arranged in a circular pattern with two or three layers of eggs. Complete clutches may have had as many as 22 eggs arranged in pairs. Sediment filled the egg-free center of the clutch and buried most but not all of the clutch. The adults squatted with their feet atop or adjacent to the eggs, their hindlimbs tightly folded and their arms pointed back and wrapping around the clutch. In this position, the body completely covers the nest. Some portions of the trunk rested directly on some eggs. This pose strongly suggests *Oviraptor* adults brooded and even incubated their eggs as modern birds do.

Troodon

Troodon, along with *Oviraptor*, belong to the Maniraptora, the theropod group that includes birds. Like *Oviraptor*, *Troodon* also suffered from misidentification. The Egg Mountain locality in the Montana Cretaceous produced the first *Troodon* eggs, and this small hill yielded a number of egg clutches on three separate horizons. Many partial skeletons and isolated bones of various age classes occurred on these horizons. These bones represented a new species of hypsilophodont, *Orodromeus makeli*. Later, when embryos were found, these were erroneously attributed to *Orodromeus*. A new find in 1993 consisted of a partial adult *Troodon* lying in contact with a clutch of "orodromeus" eggs. This suspicious discovery led to the reexamination and additional preparation of the original embryos. Three factors contributed to the original misidentification: the abundance of *Orodromeus* material on the nesting site, a lack of good *Troodon* comparative material, and the somewhat unusual teeth of *Troodon* embryos. Unlike typical *Troodon* teeth that have unusually large serrations, those of the embryos lack all signs of these denticles.

Troodon eggs belong to the Prismatoolithus variety. They have an elongate, asymmetrical, and consequently more birdlike shape, a long axis of 5 inches and a volume of roughly 27½ cubic inches. For an animal the size of *Troodon*, roughly 100 pounds, this egg is fairly large. It far exceeds the average egg produced by a 100-pound, extant reptile, but is only about half the size one would expect a similarly sized bird to lay. In relative size, *Troodon* eggs compare to those of *Oviraptor*. *Troodon* placed its eggs nearly vertical, pointy end down, within the ground. The egg tops remained exposed and unburied. Hatched clutches lack the upper portions of the eggs. This nesting style would have prohibited rotation of the eggs.

Clutches consist of up to 24 eggs and show a paired-egg pattern when viewed from below, and a closely packed pattern from above. Clutches sit in the center of an earthen nest. An excellently preserved specimen shows a nest structure, clearly recognizable from

This restoration of *Oviraptor* brooding its eggs is based exactly on the position of a skeleton preserved in brooding position atop a ring of eggs. The eggs were partly exposed and in contact with the body, whose arms were draped over many of the eggs. Feathers are limited here to reveal the eggs, but it is possible that they actually covered the entire nest. The dinosaur may have rotated its body in order to evenly incubate the eggs.

the clutch itself. The nest consists of a bowl-shaped depression surrounded by a distinct raised rim. Apparently, *Troodon* excavated this structure from soils. To date, no young or hatchlings have ever been found closely associated with a nest or egg clutch.

The egg size, shape, and microstructure of *Troodon* suggest a closer relationship to birds than found in any other dinosaur for which reproductive information exists. *Troodon* appears to have laid two eggs at a time at daily or greater intervals from two bird-like reproductive tracts. After completing the clutch, the adult gathered the eggs slightly and began incubating them. Eggs hatched synchronously, and the precocial young soon left the nest. A similar pattern of egg laying and care occurs in primitive living birds such as ratites (ostriches and their kin), galliformes (turkey, pheasant, etc.) and anseriformes (ducks, screamers). Nevertheless, some notable differences exist between birds and *Troodon*. Modern birds typically have only one functional ovary and oviduct; produce a maximum of one egg per day, although this egg is much larger; incubate their eggs free of sediment; and rotate their eggs.

Conclusions

Currently, few dinosaur species provide sufficient specimens for detailed reproductive interpretation. Those species that do present a limited but diverse picture of behavior. Even fairly closely related species, such as *Maiasaura* and *Hypacrosaurus* and *Oviraptor* and *Troodon*, display significant differences in egg size and shape, eggshell microstructure, clutch arrangements, and the altricial versus precocial nature of hatchlings. The 80 described varieties of eggshell suggest that paleontologists have only glimpsed a small portion of the potential diversity in dinosaur reproductive behavior.

Present knowledge allows for a few, somewhat tentative, generalizations about dinosaur reproduction:

- Sexual display features occur commonly in a variety of groups, and sexual selection provided a significant mechanism for dinosaur evolution.

- Apparently all dinosaurs laid hard, largely calcitic eggs. No evidence exists suggesting that any dinosaurs had live birth.

- In four of the five examples discussed above, egg clutches occur in high concentrations both stratigraphically and geographically. Some other egg localities also produce extremely rich egg assemblages. Although not likely universal, colonial nesting and site fidelity may characterize many dinosaur species.

- Rapid juvenile growth appears typical of later Upper Jurassic and Cretaceous dinosaur species.

- Clutches consisting of irregularly arranged egg concentrations occur in a variety of groups—hadrosaurs, sauropods, and at least some theropods. Although how these dinosaurs incubated their clutches largely remains unknown, these clutches may indicate that some groups possessed a more crocodilian style of reproduction. These dinosaurs would have produced larger numbers of smaller eggs from two functional reproductive tracts and laid the eggs en masse.

- Several reproductive aspects, which among living vertebrates typify birds, first evolved within theropod dinosaurs. The unknown Jurassic theropod from Portugal currently displays no avian features in its nest or eggs other than some microscopic aspects of the shell. In contrast, the two Cretaceous theropods display a number of features trending toward birds: larger and, at least in *Troodon*, asymmetric eggs; one egg produced per oviduct at a time; a loss of egg retention; eggs partially exposed in nests; and incubation by brooding.

In many ways, the study of dinosaur reproduction is currently in its infancy. Sexing individual skeletons still evades a definitive solution, and, consequently, the mating systems of dinosaurs remain only hypothetical. Identifiable eggs exist only for a handful of species, and paleontologists lack eggs or eggshell for many entire groups: prosauropods, basal sauropods, ankylosaurs, stegosaurs, hypsilophodonts, pachcephalosaurs, and ceratopsians. Additionally, dinosaur nest structures and the function of diverse types of dinosaur eggshell await new discoveries and further experimentation.

Chapter Six
Dinosaur Bioenergetics

Introduction

Consider the problem faced by a baby sauropod that has just cracked its way out of its egg into the big, wide Mesozoic world. It weighs just a few pounds but has to grow ten thousandfold to tens of tons! How is it going to do it? Among land animals, growth and metabolism appear to be closely linked. All reptiles have low metabolic rates and all grow slowly. An alligator reaches only a few pounds in a year. Only more energetic mammals and birds are able to achieve rapid growth. A 200-pound ostrich is all grown up in not much more than a year. The growth and energetics of dinosaurs were not a subject of great interest for almost a century and a half after their discovery. It was generally assumed that they were slow-growing and had low metabolic rates, like living reptiles. The matter only became subject to serious scientific scrutiny over a quarter of a century ago, when it began to be realized that dinosaurs were more similar to mammals and especially to birds in form, and, consequently, in function as well. The result has been a wave of investigation into the patterns of growth and energetics of dinosaurs.

Robert Bakker's *Dinosaur Renaissance* represented the acme of the first wave of the hypothesis of dinosaur endothermy. It attempted to overturn completely the reptilian concept of dinosaurs, arguing that they all had avian energetics, and that the small ones were even feathered. Aspects of Bakker's hypothesis would be challenged in subsequent years. The use of predator/prey ratios is as controversial today as then, and has proved difficult to verify. The way in which Bakker used bone microstructure, particularly haversion canals, turned out to be nondefinitive, and dinosaur bones do on occasion contain growth rings. It turns out, though, that mammals and even large bird bones sometimes include similar rings. On the positive side, latitudinal zonation has proved to be important not only for the mammal-like therapsids as Bakker thought, but for dinosaurs as well. In their essay and update on Australian Cretaceous dinosaurs, Tom and Patricia Rich and colleagues show that dinosaurs were living in a polar climate that, although not as harsh as today, was still bitterly cold, to the point that winter snows, hard freezes, and—as their update explains—even permafrosts occurred. Yet the dinosaurs show signs of having been active, something they could have done only if they had the ability to generate heat internally, and then retain it.

Kristina Curry Rogers tells the story of how bone microstructure demonstrates that most dinosaurs grew rapidly, in the manner of birds and mammals. So sauropods grew bigger than elephants not in the way Mesozoic crocodilians did—by gradually putting on the pounds year after year until great age resulted in great bulk—but by using the same pattern of fast growth used by elephants to become enormous while still teenagers. Bone microstructure directly records growth rather than metabolism, but by doing so it can indirectly reflect energetics, because only high-metabolic-rate endotherms have the ability to grow fast; thus, evidence for the latter is evidence for the former.

Reese Barrick has been involved in leading-edge research that could have only been dreamed of in the 1970s, using bone isotope ratios to more directly measure the thermodynamics of dinosaurs. His essay combines this chemical analysis with other lines of evidence to conclude, as have most other researchers, that Bakker was correct: dinosaurs did not have reptilian energetics, and they consumed and burned oxygen at rates far higher than seen in modern reptiles.

The icing on the energetic cake is the discovery of what are almost certainly simple but dense feather coverings on small theropod dinosaurs. They come as close to proving as it can be proved that these dinosaurs were true endotherms with levels of heat production and temperature stability not seen in any reptile. They also offer an explanation of how those little polar dinosaurs got through those long winter nights without freezing their dinosaurian booties off.

Growth Rates Among the Dinosaurs

by Kristina Curry Rogers

Dinosaurs ranged in size from that of the tiny house sparrow that frequents your back-yard feeder to the immense sauropods, like *Apatosaurus,* that wandered through Meso-zoic landscapes. With such a diverse group of animals to work with and taking into account that only a tiny specialized fraction of the group persists today (all living birds), it is no wonder that the growth rates and longevities of extinct dinosaurs are the source of heated debate among paleontologists. We wonder: How quickly did the dinosaurs grow? How long did it take for them to reach sexual maturity and adult sizes? Just how long might they have lived? Although the very name Dinosauria ("terrible lizards") evokes an image of slow-growing, cold-blooded behemoths, new discoveries and labo-ratory techniques are indicating that this early, traditional view of dinosaurs is incorrect. This essay provides a brief overview of the evidence from living and fossil animals that pertains to our new understanding of dinosaurian growth.

What Is Growth?

In a general sense, everyone knows what we mean when we talk about growth. During ontogeny (an organism's life history), plants and animals get larger. Overall body shape and size change in one or more dimensions. These changes are governed by several gen-eral rules. As an animal gets larger, its mass increases cubically. The skeleton must grow accordingly to accommodate for the new stresses and strains resulting from weight gain and redistribution. The skeleton, on the other hand, can grow only in length and girth, and, therefore, grows at a less exponential rate than the skin, muscle, fur, or feathers that surround it. While skeletal growth is most obvious in linear dimensions (i.e., getting longer or thicker), it also includes important changes in shape and proportion, fusion of different parts of the skeleton, and even changes in the microscopic cellular and vascular organization. This type of growth changes the shape of a bone at different rates in mul-tiple dimensions and is called "allometry." Allometric growth constitutes the majority of growth in all vertebrates and is thought to be important for constraining our general body form. A good example of allometry can be seen in the changes in face shape that occur as animals age. When many vertebrates are young (e.g, dogs, humans, some birds, and dinosaurs), they have large heads relative to their body size. Think of a puppy (or a juvenile dinosaur). It probably has big eyes and a blunt snout; maybe its head seems too big for its body, or it has large unwieldy paws that make it seem clumsy. As these animals grow up, their bodies change and catch up to the size of the heads or paws. The bodies of adult animals are usually well proportioned, and animals as adults are capable of smooth, well-coordinated movements.

An additional type of growth occurs when all parts of the skeleton change at the same rate, and bone shape stays constant throughout ontogeny. This type of growth is deemed "isometry," and is quite rare in vertebrate organisms.

Another important aspect of growth is that, at some point in the life history of many living vertebrates, it stops. In animals like modern elephants, birds, and even humans, the majority of growth occurs while the animal is still a juvenile. Often, growth gradually slows down after puberty and may dramatically decrease once the animal reaches adulthood. This type of growth is almost like a "finishing up" of the product that been forming throughout ontogeny. If you were to plot the weight of an animal that grows in this manner over time (or the length of its bones) versus the age in years, it would level off, or plateau, at older ages. This growth strategy is called "determinate growth" because once adult size is attained, relatively little additional skeletal growth or mass increase occurs, if the organism maintains an active lifestyle.

Other animals—especially ectotherms (cold-blooded animals such as crocodiles and turtles)—grow more slowly and steadily during life. These animals might grow throughout ontogeny, right up to the moment of death. This type of growth is called "indeterminate." Instead of a growth-curve plateau in late ontogeny, these animals just keep growing steadily, and their curve has a straightened trajectory.

Growth series record various stages of development from hatchling to adulthood, including the phenonmena mentioned above. Growth series group animals of different sizes into a continuum from small to large. If the entire animal is not available, their bones can serve as a proxy for changes over time. Growth series allow us to establish general trends in the ontogeny for a single species of animal, such as increasing limb robusticity and muscle attachments or increasing numbers of teeth. These general ontogenetic trends are useful for interpreting changing physical and biomechanical environments (e.g., changing from juvenile bipedalism to quadrupedalism as an adult).

How to Study Growth in the Fossil Record

Growth series currently exist for a number of dinosaurs. We can see that as dinosaurs grew older, they grew bigger. This comes as no great surprise because the same trend is evident in all other vertebrates. But what about rates and ages? While we can easily get the general sense of aging and size increase over time just by looking at bones from a growth series, this method doesn't allow us to measure growth increments. Looking

As these baby duck-bill *Hypacrosaurus* emerged from their mound nests they faced a daunting task: grow from the size of house cats to rhino-sized adults quickly enough to avoid being eaten before they can reproduce. The chicks are shown insulated with a downy covering they may have needed to hold in the high body temperatures that could have been the key to rapid growth.

only at the changes in external bone features leaves us in the dark with regard to how quickly dinosaurs might have grown and how old they might have been at different points in their growth series. So what are the steps by which a growth series may be turned into an actual measurement of growth rate and age?

In living animals found in the wild, one of the most common means (aside from visual clues like plumage and horns) of establishing relative age is to measure mass. We know that at certain points during a creature's ontogeny, it should weigh specific amounts. If one animal weighs significantly more than another of the same species, it may be hypothesized that the heavier animal is older. Similarly, the mass increase in modern animals is measured day to day as a proxy for growth rates. The procedure is quite simple. Animals are weighed, weights are recorded, and the amount of change from day to day indicates how quickly or slowly an animal is growing. Generally, animals grow most rapidly when they are juveniles (up to 50 percent of adult size), then begin to slow down into their subadult ages. It's certainly not as easy to measure mass increase in dinosaurs. Unlike the small birds and mammals living on the planet today (even a rhinoceros is small relative to a giant sauropod), weighing a dinosaur on a regular scale is daunting. Instead, dinosaur weight estimates are achieved in a variety of ways, and growth rates are frequently extrapolated from living animals. For example, in a groundbreaking study on growth rates and reproductive strategies of some extinct dinosaurs, Case first estimated the weights of *Hypselosaurus* (a sauropod) and *Protoceratops* (a ceratopsian). He then applied the relationship between maximum growth rate and body weight of living reptiles to *Hypselosaurus* and *Protoceratops*. From these data, Case concluded that ceratopsians may have taken up to 38 years to reach adult size, while the much larger sauropods may still have been growing at 118!

The most critical assumption in this analysis and others has been the use of reptilian growth rates in extrapolating dinosaurian ages. Crocodilians and other reptiles normally increase their body mass much more slowly than living birds and mammals do. The use of mammalian or avian growth rates reduces these ages by nearly 10 times. So which animals are the most appropriate analogues for studying dinosaurian growth? Both crocodilians and birds are modern representatives of the larger group, Archosauria, which includes dinosaurs. Thus, either one of these animals is equally likely to enlighten us on dinosaurian growth strategies. Did dinosaurs retain the slow growth rates of early archosaurs or did they grow more like birds?

Rather than introducing assumptions of prolonged mass increase and slow growth rates for dinosaurs, we should take a more direct approach. Dinosaurian soft tissues may not be readily preserved in the fossil record, but their bones are. Why not utilize the bone tissue itself to get to the bottom of this debate about dinosaurian growth? Bone histology is the study of bone tissue. Due to the unique and prevalent preservation of dinosaurian bones, we have a direct source of evidence for the record of growth in animals long extinct. It is here, inside dinosaur bones and under microscopes, that the answers lie.

Bone Histology and Growth Rates

During the dinosaur renaissance of the past three decades, it has become far more common to view dinosaurs in a biological and active context. No longer are they perceived as the slow-moving, ponderous beasts of yore. Scientists from around the world have cracked or cut open dinosaur bones and have begun to decipher the stories that are preserved inside.

In an adult human, the entire skeleton is gradually reformed every seven or so years. The microscopic internal structure of bones, its histology, reflects the effects of a rapidly or slowly growing body. Throughout the life of an animal, whether it's a bird, a crocodile, a dinosaur, or a human, bones are growing and changing. This may come as a surprise to those who think of bones only as static and solid structures. In reality, bones are dynamic; they are made up of mineral components, proteins, numerous blood vessels, nerves, and cells. Organization of bone mineral, blood vessels, and the abundance of bone-producing cells all provide clues to the speed and pattern of bone growth.

Organization of Bone Tissue

As a bone begins to grow, tiny cells called osteoblasts lay down the first bit of bone tissue, called osteoid, that is very rich in a special type of elastic protein called collagen. Bones generally grow "out" (in girth), and in length. Growth in bones is analogous to the growth observed in trees. New bone is deposited on the outside of a bone. Longitudinal growth occurs at a specialized region of the bone shaft called the metaphysis, or "growth plate." If a bone is growing very quickly, the collagen has no time to be laid down in an organized fashion, and is called woven, or fibrous, bone. Alternatively, when bones are growing very slowly, the collagen fibers have plenty of time to become organized, and with each pulse of bone growth, the fibers arrange themselves in parallel. This type of bone tissue is commonly known as lamellar bone because each increment of bone growth is called a lamella. As time passes, the bone mineral hydroxyapatite precipitates on these strands of collagen in the same direction as the collagen fibers themselves. In most fast-growing animals, like mammals and birds, the two types of bone tissue combine. This combination tissue is called the "fibrolamellar complex." Under a microscope, a special sort of polarized light allows these different fiber patterns to be distinguished. Woven bone tissue will change color all at once as you turn the stage of the microscope, making it easy to distinguish between fast- and slow-growing regions of the bones.

Organization of Bone's Blood Supply

Other distinctive features of modern bones are the arrangements of the vascular canals. Vascular canals are spaces in the living bones that house blood vessels and nerves. Generally, when bones are growing quickly and need abundant blood and nutrient supplies, there are numerous vascular canals that course through the bone tissue. Different vascular patterns indicate qualitative differences in the speed of growth, with more canals in a greater variety of directions indicative of faster growth. In the slowest-growing bones, vascular canals may be absent or are predominately longitudinal, circular, or radial. In contrast, when an animal is growing very quickly, the various canal types course through the bone in complex, interweaving patterns. Each of the different types of combinations is given a descriptive name: "plexiform vascularity" includes all types of canals mentioned above; "reticular vascularity" is characterized by oblique interweaving canals; Another combined vascular pattern is called "laminar"; laminar vascularity occurs when longitudinal and circular canals interweave. Recent studies have shown that laminar and reticular vascular patterns actually occur in the fastest-growing bones in modern birds and mammals, with plexiform bone deposits occurring at slightly lower bone-growth rates.

Stop and Go: The Periodicity of Bone Growth

In some modern animals, especially those living in rapidly fluctuating environments (e.g., manatees or sea turtles) or those that are ectothermic (e.g., crocodilians), an additional bone tissue pattern is observed. Conspicuous "growth rings," or zones, result from the periodic slowing or cessation of growth. Much like the rings that are deposited in trees in the winter, commonly, growth rings in bones are yearly or seasonal events, and often record ages in living amphibians and reptiles. The areas of slowest growth appear as dense bands known as annuli, while complete cessation of growth is indicated by features called "lines of arrested growth" (LAGs; also called "arrest lines"). LAGs indicate the position of the bone's external surface at the time that the LAG was deposited. Bone that is characterized by the presence of LAGs, annuli, cycles, or a combination of these, is typically termed "zonal bone." Bone occurring in zones may be fibrolamellar, with abundant vascularity, telling the tale of fast growth for at least part of a year (e.g., polar bears). Alternatively, bone within zones may be lamellar and poorly vascularized if an animal deposits bone slowly throughout its active growth period (e.g., wild crocodiles and alligators).

Cycles and/or LAGs are frequently used in living amphibians and reptiles to determine the age of animals trapped in the wild. Simply counting the observable cycles isn't enough for several reasons. First, bone remodels throughout ontogeny, and because bone remodeling forces the loss of original bone organization, LAGs that are deposited early in ontogeny are often lost and can cause the miscounting of cycles. For this reason, cycles are not always the most reliable indicators of age, even in living animals. In addition, while there is a general correlation between seasonality and deposition of LAGs, few studies have been conducted on living animals to test this hypothesis. In most living animals with a wide histological sample, LAGs and cycles do not occur at the same frequency among different bones, even in the skeleton of a single animal. This means that if we examine only one bone or a few bones, we might miscount (there may be 10 cycles present in a single bone with a unique ontogenetic development or function, while in another bone from the same animal, none are present). Histological sampling must encompass as many bones, from as many individuals, at as many stages of development, as possible.

How Fast Is Fast?

Bone organization at a microscopic level can certainly tell us a lot about the qualitative relationships of rapid or slow growth in living animals. But what about quantification? What numbers can be placed on fast growth, and how different is it from slow growth? More important, do our observations about the relative rapidity of woven, highly vascularized bone hold true in experimental studies?

There is a stark lack of data on living vertebrate bone-depositional rates, and virtually no experimental quantification. However, several groundbreaking new experiments in modern duck, emu, and ostrich indicate that our thoughts on the relative speeds of bone growth are right on the mark (with a few exceptions). Experiments are conducted by injecting living birds of a known age and weight with fluorescent markers at regular intervals. The fluorescent label is absorbed by the animal's bones as they grow. Animals are then sacrificed, or die naturally, and their bones are prepared for histological study. Very thin slices of the bones are made and polished smooth. They are viewed under fluorescent light, making the markers visible. Since the amount of time that passed between injections is known, we can calculate how much bone grew in a given number of days, just by measuring, and we can then figure out a daily rate of growth. The bone's vascular and fibrillar organization between markers are also recorded, thus providing us with a qualitative growth hypothesis to test. It turns out that in living ostrich, emu, duck, and mink, fibrolamellar bone with laminar and reticular vascular patterns grows very quickly. At long last, our hypotheses are being tested in living animals. Such crucial hypothesis testing is the only way to begin to understand and make interpretations of the extinct dinosaurs.

Bone Histology and Growth Rates in Extinct Dinosaurs

Now that we have a basic understanding of the microscopic structure of bones in living animals, we can apply it to the dinosaurs. Fortunately, bones are among the most common fossils in the dinosaurian fossil record. Even more fortuitous, the original arrangement of bone mineral is often preserved in dinosaur fossils and allows us to interpret those microstructural features in terms of growth rates just as we do in living vertebrates.

Unfortunately, dinosaur bone tissue has proved to be just as diverse as the bone of living mammals, birds, and crocodiles, so few generalizations are possible. The data available on dinosaurian growth rates are much more extensive than just a few years ago but remain tentative primarily due to the persistent lack of detailed sampling. In no case described below where LAGs have been used to age dinosaurs has the sample included both a variety of bones and a group of individuals at different ages. In one, I present a case study of recent growth-rate analysis for the sauropod dinosaur *Apatosaurus*. This study serves as an example of histological methodology and highlights the importance of bone microstructure for getting to the bottom of growth strategy in one of the "super-sauropods."

Case Study: Apatosaurus Histology and Growth Rate

I recently investigated the bone microstructure of scapulae (shoulder blade), radii, and ulnae (two lower armbones), in juvenile, subadult, and adult *Apatosaurus*. All three bones show that *Apatosaurus* grew relatively rapidly: until 91 percent of growth had occurred (imagine a 6-foot-long hatchling growing to be nearly 90 feet long). The bone tissue is fibrolamellar and exhibits laminar-reticular vascularity. At about 91 percent of adult size, *Apatosaurus* began to lay down slow-growing bone punctuated by LAGs and very narrow, avascular, or longitudinally vascularized zones. Laminar, fibrolamellar bone is only rarely known from

An African elephant bull grows from 200 lbs. at birth to 7 tons, which is an eighty-fold increase in two to three decades. The largest sauropods such as *Argentinosaurus* had an even greater challenge growing from few pounds at hatching to 100 tons, a fifty thousand-fold increase, which evidence indicates they achieved in just a few decades.

living crocodilians, but it is abundant in living mammals and birds. In general, this means that *Apatosaurus* grew more like mammals and birds than it did like crocodilians. Just as in birds and mammals, the occurrence of tightly-spaced LAGs and zones at the external margin of *Apatosaurus's* bones indicate an extreme growth-rate decrease that is consistent with attaining adult size. It looks as though *Apatosaurus,* and probably lots of other large sauropods, may have stopped growing when they got to some maximum size, just as people, birds, and mammals do.

Apatosaurus maintained a high rate of growth until it reached adult size. Cycle counts in two scapulae in the study yielded age estimates of five years for a half-grown *Apatosaurus*, and between 10 and 16 years for a nearly full-grown individual. These estimates take into account cycles that may have been eroded during ontogeny.

Next, the thickness of individual cycles in the scapulae were measured. Because cycles in living animals most frequently occur on a yearly or seasonal basis, Curry assumed that each cycle took about 377 days to be laid down (377 was the number of days per year in the Jurassic period when *Apatosaurus* was alive. The number of days per year has shortened during the history of life on Earth due to tidal forces). The growth rate obtained for *Apatosaurus* is well within the range known for living birds and mammals (it is actually at the low end of the spectrum), but is far higher than that observed in living crocodilians and lizards. Instead of taking over a century to reach adulthood, *Apatosaurus* may have reached its adult size in less than 20 years.

Current Status: What We Know About Dinosaur Growth

Histological details in the following descriptions are taken from recent works on dinosaur bones.

Ornithopods

Dryosaurus

Dryosaurus femora indicate that this ornithopod grew quickly throughout ontogeny without any pauses in bone deposition. Fibrolamellar bone predominates. Anusuya Chinsamy suggests an indeterminate growth pattern for *Dryosaurus*, a pattern commonly observed in some living reptiles. There is no late-stage ontogenetic decrease in bone deposition.

Rhabodon

LAGs have been observed in several elements of an ornithopod called *Rhabodon*. The LAGs are very much like those observed in living crocodiles. They are well developed and regularly spaced, and zones decrease in thickness toward the external margin of the bone. Such a decrease is to be expected if the animal was growing toward the attainment of a maximum size. Following the strong decline in zonal thickness, a change in zonal tissue is observed, from fibrolamellar and well vascularized to a lamellar matrix with extremely low vascularity. This observation is also consistent with the attainment of adult size. As many as 12–17 LAGs may have been present in this femur, and I estimated that near-adult size would have been reached by age 12.

Hadrosaurs

Hypacrosaurus

Histological thin sections taken from the holotype of *H. stebingeri* indicate variation among bone tissue throughout the skeleton. Overall growth, although periodically interrupted during ontogeny, was relatively rapid until the attainment of adult sizes. The presence of plexiform and longitudinal vascularity indicates a slower growth rate than that commonly observed in living birds, and may more closely approximate that of extant cows and sheep.

Maiasaura

A growth series including six distinct ontogenetic stages was sampled for *Maiasaura*. Multiple elements from the forelimb, hindlimb, vertebrae, and ribs were studied to reveal that growth rates in *Maiasaura* varied ontogenetically. Juvenile growth rates were high, as indicated by the predominance of highly vascularized woven bone. Subadults developed laminar fibrolamellar bone. Femora exhibit LAGs, and indicate that *Maiasaura* may have reached a length of over 9 feet in its first year of growth. This growth rate is consistent with that observed in extant flightless birds (e.g., ostrich), and is significantly higher than the growth rates known from "typical" reptiles.

Ceratopsians

Psittacosaurus

A growth series spanning juvenile to adult developmental stages form the basis for conclusions on *Psittacosaurus* growth strategy. Adult sizes were attained between 8 and 9 years of age (as counted from LAGs and measured). An S-shaped growth curve illustrates the results, which indicate that *Psittacosaurus* grew seven times faster than most living reptiles and two times faster than most marsupials. Bodily growth in *Psittacosaurus* was two to four times slower than most members of the mammalian and avian groups.

Prosauropods

Massospondylus

A comparison of the amount of vascularization among *Massospondylus*, the secretary bird, ostrich, and Nile crocodile indicate that the bones of *Massospondylus* were even better vascularized than the avian taxa. Chinsamy found that *Massospondylus* took about 15 years to reach adult size, and may have had an indeterminate growth pattern. The hypothesized growth rate—29 years to body size of 600-plus pounds is similar to that observed in living crocodilians. This hypothesis is problematic due to the inconsistent histological data.

Sauropods

Lapparentosaurus

Humeri of *Lapparentosaurus* indicate a growth pattern nearly identical to that observed in *Apatosaurus* scapulae. Discreet cycles of bone deposition indicate that yearly variation in bone vascularity occur, but always in the context of fibrolamellar bone. In *Lapparentosaurus*, just as in *Apatosaurus*, no true LAGs occur until adult size is reached, and laminar/plexiform fibrolamellar bone predominates until late ontogeny. The authors propose a bone appositional rate similar to rates observed in living cows and other large mammals.

Theropods

Syntarsus

Syntarsus femora are characterized by zonal bone consisting of fibrolamellar bone with sparse longitudinal vascular canals. These zones are punctuated by LAGs. At least one LAG is viewable in the youngest animals sampled. Secondary reconstruction is also pervasive, and peripheral rest lines occur in the outer cortex, indicating the attainment of a maximum size. *Syntarsus LAG* counts provide an estimate of seven years to reach an adult size of approximately 45 pounds. This rate is as rapid as some maturing mammals of similar size (e.g., marsupials and primates).

Troodon

Varricchio observed metatarsals and tibiae of an ontogenetic series of *Troodon*. Bone passed through three organizational stages: rapid fibrolamellar bone deposition with laminar vascularity characterized juveniles; lower vascularized lamellar-zonal bone occurred in subadults; and avascular lamellar bone was present in adults. Varricchio tentatively assigned an age of 3–5 years to *Troodon* adults. By modern standards, *Troodon* was a large predator, with mass similar to that of modern humans (at least 110 pounds), and such a growth rate exceeds that known in at least some living mammals.

Paleoecological Implications of Dinosaur Growth

So why did some dinosaurs grow more quickly than others? How did they find enough to eat from the seemingly meager food supply during the Mesozoic and with less than optimally constructed feeding structures, enabling them to grow as quickly as modern birds and mammals? And, if living birds and mammals are growing at the same rates, or even faster, than dinosaurs, why don't they get as big as the biggest sauropods? All of these questions can be examined in light of our knowledge of dinosaurian life history strategy (the way animals behave, grow, and live), but we still have much to learn about dinosaur growth.

Dinosaurs probably had different growth strategies, just as modern animals do, because they were doing a variety of different things. Some, like the enormous sauropods, wandered the Earth eating constantly. Other dinosaurs, namely the theropods, were faster,

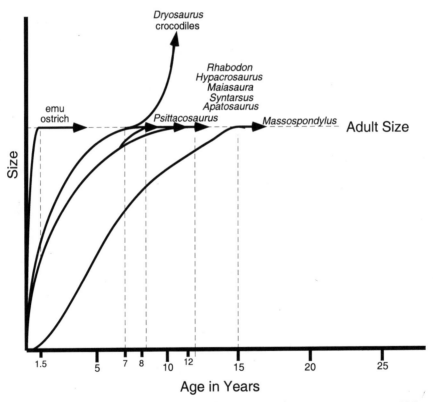

Growth curves for ostrich, emu, alligator, *Dryosaurus, Psittacosaurus, Rhabodon, Hypacrosaurus, Maiasaura, Syntarsus, Apatosaurus,* and *Massospondylus.* It is interesting to note that most dinosaurs have the more "S-shaped" growth curve typical of most living mammals and birds. Such a growth curve indicates slowing growth rate into adulthood, with a dramatic slow down once adult size is achieved. (Figure is modified from Weishampel and Fastovsky, 1997.)

smaller, and predatory. Growth spans of dinosaurs like the sauropods put limits on their growth rates. Predation, intraspecific combat, disease, injury, and starvation limited lifespans. Dinosaurs living in isolation from dangerous predators and relatively free from new diseases, might have lived up to 100 or 180 years. Heavily armored animals and camouflaged animals may have a better chance at survival. Most other animals, though, probably do not live over more than 60–100 years. Lifespans must be kept to a century and a half or less to minimize competition with juvenile successors. Animals must start to reproduce by around age 20, or the chances for dying before leaving a gene pool contribution gets higher and higher. Estimates that giant sauropods took many decades to reach sexual maturity are thus unreasonable. Sauropods and other dinosaurs lived on the planet for nearly 200 million years. Such a temporal range is not one expected for a group ill-equipped to deal with its surroundings. Instead of prolonged slow growth, it is more reasonable to invoke rapid juvenile growth rates to account for the large size of sauropods and other dinosaurs.

The conclusion to the debates on dinosaurian growth stategy is not yet final. It is obvious that dinosaurs followed the same rules for growth that all other vertebrates do. Our hope for getting to the bottom of the remaining questions for dinosaurs lies with a better knowledge of growth processes in their living representatives, the birds, and their cousins, the crocodiles.

The Thermodynamics of Dinosaurs

by Reese Barrick

Three reasons that dinosaurs are so fascinating are that they are extinct, some of them grew to enormous size, and they seem unlike any animal that is still with us today. These facts lead us to want to know more about their lives. For example, we want to know how fast they grew, how much they ate, if they had long migrations, how they reproduced, and how strong and fast and smart they were.

While studying modern animals, biologists have discovered that the two most important features in understanding the life history of an animal are body size and metabolism. Metabolism is essentially the amount of energy or calories burned by an animal (as measured while resting and while active). The burning of these calories produces heat. So animals with high metabolic rates produce more heat than do animals with low metabolic rates. If we were to know the body size and metabolism of each dinosaur species, it would be possible to predict many of the aspects of the day-to-day lives of dinosaurs that kids and scientists alike want to know. Fortunately, several good methods of estimating the body weights of dinosaurs make it relatively easy for paleontologists to estimate this important feature of dinosaur biology. Because we know the body weights of most dinosaurs, a great deal of attention has been given to deciphering the metabolic rates of dinosaurs, and this is why the study of dinosaur metabolism and temperature regulation has become so contentious and interesting.

Metabolic Rates: Endothermy and Ectothermy

In living species, the terms endothermy (warm-bloodedness) and ectothermy (cold-bloodedness) describe two hypothetical extremes of a continuum of metabolic strategies. Endotherms are animals that maintain constant body temperatures through internally produced heat from high metabolic rates. Ectotherms depend upon external heat sources for regulating body temperature. Thus, ideally, endotherms have high metabolic rates and ectotherms have low metabolic rates. Discussions of endothermy generally presume constant body temperatures and high metabolic rates (equal to those of modern mammals and birds). Discussions of ectothermy generally presume fluctuating body temperatures and low metabolic rates similar to those of modern reptiles. The pure application of either strategy throughout an individual's entire lifetime is rare. For example, baby birds are ectothermic for up to a week after they hatch. Nevertheless, in species that are primarily ectothermic, metabolic rates rise and fall in a relatively consistent relationship with environmental (air or water) temperature. Most living ectotherms (reptiles, amphibians, and fish) have adopted various strategies to minimize the effects of fluctuating ambient temperature on their metabolic processes. For example, they may limit themselves to relatively narrow temperature conditions by moving to different environments throughout the day or year. Food supply and environmental temperature are the most important factors controlling growth efficiency in ectotherms. These in turn can affect both the animals' ability to survive and their potential for successful reproduction.

Metabolic rates of primarily endothermic species (mammals and birds) are also closely linked to temperature. These species have adopted one or many strategies for generating heat within their body tissues to maintain their body temperature in a relatively narrow range. This body temperature is generally above the ambient environmental temperature. Because most mammals and birds are relatively small (when compared to many dinosaurs), this heat generation is usually combined with mechanisms to retard heat loss, such as having fur, feathers, or layers of fat. Endotherms must consistently eat large amounts of food to support high rates of metabolism and to maintain a constant body temperature. Doing so allows endotherms to successfully exploit a much greater range of environmental conditions than can be exploited by ectotherms.

Ectotherm metabolism, on the other hand, adjusts up, to take advantage of a plentiful food supply, or down, to prolong survival during periods of scarce resources. Endotherms generally have metabolic rates that are 10 times higher than do ectotherms. These two metabolic modes dictate important contrasts in life strategies, including rates and amounts of food consumption and growth rates. Metabolic rates change with body size in both endotherms and ectotherms. The metabolic rate (energy burned and heat produced) of each individual cell in the body is much lower for large animals than it is for small animals. Thus, small animals have a higher relative metabolic rate (higher metabolic rate per cell or pound of flesh) than large animals. However, because the largest animals have a greater total number of cells, they, therefore, have the highest metabolic rates in both ectotherms and endotherms.

When we discuss the fact that mammals and birds have metabolic rates that are 10 times higher than reptiles, we must keep in mind that we are comparing similar-sized animals. Large mammals may actually have lower relative metabolic rates per cell than small reptiles, and large reptiles could have higher total metabolic rates than small mammals or birds, due to the great difference in total number of cells. When comparing similar-sized mammals or birds to reptiles, however, the endothermic animals always have the higher metabolic rates. It has been supposed that the distinction between the metabolic strategies of endothermy and ectothermy are greatest at small body size (tenfold difference in metabolic rates) and least (zero to threefold difference in metabolic rates) at extremely large body sizes. This supposition assumes (with some supporting evidence) that the metabolic rate per cell decreases with size at a more rapid rate in endotherms than in ectotherms. However, it is incorrectly assumed that the sizes of *typical dinosaurs* are in the range where the distinction between endothermy and ectothermy is subtle, if not obscure. This is perhaps true of large sauropods, but most dinosaurs had adult sizes between that of a chicken and that of a modern male African elephant.

Advantages to Ectothermy and Endothermy

There are two main advantages to being an ectotherm. The primary advantage is the ability to survive conditions of widely fluctuating or limited food supplies. Because of the low metabolic rates of reptiles, little energy is burned to maintain body temperature or to maintain active physiologic functions. Therefore, more of the energy obtained from food may be put directly into growth. Thus, ectotherms are much more efficient than endotherms in transferring food into body growth.

While reptilians may be more efficient in their ability to turn food into flesh, endotherms generally grow and mature much faster than ectotherms. Keeping body temperatures within narrow limits allows the maximum efficiency of thousands of enzymes, which subsequently allows relatively continuous growth independent of environmental (air/water) temperature. Reptiles grow more rapidly at warmer temperatures than colder temperatures, and growth shuts down below 8°F. Thus, while the conversion of food to flesh is efficient in ectotherms, it is seasonal and slow when the animal is not at optimum temperatures. The conversion of food to flesh may be inefficient in endotherms, but growth is continuous and rapid until adult size is reached. The need to continuously feed the high metabolism in endotherms is supported by the ability of these animals to have high rates of continuous activity. This means that endotherms can spend longer periods of time searching greater areas for food or mates than can ectotherms. Increased growth rates mean that endotherms reach their reproductive size at much earlier ages than reptiles.

What would ectothermic dinosaurs be like?

Many reptiles are capable of great feats of strength and may be just as fast as similarly sized mammals. However, they typically rely upon anaerobic metabolism for high energy activities. Anaerobic metabolism is sustainable only for short bursts, causing animals to fatigue rapidly due to the buildup of lactic acid. The effects of lactic acid buildup also occur in endotherms, as can be seen in sporting events, especially in hot weather, where atheletes' muscles cramp up. Thus, ectothermic dinosaurs would be limited to slow, sustained movements with few bursts of high speed and energy. Chase scenes between tyrannosaurs and hadrosaurs (duck-billed dinosaurs) would be very brief indeed. Hypothesized long annual migrations of hundreds to thousands of miles by herds of hadrosaurs would also be unlikely, as they would require long hours of continuous activity each day for days, weeks, or months.

What are the food consumption needs of ectothermic dinosaurs? Low metabolic rates require relatively low inputs of food. Some crocodilians in Africa, for example, may survive on one large meal of wildebeest per year. An ectothermic 9-ton *Tyrannosaurus rex* would have required 12-plus pounds of meat per day to survive. This would amount to almost 4,500 pounds of meat per year, which is the equivalent of a medium-sized hadrosaur. Given the size of a tyrannosaur's stomach, it would only take about six to eight good meals per year to sustain an adult tyrannosaur if it were ectothermic. Similarly, plant-eating dinosaurs would have needed relatively little food for survival. A 2¼-ton hadrosaur would eat the same amount of food as a small (110-pound) deer, while a 7–9-ton *Triceratops* would have eaten quanities of food similar to an adult horse. Small plant-eating dinosaurs like hypsilophodonts or protoceratopsians would have eaten similar volumes as rabbits. Small carnivores such as *Troodon* or *Velociraptor* could have easily survived on one rat-sized mammal per week. Small dinosaurs would have taken at least five to six years to reach sexual maturity and adult body size, while medium-to-large dinosaurs would have taken 15 to 25 years; and truly great-sized sauropods would have taken over 50 years to reach adult body size.

What would endothermic dinosaurs be like?

Dinosaurs with metabolic rates similar to modern mammals and birds would have been extremely active. Herds of 2½–5-ton duck-billed dinosaurs that numbered in the thousands to tens of thousands would have decimated local vegetation, probably requiring long annual migrations. A 7-ton *Tyrannosaurus* would have required 220 pounds of food per day. This would have required it to eat large meals every three to four days. Such food intake requirements would have demanded prolonged periods of high activity in the search for food. These necessities would also likely have made carnivorous dinosaurs highly territorial, and perhaps led the large tyrannosaurs to follow the migrations of the large herbivorous hadrosaurs in order to maintain suitable food availability. Growth rates would have been relatively rapid for endothermic dinosaurs, with small species advancing to adult size in two to three years and large species in six to seven years, while the giant sauropods would have reached sexual maturity and adult body size in 10 to 15 years.

The Case for Ectothermy

Several lines of evidence have led scientists to suggest that dinosaurs were cold-blooded ectotherms with typical reptilian metabolic rates. The first line of evidence is that dinosaurs are closely related to the crocodilians that are cold-blooded. A second line of evidence comes from experiments that model the effects of high and low metabolic rates on large dinosaurs. Such modeling experiments suggest several problems for large dinosaurs with high metabolic rates. The first is the need for these animals (especially plant-eaters) to eat enough food to continuously feed high metabolic rates. Complex dentition in plant-eating dinosaurs did not evolve until the last few million years of dinosaur time, and the largest plant-eating dinosaurs had simple dentition as well as tiny heads relative to their bodies. Even eating 24 hours a day, it would have been difficult for these animals to have ingested enough to survive. Second, large animals lose heat to the environment much more slowly than small animals due to their large volume relative to the surface area of their skin. This suggests that the larger dinosaurs would have had tremendous problems disposing the heat created by their metabolism if they were endothermic, and they could have easily overheated, resulting in death.

Two separate lines of evidence for ectothermy come from the respiratory system of dinosaurs. One comes from the study of the nasal respiratory passages of dinosaurs. Mammals and birds have bone or cartilage or protrusions called respiratory turbinates in the front of the nasal region. These turbinates are covered with moist membranes that exchange heat and water with respired air. As it is inhaled, cool, dry air is warmed and moisturized across the membranes before entering the lungs. Warm moist air exhaled from the lungs is cooled, and the moisture is trapped prior to leaving the nose. This is important because mammals and birds have very high rates of breathing and, without the turbinates, could lose significant amounts of water while respiring, thereby dehydrating their lungs and bodies. To date, no definite respiratory turbinates have been found in dinosaur nasal passages. There has been evidence that some hadrosaurs may have had turbinates present; furthermore, the large nasal passages of ceratopsians and sauropods suggest the possibility of structures that functioned similarly to turbinates present in these dinosaurs as well. The second aspect of the respiratory system that has suggested ectothermy has come from evidence for the structure of the lungs. Several researchers have suggested that dinosaurs must have used their ribs to expand and contract their lungs to allow breathing rather than using a diaphragm, as do mammals, or an air sac system, as birds use. This would have become a problem for large dinosaurs, especially sauropods, in that they had such narrow, long air passages from their throats to their lungs; the inefficiency of "rib breathing" would create a poor exchange of inhaled and exhaled air. This means that these animals would not have been able to get enough fresh air and oxygen into their lungs to support a high metabolism. (See Gus

Leahy's essay "Noses, Guts, and Lungs" in Chapter 2 for further discussion of nasal passages and lungs.)

Evidence for Endothermy

Some types of evidence proposed for ectothermy have also been used to support warm-blooded interpretations for dinosaurs. Many, if not most, paleontologists accept that birds are descendants of dinosaurs. As birds are endotherms, therefore, it has been suggested that their ancestors would also have had the high metabolic rates of their descendents. Dinosaurs had long limbs and upright postures relative to ectothermic reptiles, suggesting high rates of activity. It has been argued that high activity rates, especially in large dinosaurs, must have been supported by high metabolic rates, as they are in birds and mammals. Problems of overheating would have been compensated for by modeling relatively slim dinosaurs and many of their unique body features (for example, the long necks and tails of sauropods, the plates of stegosaurs, the frills of ceratopsians and the sailbacks of spinosaurs) as adaptations for getting rid of excess heat created by their high metabolic rates. It is also valid to note that only the large sauropods grew to be significantly larger than the largest mammals in the fossil record.

Bone histology has also been used to suggest high metabolic rates for dinosaurs. Most dinosaurs, especially those from the Cretaceous period (144 million to 65 million years ago), have little or no slow-growth zonal bone that is typical of modern reptiles. Growth lines may be present in some bones but are not ubiquitous. Dinosaur bones also have a type of fast-growing bone called fibrolamellar that is common in modern endotherms but only present intermittently in some crocodiles and turtles. (See Kristina Curry Rogers' essay "Growth Rates Among the Dinosaurs" in Chapter 6.) It has also been demonstrated that modern mammals can have growth lines present in their bones, suggesting that their presence is not definitive for cold-blooded interpretations. The spacing of growth lines in Jurassic and Cretaceous dinosaurs has been utilized to determine the growth rates of these dinosaurs. Sauropods have been shown to reach adult size (appropriately 80 percent of maximum size) in eight to 12 years, while hadrosaurs reach adult size in five to eight years. These rapid growth rates for large animals are strongly correlated to endothermy rather than ectothermy. While hadrosaurs and ceratopsians had complex and efficient dentition for processing large amounts of food to support high metabolic rates, sauropods did not. For sauropods, there is evidence that they did not chew their food but swallowed foliage whole and used gizzard stones to process the vegetation into easily digestible parts. This allowed sauropods to eat large quantities of vegetation without chewing.

Small endotherms have the problem of retaining their metabolic heat, as opposed to trying to eliminate it as do large endotherms. For example, elephants use their large, thin ears to expel excess heat and have little hair or fur that would retain heat. Small mammals and birds—or those that live in cold climates—have abundant fur or feathers to help retain heat. The relatively recent discoveries of small birdlike dinosaurs with feathers has supported not only the relationship between dinosaurs and birds but also their use of feathers as thermal covers to retain heat, a trait found only in endotherms. We also know that dinosaurs inhabited every continent, including Antarctica. Even though the continents have moved around the globe through time, it appears that Antarctica and Alaska were located at near-polar latitudes during the Mesozoic era, during which time dinosaurs were inhabiting these areas. While climates were warmer in these regions than they are today, there is abundant evidence that they were too cold for ectotherms to survive. Thus, the abundance of near-polar dinosaurs supports warm-blooded endothermic interpretations.

Another line of evidence against ectothermy in dinosaurs has come from the chemistry of dinosaur bones. Oxygen is one of the most abundant elements in the Earth's crust, atmosphere, and hydrosphere, as well as in animal blood, tissue, and bone. Atoms of oxygen may contain between 16 and 18 neutrons in their nucleus, though the number of protons remains constant. This means there are two different masses or sizes of oxygen atoms. This can be analogized comparing large Delicious apples with smaller MacIntosh apples; they are both apples but of a different size and mass. The differences in the mass of these isotopes of oxygen result in their having slightly different chemical properties.

Differences in these properties can lead to isotope effects in chemical reactions. One of the most useful of these mass-related isotope effects is the dependence upon temperature of the exchange between oxygen atoms in fluids and those in minerals. This means that, during bone formation, the ratio of the oxygen-18 to oxygen-16 atoms in the bone will vary, depending upon body temperature. Therefore, changes in the oxygen ratios within bones represent fluctuations in body temperature. This gives scientists a thermomether with which to measure body temperature variations in extinct animals. In warm-blooded animals, the core body temperature does not vary much (less than 40°F) while in cold-blooded animals, core temperatures fluctuate more widely despite behavioral attempts to reduce the total body temperature variation. Most of the dinosaurs studied with this methodology have indicated temperature variations more similar to modern endotherms than ectotherms. These studies have included both large (22 tons) and small (44 pounds) dinosaurs. While very large dinosaurs could maintain constant body temperatures with ectothermic metabolic rates simply because their large size would act as a heat-conserving thermal layer, like fur, the small dinosaurs could not, nor could the juveniles of the large dinosaurs. Even the large dinosaurs would have had a difficult time maintaining summer body temperatures that were within three or four degrees of winter body temperatures. This isotopic method therefore indicates that most dinosaurs of the

Cretaceous and Late Jurassic periods were not ectotherms but utilized relatively high metabolic rates to maintain stable body temperatures.

Finally, a unique fossil occurrence has recently been described in which a mineralized concretion (a hard compact material that is different in composition from the surrounding sediment) within the chest cavity of a small dinosaur appears to have preserved the structure of its heart. CT scans apparently show two separated ventricle cavities, indicating a four-chambered heart. This extraordinary find also indicates the presence of only one systemic arch (aorta) similar to modern birds, rather than two arches similar to all modern reptiles. This find supports the suggestion of Robert Reid that such an

The multitude of dinosaurs had to cope with a wide variety of climates. In southern Africa Early Jurassic *Syntarsus* and *Massospondylus* lived in true deserts, where intense daytime heat followed by chilly nights were the primary thermodynamic problems. The little theropod may have relied upon a covering of feathers to screen out the searing Sun, and then hold in body warmth at night. Better protected by its bulk from thermal extremes, the prosauropod was able to do without a coat of insulation.

advanced double-pump heart was the norm for dinosaurs. His speculation was based on the need for high blood pressures to reach the great vertical heights of large dinosaurs. However, it is likely that the primary advantage for this advance heart was to improve oxygen delivery to the body that would efficiently support higher metabolic rates and longer periods of high activity. An advanced four chambered double pump heart is a major anatomical difference between dinosaurs and all modern reptiles and indicates that the physiology and metabolic rates of dinosaurs were different from all modern cold-bloods.

Where Does the Evidence Lead?

Evidence from heart, isotopes, and bone histology clearly indicates that dinosaurs were not truly reptilian in their physiology. Bone histology and the lack of evidence for turbinates also seem to indicate that dinosaurs were not mammalian or avian in their metabolism. Is there another possibility? Many scientists are now suggesting that dinosaurs were unique in their metabolic structure and were intermediate in metabolism between modern crocodilians and birds. Some authors have proposed that dinosaurs were supercharged reptiles with temperature stabilized by bulk, with high activity rates but low resting rates of metabolism. Reid suggests that dinosaurs were "failed" endotherms that relied on size to maintain body temperature. He follows the suggestion of earlier researchers that endothermy evolves at small body sizes and notes that the trend in early dinosaurs was toward gigantism rather than miniaturism. However, he does suggest that dinosaurs had somewhat increased metabolic rates from reptiles and might be considered intermediate animals. Reid envisions small dinosaurs as animals with high activity metabolic rates that probably could not maintain homeothermy (constant body temperatures) while resting. James Farlow also suggests that dinosaurs were intermediate in their metabolic stature with the ability to raise and lower their metabolic rates daily and seasonally. Such a scenario also implies daily or seasonal inability to maintain stable body temperatures for small dinosaurs. This point is certainly countered by the evidence from oxygen isotopes. While isotopic data suggesting stable body temperatures from large adult dinosaurs (4½–44 tons) could be interpreted as resulting from their large bulk, data from juveniles and small adult dinosaurs (45–660 pounds) also suggest stability of core body temperatures. Thus, it is likely that temperature stability in most small dinosaurs was a constant, rather than daily or seasonal feature.

Intermediate Metabolic Rates (Mesometabolism)

At any given body size, mammals have metabolic rates 10 times higher than lizards. Bird metabolic rates are higher than those of mammals at very small body sizes, but lower at larger body sizes. When you extend the correlation lines to body sizes to include the largest dinosaurs, bird metabolic rates begin to approach lizard metabolic rates, while the mammalian line does not. A metabolic rate five times higher than that of lizards has been proposed as the minimal metabolic rate that could support endothermy. If we propose this intermediate metabolic rate, or mesometabolism, as appropriate for dinosaurs, it is possible to test whether it supports all of the data that has been accumulated on dinosaurian physiology. When a line at this metabolic level is added to the plot with those of mammals, birds, and lizards, it is interesting to note that the bird and dinosaur lines intersect between 6½ and 9 tons, which is in the size range of large dinosaurs, excluding the sauropods. This leads one to consider two possible scenarios for dinosaurs. One is that they had avian- rather than mammalian-style endothermy. Such a scenario is attractive in that the smallest adult species and the babies of the large species of dinosaurs (9–44 pounds) would have had a very high metabolic rate (10 times that of reptiles) with supporting evidence coming from the thermal feather coverings discovered in some small theropods. Medium-sized (220–2,200-pound) adult and juvenile dinosaurs would have slightly lower metabolic rates (six to eight times that of reptiles), with large dinosaurs (4½–11 tons) having metabolic rates only five times greater than the reptilian rate. The largest adult sauropods (22–88 tons) would have the lowest metabolic rates (only three to four times greater than the reptilian rate). A second option is to model dinosaurs with intermediate (five times greater than reptiles) metabolic rates for all sizes, in parallel to the mammal and lizard line. This intermediate metabolic rate would support evidence for stable body temperatures in small, medium, and large dinosaurs noted from oxygen isotope studies. This much lower metabolic rate relative to mammals would also not contradict the lack of nasal turbinates or similar structures in many dinosaurs.

What about the histologic evidence? An intermediate metabolic rate would allow greater efficiency in turning food into growth than seen in true endotherms, as well as the constant growth and efficient enzyme activity associated with maintenance of stable body temperatures. Thus, mesometabolism supports the rapid growth features of dinosaur bones. This intermediate metabolic rate would also allow dinosaurs to survive on more seasonally fluctuating food sources than mammals or birds and to shut down growth when resources were scarce. This would support the evidence of the reptilian "growth ring" characters of dinosaur bone. Efficient double-pump hearts do not evolve because of any "need" to have increased metabolic rates; rather, the evolution of such a heart allows the evolution of increased metabolic rates because it supports the circulation and oxygenation necessary to support increased metabolic rates. Thus, the discovery of a fully four-chambered heart in dinosaurs also supports interpretations of intermediate and/or high metabolic rates.

Other Support for Mesometabolism

Dinosaurs with a fivefold increase in metabolism over reptiles would have had difficulty surviving at very small body sizes (⅓ ounce to 2½ pounds) where there are numerous species of mammals, birds, and lizards. Reptiles are able to survive at such small sizes because of their ectothermy, which allows wide fluctuations in body temperature. However, endotherms, with rare exceptions, cannot suffer greatly lowered body temperatures for extended periods without fatality. Thus, very small endotherms have very high metabolic rates to maintain stable body temperatures. Birds, which have lower field metabolic rates than mammals at body sizes greater than 9 pounds actually have higher metabolic rates than mammals with body sizes smaller than 2¼ pounds. It might be hypothesized that this relation reflects the fact that most tiny mammals are fossorial and use burrows to help moderate environmental temperature extremes, whereas birds generally do not, thus requiring higher metabolic rates to maintain temperature stability. If dinosaurs were mesometabolic, they could not have survived as adults at these small body sizes (less than 2 pounds) as they would not have been able to maintain temperature stability through metabolism. They were also not fossorial, and, unlike ectotherms, they could not have survived prolonged periods of low body temperatures. Indeed, there are no species of dinosaurs that fit into this size range (less than 2 pounds), and there is now evidence that the small species of dinosaurs related to birds in the 2¼–22-pound size range also had a thermal covering of feathers to conserve metabolic heat. Evidence of rapid growth in dinosaurs also suggests the ability of hatchling dinosaurs to grow rapidly past this minimum body size threshold.

It is also valid to look at the climatic environments in which dinosaurs lived and evolved. Much of the Mesozoic period when dinosaurs ruled the land was typified by much higher levels of greenhouse gases (such as carbon dioxide) than are present today. This means that dinosaurs generally lived in warmer environments that were perhaps less seasonal with respect to temperature relative to today. In the Cretaceous, seasonal and mean annual temperatures in Montana would have been more similar to those presently in North Carolina or Louisiana, while those in Louisiana would have been similar to those in southern Mexico. Thus, dinosaurs would not have necessarily required the evolution of modern mammalian metabolic rates to maintain stable body temperatures and higher rates of activity than seen in the other reptiles with which they were competing. The optimum "endothermic" metabolic rates may indeed have been much lower than would be required in the "icehouse" world of today or even at the end of the Paleozoic. Mesometabolism that supported higher activity and growth to large sizes would have mitigated the need for the very high metabolic rates seen in modern birds and mammals. Late Jurassic through Early Cretaceous climates were much cooler than in the Late Triassic through Early Jurassic or in the Late Cretaceous, with some evidence suggesting polar icecaps. Cooler climates would have supported the evolution of mesometabolic giants with less stress of overheating than in preceding or subsequent greenhouse periods. It was, in fact, during the Late Jurassic and Early Cretaceous when the largest sauropod dinosaurs dominated the terrestrial landscape, and when some carnivorous dinosaurs reached sizes even greater than the well-known *Tyrannosaurus*.

What would Mesometabolic dinosaurs be like?

We can use the "dinosaur line" to better understand several aspects of dinosaur biology. First, we choose the mass of a dinosaur—for example, *Triceratops*. Knowing that a *Triceratops* adult individual weighed about 6½ tons, we can take the metabolic rate and correlate it with this mass. We do this by finding the same metabolic rate in another mammal, and then finding the equivalent mammalian body size. This process allows us to find a mammal whose food needs were similar to those of a specific dinosaur. A 6½-ton *Triceratops* would have required as much food as an 1,800-pound mammal, such as a large bison. A 2–4-ton duck-billed dinosaur would have eaten like a small-to-medium-sized bison. This makes it much easier to envision the ecologic damage done by herds of thousands of hadrosaurs as suggested by bonebeds and nesting sites in Montana than if they had the very high metabolic rates and food needs of 2¼–4½-ton mammals such as elephants. Similarly, giant (33–55-ton) sauropods would have had equivalent food requirements as modern (5½–7½-ton) elephants. On the other hand, if they were truly ectotherms, these sauropods would have eaten similar amounts of food as bison, and hadrosaurs would have had similar food requirements as sheep. What about carnivores? An adult mesometabolic *T. rex* would have required approximately 40 pounds of meat per day. Thus, the food requirements necessary to support high, stable body temperatures are low, but still support many of the advantages of endothermy. At this rate of consumption, a *T. rex* would have needed to eat a large meal every one to two weeks rather than two to three days if it had a mammalian metabolic rate, or every two months if it maintained only a reptilian metabolic rate. The combination of core temperature stablity with the relatively low metabolic needs of mesometabolic dinosaurs could have produced growth rates at least as rapid as those seen in modern endotherms. This interpretation is supported by the rapid growth rates indicated by the bone histology of hadrosaurs and sauropods noted previously.

Were all dinosaurs created equal?

Not all mammals nor all reptiles of a similar size have the identical metabolic rate. There is historical evidence that the earliest dinosaurs had lower metabolic rates than did the latest dinosaurs from the Cretaceous. Also, advanced dental batteries that allowed the processing of large amounts of food for herbivorous dinosaurs did not show until the Late Cretaceous. Trends in limb length and construction that indicate higher activity levels as well as increasing brain size all point to increasing metabolic rates throughout the Mesozoic. There is also evidence that large Cretaceous ankylosaurs may have had lower metabolic rates than other dinosaurs. For example, these dinosaurs had poor dentition for processing food, heavy protective armor, and low mobility. Oxygen isotope evidence suggests they underwent seasonally variable core body temperatures to a much greater extent than ornithopods, ceratopsians, or theropods. Thus, throughout the Mesozoic, species of dinosaurs may have occupied much of the area between the mammalian and reptilian metabolic rates. Some dinosaurs, such as the small theropods that were closely related to birds, probably had metabolic rates close to modern avian levels. Many Jurassic and Cretaceous dinosaurs probably had metabolic rates five times the reptilian rate, and many Triassic and certain Jurassic and Cretaceous forms fall between the reptilian and proposed dinosaurian rates.

Conclusions

Understanding the metabolic strategies employed by dinosaurs is vital if we are to reconstruct their life histories and the ecologic roles they filled throughout the Mesozoic. Knowing both metabolic rate and body size for dinosaur species, we can derive accurate pictures of their food requirements and thus the effects they had on their environment, including prey species and vegetation. We can also begin to understand their growth rates, lifespans, population densities, and reproductive strategies with much greater confidence. While the picture of dinosaurs as endotherms or ectotherms has shifted back and forth in the past, a new consensus picture of dinosaurs having intermediate metabolic rates is beginning to emerge. This picture makes these animals even more interesting, as there are no vertebrates alive today employing such a strategy. Direct evidence from bone histology (including growth rates), oxygen isotopes (body temperature stability), and the discovery of an advanced birdlike heart all definitively point to dinosaurs having metabolic rates well elevated above those of modern reptiles. There is also much indirect evidence supporting this interpretation. An intermediate metabolic rate is supported by the lack of modern nasal turbinates and bone histology. Over the 150 million years of dinosaur evolution, it is likely that dinosaurs exhibited metabolic strategies from only slightly elevated rates above those of modern reptiles to near or equal to those of modern birds. For dinosaur species that exhibit core temperature stability in juvenile as well as adult stages, a hypothesis of an intermediate metabolic rate five times greater than modern ectotherms is a good starting point for studying food requirements and other ecologic parameters. The fossil record is continuously proving to us that dinosaurs were truly unique among terrestrial vertebrates in the history of life. Our fascination with them will not end soon.

Australia's Polar Dinosaurs

by Thomas Hewitt Rich, Patricia Vickers-Rich, Anusuya Chinsamy, Andrew Constantine, and Timothy Flannery

In the Early Cretaceous period, about 100 million years ago, Australia lay alongside Antarctica, which straddled the South Pole, as it does today. Australia's southeastern corner, now the state of Victoria, lay well inside the Antarctic Circle. At that time, the region hosted an assemblage of animals and plants that lived under climatic conditions having no modern analogue. The average temperature appears to have ranged from frigid to low-temperate. Through the long winter, the sun did not shine for weeks or months at a time.

Many dinosaur lineages survived in this strange environment after they had died out in other places. At least one member of the group evolved an adaptation to the cold and to the dark that is interesting both in itself and for what it tells of the passing of a biological epoch. If global cooling indeed killed the dinosaurs, as many paleontologists have suggested, then Australia's species were the ones most likely to have survived the longest. Did their adaptations to an already marginal climate help them survive a sharp cooling trend, one that caught species living on other continents unprepared?

Although the Cretaceous fossil plants of southeastern Australia have been studied for more than a century, the animals remained mostly undiscovered until recently. In 1903, the geologist William Hamilton Ferguson found two bones that have had a bearing on later paleontological work—the tooth of a lungfish and the claw of a carnivorous dinosaur, assigned to the theropod genus *Megalosaurus*. For the next 75 years, as no further finds joined them, these bones lay neglected in a cabinet in Museum Victoria. Then, in 1978, two graduate students at Monash University, Tim F. Flannery and John A. Long, discovered near Ferguson's original site the first specimens of a trove of dinosaur bones embedded in hard sandstones and mudstones from the Early Cretaceous.

These discoveries—only an hour-and-a-half drive southeast of Melbourne—encouraged paleontologists to prospect other coastal sites. In 1980, a rich lode was struck in the Otway Ranges, which the Victorian government, at Tom Rich's suggestion, has since named Dinosaur Cove. There, for a decade, with the help of Earthwatch and other volunteers—along with the National Geographic Society—the Australian Research Council and Atlas Copco, a manufacturer of mining equipment, devoted three months out of every year to chiseling, hammering and, on occasion, blasting tunnels into the fossil-bearing strata. With the work at Dinosaur Cove completed in 1994, effort has since been concentrated at a site about 185 miles east of there, Flat Rocks. The rocks at that locality are about 10 million years older than those at Dinosaur Cove. This project has been at the center of Thomas and Patricia Rich's lives, as well as their children's, their coworkers', and even their parents' (two of whom are paleontologists).

Dinosaur Cove, Flat Rocks, and other sites of similar character were formed when violent, seasonal streams swept broad floodplains of their accumulated bones and plant life, depositing this flotsam and jetsam at the bottom of shallow channels. These deposits are restricted to the southern Victorian coast because only there could gnawing waves expose the sediments laid down in the rift valley that formed when Australia and Antarctica went their separate ways, as did the other fragments of Gondwana, the ancient super-continent. Only two fossil sites from the same period have been found inland, one in sediments laid down under far quieter conditions at the bottom of an ancient lake. The inland site has, therefore, yielded some uncommonly well preserved specimens.

It must be noted that southeastern Australia's dinosaurs are known from a mere 7,000 individual bones and two partial skeletons. Just a few hundred of the bones can be assigned to a given species or genus. What they lack in number, however, they make up in scientific interest.

All efforts at interpretation revolve around the estimation of temperature, for which three methods have been tried. Robert T. Gregory of Southern Methodist University and his associates infer Australian paleoclimate from the ratio of oxygen-18 to oxygen-16 trapped in concretions in these ancient rocks. They find that mean annual temperatures probably approached 32°F, but might have reached as high as 43–46°F. Such values occur today in Hudson Bay, Saskatchewan (32°F) and in Minneapolis and Toronto (46°F).

Andrew Constantine's work on structures preserved in the rocks in which the dinosaur bones are buried revealed evidence for the former existence of permafrost, ice wedging, patterned ground, and hummocky ground. Such features are formed today in regions with mean annual temperatures of 27°–37°F. These structures are not as commonly encountered as the concretions. However, one of the places where they are most obvious is less than 10 feet stratigraphically below the Flat Rocks locality where dinosaurs, mammals, and associated fauna have been found. Evidence for the occurrence of permafrost has never been previously reported in association with dinosaurs.

Robert A. Spicer of the University of Oxford and Judith Totman Parrish of the University of Arizona deduce temperature using another method—the structure of ancient plants—arriving at somewhat high mean annual temperature of 50°F. Their research and that of Jack Douglas, Barbara Wagstaff, Andrew Drinnan, and Jenny McEwan-Mason has demonstrated that polar Australia supported conifers, ginkgoes, ferns, cycads, bryophytes, and horsetails, but only a few angiosperms, or flowering plants, identifiable by sprinking of pollen. The angiosperms were then just beginning to spread into new niches. Perhaps they got their start by exploiting weedy ecological systems in the rift valleys that formed as the supercontinent split apart.

Spicer and Parrish noticed that evergreens, which provided forage in all seasons, had thick cuticle and other structural features, indicating adaptations to cold or scarcity of water at times during the year (perhaps drought brought on by winter freezing). Deciduous plants offer another climatic clue: they seem to have lost all their leaves at once. These mass falls may have been triggered by darkness or cold. Drought, however, probably did not serve as a constant cue—the sedimentary record and the abundance of ferns and bryophytes argue for conditions that were moist in all seasons except perhaps winter.

If the higher estimate of mean temperature is correct, Australia was both temperate and subject to a period of continuous darkness every year—a combination with absolutely no modern counterpart. The winter night lasted between six weeks and four and a half months depending on the true paleolatitude. Because the lower extreme of temperature would then have fallen well below the mean, most of the vertebrates preserved as fossils must have lived quite close to their thermal limits. Some, such as lungfish, cannot now breed in waters colder than 50°F.

If, on the other hand, the lower mean temperature is correct, it becomes more than a normal scientific challenge to understand how this paleocommunity functioned at all. Before seriously attacking this problem, scientists will first have to demonstrate that it exists. To refine the estimate of the average annual temperature, a multidisciplinary team is comparing floral, geochemical, and other forms of evidence.

Nothing in this fauna is quite so peculiar to the region as the koala is today, for although the species and genera were local, they belonged to cosmopolitan families. Yet their adaptations are striking, as is the fact that some survived beyond the time of demise for their families elsewhere.

Among such anachronisms—or relicts—are the labyrinthodont amphibians, ancestors of modern amphibians and reptiles. Most paleontologists had thought this group was extinct by the Jurassic, some 160 million years ago. In the past 15 years, however, Michael Cleeland and Lesley Kool of Monash University found three jaws from this group in Victorian sediments dating from the Early Cretaceous. Two of the jaws were unmistakable, because their teeth had the labyrinthine infolding of the enamel that gives this group its name. At least one large species of labyrinthodont, *Koolasuchus cleelandi*, lived in polar Australia 115 million years ago, several million years after the group had died out elsewhere.

How did they survive? We suspect that the cool weather preserved the animals from competition with crocodiles, which were probably poorly adapted to the conditions prevailing in southeastern Australia until the onset of climatic warming the last 5 million years of the Early Cretaceous. The hypothesis rests on the fact that contemporary crocodilians now live in waters no colder than 50°F, whereas some modern frogs and salamanders can be active in meltwater from snow.

Another late survivor was a close relative of the familiar *Allosaurus*, a carnivorous theropod. Elsewhere in the world, this animal ranged up to 16½ in height, but the southeastren specimen stood no more than 6½ feet high—hardly taller than a human. This "pygmy," presumably a juvenile, is the latest-surviving allosaur that has yet been found. It remains unclear whether this species also owed its longevity to its occupation of some niche that cold climate may have provided for it. The discovery of juvenile forms (but no eggshell, as yet) does suggest that these dinosaurs were not just casual visitors but lived near the pole for much of the year, using the area as a nursery during the period of maximum sunlight.

Unlike the allosaurs, many dinosaurs of Australia were not the last in their lineage; some may have been the first. At least two and perhaps as many as four families of dinosaurs have been recognized that include forms that are either the oldest or among the oldest of their kind. For instance, the ornithomimosaurs, carnivores of ostrichlike size and appearance, are manifestly primitive and among the oldest within this group. The elongated, slender hindlimbs of the Australian species made them the gazelles of the dinosaur world, able to escape from predators and to run down prey. The ornithomimosaurs may have originated in Gondwana and spread northward to join the later Cretaceous faunas of North America and Eurasia, where they enjoyed wide success.

Two very small theropods remain unidentified, but one seems to resemble an egg-eating oviraptorosaur, known until now exclusively from the younger Cretaceous rocks of North America and Asia. These groups may also have an origin in Gondwana. Yet another dinosaur group is the neoceratopsians, or horned dinosaurs. Identification is tentative, as it is based on just two ulnae (part of the lower arm), but the similarity to *Leptoceratops*, a sheep-sized browser, is uncanny. Previously, all neoceratopsian records dated from the Late Cretaceous and, with the exception of a few bones from Argentina, came from the Northern Hemisphere. Recently, there have been reports of Early Cretaceous neoceratopsians from China and Utah. This dinosaur family may also have arisen in the southern supercontinent.

In addition to some dinosaurs, there are mammals that may be among the earliest members of their group. The miniscule *Ausktribosphenos* resembles the living spineless hedgehog *Neotetracus*. This animal may possibly have been a placental. If so, it is as old as the oldest reported from the Northern Hemisphere and twice the age of the oldest marsupial fossils yet found in Australia. This is surprising because the domination of Australia by marsupials has been conventionally explained by the apparent fact that land-dwelling placentals first reached the continent long after the marsupials.

Another mammalian group whose presence is no surprise at all is the monotremes. An isolated limb bone of one of them has a structure suggestive of a more upright stance than either the echidna or platypus. A second species is by far the smallest monotreme yet known, weighing about only 1 percent as much as any other known living or fossil member of the group.

The Australian Early Cretaceous also reshaped forms that continued to flourish in other regions. By far the most successful such group consisted of the hypsilophodontid dinosaurs. These animals, most of them hardly larger than a chicken, were bipeds built for speed, with large hindlegs, small but well-developed hands, substantial tails and—for the most part—herbivorous habits. They thus resembled wallabies in both shape and ecological role.

Acute night vision is suggested by the eyes and brain of *Leallynascura amicagraphica*, a hypsilophodontid. The large eyes were common to all hypsilophodontids and may have helped the group dominate in an environment marked by seasonal darkness.

The family Hypsilophodontidae was common throughout the world from the Middle Jurassic to Late Cretaceous times, but its prominence reaches an absolute and relative peak in the Victorian sediments. Not only do hypsilophodontids constitute the bulk of dinosaur remains, they are also represented by four to five genera, depending on the taxonomic criteria one uses, and five to six species. Other geographic areas, some much more richly endowed with dinosaur species, never harbored more than three kinds of hypsilophodontids at a time. Something clearly favored the diversification of this group in polar Australia.

A particularly intriguing adaptation of at least one species of polar hypsilophodontid is suggested by the magnificently preserved brain cast of *Leaellynasaura amicagraphica* (named after the Rich's daughter, friends of the Museum of Victoria and the National Geographic Society). The brain, unusually large for a dinosaur of this size, bears the marks of optic lobes whose relative size is easily the greatest ever documented in a hypsilophodontid. How is one to interpret these enlarged lobes? We have hypothesized that they enhanced the animals' ability to see in the dark, enabling them to forage effectively during the long winter months. There would have been no lack of food then, for those capable of seeing it: the herbivores could have lived off evergreens and deciduous leaf mats.

327

This hypothesis also explains why the hypsilophodontids came to dominate the polar environment in the first place. Hypsilophodontids everywhere in the world had large eyes and, presumably, acute vision. That trait could have given them their foothold in polar Australia. Once established in this "protected" environment, the hypsilophodontids could have competed with one another to produce the observed diversity of genera and species, perhaps all sharing hypertrophied optic lobes. If the animals foraged at night, they must have been active at freezing or subfreezing temperatures. This feat goes far beyond the cold tolerance of any modern reptiles, even the New Zealand tuatara, *Sphenodon punctatus,* which can remain active at 41°F provided it can sun itself. *Leaellynasaura* could have survived solely by maintaining a constant body temperature, eating frequently, as birds do in wintertime.

That the hypsilophodontids were active through the winter night is attested to by the microscopic structure found in their bones by Anusuya Chinsamy. Lines of Arrested Growth, or LAGs, are formed when terrestrial vertebrates markedly slow down their growth or cease growing altogether. These lines can be formed during a period of lack of food or water, or they can be formed when an animal aestivates or hibernates. Lacking LAGSs, an animal evidently did none of those things. LAGs appear as dark lines of dense bone against a background of lighter bone. The hypsilophodontids from polar southeastern Australia and elsewhere all lack LAGs, unlike the majority of dinosaurs examined. So they seem to have been a group that maintained a relatively uniform metabolic activity year-round, a behavior pattern they evidently continued at high latitude despite the cold and prolonged darkness of the polar winter.

Pterosaurs, flying reptiles, and ankylosaurs, heavily armored dinosaurs, also appear in southern Victorian fauns, but are so fragmentary that they tell us little of the animals' lives. Much can be gleaned from one handful of teeth, however, for they come from plesiosaurs. These long-necked reptiles, not themselves dinosaurs, generally paddled the seas, but in southern Victoria, they inhabited the freshwaters of the ancient valley between Australia and Antarctica. They are reminiscent of the Ganges River dolphin, one of the few cetaceans that live in freshwater.

The sauropods are one of the few major groups of dinosaurs that are absent. These giants, familiar from the example of *Apatosaurus* (or *Brontosaurus,* as it is more popularly known), lived at that time in Australia's lower latitudes. None, however, has been found further south nor, indeed, in any of the nine Cretaceous polar dinosaur sites so far identified in both hemispheres. The only polar sauropod yet discovered is the much older (Early Jurassic) *Rhoetosaurus* from northeastern Australia.

The apparent restriction of these large dinosaurs to lower latitudes in the Cretaceous of Australia may be real or merely an artifact of sampling. We worry about this question because the floodwaters that broke out of rain-swollen rivers would have collected small and medium-size bones but left large ones behind. The body of a sauropod would have stayed put, rather than floating to a place where many specimens were concentrated in the small flood channels, which were no more than 16–33 feet in width and 8–12 inches in depth.

Yet we suspect there was an underlying tendency toward small body size in these polar environs. None of the hypsilophodontids, it must be remembered, stood taller than a human, and most were barely knee-high. The dwarf *Allosaurus* matches the smallest we have examined in the North American collections. The ornithomimosaur is equally unprepossessing, and the protoceratopsid and the ankylosaur are no bigger than sheep. A single fragment of a claw constitutes our sole record of a form—a carnivore, apparently similar to *Baryonyx* of England—which may have measured up to 26 feet in length. This pattern contradicts the scaling laws that Bergmann and Allen

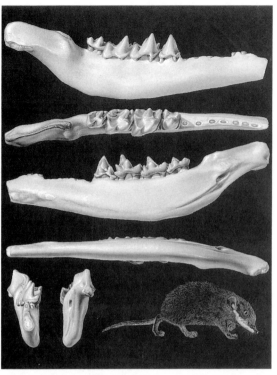

When alive during the Cretaceous, *Ausktribosphenos* from southeastern Australia may have resembled this spineless hedgehog from China. The jaw superimposed on the drawing of the living animals shows what is known of the fossil.

formulated in the nineteenth century. According to these laws, animals in a given lineage tend to become larger and more compact as the average temperature of their environment falls. This trend is exemplified by the comparison of mountain lions in Canada with pumas of Central America and of human populations in the subarctic and tropical zones.

Other factors also determine body dimensions, especially the size of the territory in which a population lives. Individuals found on islands are often smaller than their mainland counterparts. For example, there were dwarf elephants on the ancient Mediterranean islands, and pygmy mammoths have been found in 4,000-year-old sediments on islands off the north coast of Siberia. Dwarfism may be a response to selective pressure to increase the number of individuals so as to ensure a gene pool diverse enough for the species to survive in a restricted area. This effect has also been noted on peninsulas—and ancient southeast Australia was a peninsula of the Gondwana landmass.

The dinosaurs on that peninsula were trapped virtually at the end of the Earth. Their direct path north was blocked by a vast inland sea, which they could have passed only by going hundreds of miles to the west before wheeling about to the north. At the end of such labors, they would have been able to catch, at most, an hour of sun a day in winter. Migration would have made little sense for such small animals.

Formidable barriers did not seal in the dinosaurs of the other polar site that has yielded large quantities of fossils: the North Slope of Alaska. The dinosaurs there had a clear north-south corridor along which they could migrate with ease. It is significant that those dinosaurs were big—at least equal in size to caribou, wildebeest, and other modern animals that migrate.

One must question whether animals so superbly adapted to the cold and the dark could have been driven to extinction by an artificial winter, such as is supposed to have followed a cataclysmic event at the boundary between the Cretaceous and Tertiary formations. It is proposed that the cataclysm, perhaps a collision with a comet or asteroid or a series of volcanic eruptions, suffused the atmosphere with a blanket of dust, excluding sunlight, and freezing or starving most animals to death.

We suspect, however, that no such artificial winter could have killed the dinosaurs unless it lasted for a long time, certainly more than a few months. Otherwise, at least a few of the polar dinosaurs might well have survived the cataclysm. Of course, it is possible that some other development had already ended the reign of southern Australia's dinosaurs by the end of the Cretaceous.

Arthur Conan Doyle once dreamed of a plateau in South America that time forgot, where dinosaurs continued to reign. Reports in 1993 that dwarf mammoths survived to early historical times, on an island off the coast of Siberia, give force to such speculation. If dinosaurs found a similar haven in which they outlived the rest of their kind, then we think polar Gondwana, including southern Australia, is a likely place to look for it.

Dinosaur Renaissance

by Robert T. Bakker

The dinosaur is for most people the epitome of extinctness, the prototype of an animal so maladapted to a changing environment that it dies out, leaving fossils but no descendants. Dinosaurs have a bad public image as symbols of obsolescence and hulking inefficiency; in political cartoons they are know-nothing conservatives that plod through miasmic swamps to inevitable extinction. Most contemporary paleontologists have had little interest in dinosaurs; the creatures were an evolutionary novelty, to be sure, and some were very big, but they did not appear to merit much serious study because they did not seem to go anywhere: no modern vertebrate groups were descended from them.

Recent research is rewriting the dinosaur dossier. It appears that they were more interesting creatures, better adapted to a wide range of environments and immensely more sophisticated in their bioenergetic machinery than had been thought. This essay presents some of the evidence that has led to reevaluation of the dinosaurs' role in animal evolution. The evidence suggests, in fact, that the dinosaurs never died out completely. One group still lives. We call them birds.

Ectothermy and Endothermy

Dinosaurs are usually portrayed as cold-blooded animals, with a physiology like that of living lizards or crocodiles. Modern land ecosystems clearly show that in large animals cold-bloodedness (ectothermy) is competitively inferior to warm-bloodedness (endothermy), the bioenergetic system of birds and mammals. Small reptiles and amphibians are common and diverse, particularly in the tropics, but in nearly all habitats, the overwhelming majority of land vertebrates with an adult weight of 22 pounds or more are endothermic birds and mammals. Why?

The term cold-bloodedness is a bit misleading: on a sunny day, a lizard's body temperature may be higher than a man's. The key distinction between ectothermy and endothermy is the rate of body-heat production and long-term temperature stability. The resting metabolic heat production of living reptiles is too low to affect body temperature significantly in most situations, and reptiles of today must use external heat sources to raise their body temperature above the air temperature—which is why they bask in the sun or on warm rocks. Once big lizards, big crocodiles, or turtles in a warm climate achieve a high body temperature they can maintain it for days because large size retards heat loss; but they are still vulnerable to sudden heat drain during cloudy weather or cool nights or after a rainstorm, and so they cannot match the performance of endothermic birds and mammals.

The key to avian and mammalian endothermy is high basal metabolism: the level of heat-producing chemical activity in each cell is about four times higher in an endotherm than in an ectotherm of the same weight at the same body temperature. Additional heat is produced as it is needed, by shivering and some other special forms of thermogenesis. Except for some large tropical endotherms (elephants and ostriches, for example), birds and mammals also have a layer of hair or feathers that cuts the rate of thermal loss. By adopting high heat production and insulation, endotherms have achieved the ability to maintain more nearly constant high body temperatures than their ectothermic competitors can. A guarantee of high, constant body temperature is a powerful adaptation because the rate of work output from muscle tissue, heart, and lungs is greater at high temperatures than at low temperatures, and the endothermic animal's biochemistry can be finely tuned to operate within a narrow thermal range.

The adaptation carries a large bioenergetic price, however. The total energy budget per year of a population of endothermic birds or mammals is from 10 to 30 times higher than the energy budget of an ectothermic population of the same size and adult body weight. The price is, nonetheless, justified. Mammals and birds have been the dominant large and medium-sized land vertebrates for 60 million years in nearly all habitats.

In view of the advantage of endothermy, the remarkable success of the dinosaurs seems puzzling. The first land vertebrate communities, in the Carboniferous and Early Permian periods, were composed of reptiles and amphibians generally considered to be primitive and ectothermic. Replacing this first ectothermic dynasty were the mammal-like reptiles (therapsids), which eventually produced the first true mammals near the end of the next period, the Triassic, about when the dinosaurs were originating. One might expect that mammals would have taken over the land-vertebrate communities immediately, but they did not. From their appearance in the Triassic until the end of the Cretaceous, a span of 140 million years, mammals remained small and inconspicuous while all the ecological roles of large terrestrial herbivores and carnivores were monopolized by dinosaurs; mammals did not begin to radiate and produce large species until after the dinosaurs had already become extinct, at the end of the Cretaceous. One is forced to conclude that dinosaurs were competitively superior to mammals as large land vertebrates. And that would be baffling if dinosaurs were cold-blooded. Perhaps they were not.

Measuring Fossil Metabolism

In order to rethink traditional ideas about Permian and Mesozoic vertebrates, one needs bioenergetic data for dinosaurs, therapsids, and early mammals. How does one measure a fossil animal's metabolism? Surprising as it may seem, recent research provides three independent methods of extracting quantitative metabolic information from the fossil record. The first is bone histology. Bone is an active tissue that contributes to the formation of blood cells and participates in maintaining the calcium-phosphate balance, vital to the proper functioning of muscles and nerves. The low rate of energy flow in ectotherms places little demand on the bone compartment of the blood and calcium-phosphate system, and so the compact bone of living reptiles has a characteristic "low activity" pattern: a low density of blood vessels and few Haversian canals, which are the site of rapid calcium-phosphate exchange. Moreover, in strongly seasonal climates, where drought or winter cold forces ectotherms to become dormant, growth rings appear in the outer layers of compact bone, much like the rings in the wood of trees in similar environments. The endothermic bone of birds and mammals is dramatically different. It almost never shows growth rings, even in severe climates, and it is rich in blood vessels and frequently in Haversian canals. Fossilization often faithfully preserves the structure of bone, even in specimens 300 million years old; thus it provides one window through which to look back at the physiology of ancient animals.

The second analytic tool of paleobioenergetics is latitudinal zonation. The present continental masses have floated across the surface of the globe on lithospheric rafts, sometimes colliding and pushing up mountain ranges, sometimes pulling apart along rift zones such as those of the mid-Atlantic or East Africa. Paleomagnetic data make it possible to reconstruct the ancient positions of the continents to within about 5 degrees of latitude, and sedimentary indicators such as glacial beds and salt deposits show the severity of the latitudinal temperature gradient from the Equator to the poles in past epochs. Given the latitude and the gradient, one can plot temperature zones, and such zones should separate endotherms from ectotherms. Large reptiles with a lizardlike physiology cannot survive cool winters because they cannot warm up to an optimal body temperature during the short winter day, and they are too large to find safe hiding places for winter hibernation. That is why small lizards are found today as far north as Alberta, where they hibernate underground during the winter, but crocodiles and big lizards do not get much farther north than the northern coast of the Gulf of Mexico.

The third meter of heat production in extinct vertebrates is the predator-prey ratio: the relation of the "standing crop" of a predatory animal to that of its prey. The ratio is a constant that is a characteristic of the metabolism of the predator, regardless of the body size of the animals of the predator-prey system. The reasoning is as follows: The energy budget of an endothermic population is an order of magnitude larger than that of an ectothermic population of the same size and adult weight, but the productivity—the yield of prey tissue available to predators—is about the same for both an endothermic and an ectothermic population. In a steady-state population, the yearly gain in weight and energy value from growth and reproduction equals the weight and energy value of the carcasses of the animals that die during the year; the loss of biomass and energy through death is balanced by additions. The maximum energy value of all the carcasses a steady-state population of lizards can provide its predators is about the same as that provided by a prey population of birds or mammals of about the same numbers and adult body size. Therefore, a given prey population, either ectotherms or endotherms, can support an order of magnitude greater biomass of ectothermic predators than of endothermic predators, because of the endotherms' higher energy needs. The term standing crop refers to the biomass, or the energy value, contained in the biomass, of a population. In both ectotherms and endotherms, the energy value of carcasses produced per unit of standing crop decreases with increasing adult weight of prey animals: a herd of zebra yields from about a fourth to a third of its weight in prey carcasses a year, but a "herd" of mice can produce up to six times its weight because of its rapid turnover, reflected in a short lifespan and high metabolism per unit weight.

Now, the energy budget per unit of predator standing crop also decreases with increasing weight: lions require more than 10 times their own weight in meat per year, whereas shrews need 100 times their weight. These two bioenergetic scaling factors cancel each other, so that if the adult size of the predator is roughly the same as that of the prey (and in land-vertebrate ecosystems it usually is), the maximum ratio of predator standing crop to prey standing crop in a steady-state community is a constant independent of the adult body size in the predator-prey system. For example, spiders are ectotherms, and the ratio of a spider population's standing crop to its prey standing crop reaches a maximum of about 40 percent. Mountain boomer lizards, about 3½ ounces in adult weight, feeding on other lizards would reach a similar maximum ratio. So would the giant Komodo dragon lizards (up to 330 pounds in body weight) preying on deer, pigs, and monkeys, Endothermic mammals and birds, on the other hand, reach a maximum predator-prey biomass ratio of only from 1 to 3 percent—whether they are weasel and mouse or lion and zebra.

Some fossil deposits yield hundreds or thousands of individuals representing a single community; their live body weight can be calculated from the reconstruction of complete skeletons, and the total predator-prey biomass ratios then can be easily worked out. Predator-prey ratios are powerful tools for paleophysiology because they are the direct result of predator metabolism.

The Age of Ectothermy

The paleobioenergetic methodology outlined here can be tested by analyzing the first land-vertebrate predator-prey system, the early Permian communities of primitive reptiles and amphibians. The first predators capable of killing relatively large prey were the finback pelycosaurs of the family Sphenacodontidae, typified by *Dimetrodon*, whose tall-spined fin makes it popular with cartoonists. Although this family included the direct ancestors of mammal-like reptiles and, hence, of mammals, the sphenacodonts themselves had a very primitive level of organization, with a limb anatomy less advanced than that of living lizards. Finback bone histology was emphatically ectothermic, with a low density of blood vessels, few Haversian canals, and the distinct growth rings that are common in specimens from seasonally arid climates.

One might suspect that finbacks and their prey would be confined to warm, equable climates, and Early Permian paleogeography offers an excellent opportunity to test this prediction. During the early part of the period, icecaps covered the southern tips of the continental landmasses, all of which were part of the single southern supercontinent Gondwanaland, and glacial sediment is reported at the extreme northern tip of the Permian landmass in Siberia by Russian geologists. The Permian Equator crossed what are now the American Southwest, the Maritime Provinces of Canada, and western Europe. Here are found sediments produced in very hot climates: thick-bedded evaporite salts and fully oxidized, red-stained mudstones. The latitudinal temperature gradient in the Permian must have been at least as steep as it is at present. Three Permian floral zones reflect the strong poleward temperature gradient. The Angaran flora of Siberia displays wood with growth rings from a wet environment, implying a moist climate with cold winters. The Euramerian flora of the equatorial region had two plant associations: wet swamp communities with no growth rings in the wood, implying a continuous warm-moist growing season, and semiarid, red-bed-evaporite communities with some growth rings, reflecting a tropical dry season. In glaciated Gondwanaland, the peculiar Glossopteris flora dominated, with wood from wet environments showing sharp growth rings.

The ectothermy of the finbacks is confirmed by their geographic zonation. Finback communities are known only from near the Permian Equator; no large Early Permian land vertebrates of any kind have been found in glaciated Gondwanaland. (One peculiar little fish-eating reptile, *Mesosaurus*, is known from southern Gondwanaland, and its bone has sharp growth rings. The animal must have fed and reproduced during the Gondwanaland summer, then burrowed into the mud of lagoon bottoms to hibernate, much as large snapping turtles do today in New England.)

Excellent samples of finback communities are available for predator-prey studies, thanks largely to the lifework of the late Alfred Sherwood Romer of Harvard University. To derive a predator-prey ratio from a fossil community, one simply calculates the number of individuals, and thus the total live weight, represented by all the predator and prey specimens that are found together in a sediment representing one particular environment. In working with scattered and disarticulated skeletons, it is best to count only bones that have about the same robustness, and hence the same preservability, in both predator and prey. The humerus and the femur are good choices for finback communities: they are about the same size with respect to the body in the prey and the predator, and should give a ratio that faithfully represents the ratio of the animals in life.

In the earlier Early Permian zones, the most important finback prey were semiaquatic fish-eating amphibians and reptiles, particularly the big-headed amphibian *Eryops* and the long-snouted pelycosaur *Ophiacodon*. As the climate became more arid in Europe and America, these water-linked forms decreased in numbers, and the fully terrestrial herbivore *Diadectes* became the chief prey genus. In all zones from all environments, the calculated biomass ratio of predator to prey in finback communities is very high: from 35 to 60 percent, the same range seen in living ectothermic spiders and lizards.

All three of the paleobioenergetic indicators agree: the finback pelycosaurs and their contemporaries were ectotherms with low heat production and a lizardlike physiology that confined their distribution to the tropics.

Therapsid Communities

The mammal-like reptiles (order Therapsida), descendants of the finbacks, made their debut at the transition from the Early to the Late Permian and immediately became the dominant large land vertebrates all over the world. The three metabolism-measuring techniques show that they were endotherms.

The earliest therapsids retained many finback characteristics but had acquired limb adaptations that made possible a trotting gait and much higher running speeds. From early Late Permian to the Middle Triassic, one line of therapsids became increasingly like primitive mammals in all details of the skull, the teeth, and the limbs, so that some of the very advanced mammal-like therapsids (cynodonts) are difficult to separate from the first true mammals. The change in physiology, however, was not so gradual. Detailed studies of bone histology conducted by Armand Riqles of the University of Paris indicate that the bioenergetic transition was sudden and early: all the finbacks had fully ectothermic bone; all the early therapsids—and there is an extraordinary variety of them—had fully endothermic bone, with no growth rings and with closely packed blood vessels and Haversian canals.

The Late Permian world still had a severe latitudinal temperature gradient; some glaciation continued in Tasmania, and the southern end of Gondwanaland retained its cold-adapted Glossopteris flora. If the earliest therapsids were equipped with endothermy, they would presumably have been able to invade southern Africa, South America, and the other parts of the southern cold-temperature realm. They did exactly that. A rich diversity of early therapsid families has been found in the southern Cape District of South Africa, in Rhodesia, in Brazil, and in India—regions reaching to 65 degrees south Permian latitude. Early therapsids as large as rhinoceroses were common there, and

Hairy therapsids, mammal-like reptiles of the late Permian period some 250 million years ago, confront one another in the snows of southern Gondwanaland, at a site that is now in South Africa. *Anteosaurus*, weighing about 1,300 lbs., had bony ridges on the snout and brow for head-to-head contact in sexual or territorial behavior. Pristerognathids weighed about 110 lbs. and represent a group that included the direct ancestors of mammals. The reconstructions were based on fossils and the knowledge, from several kinds of data, that therapsids were endothermic, or "warm-blooded"; those adapted to cold would have had hairy insulation. The advent of endothermy, competitively superior to the ectothermy ("cold-bloodedness") of typical reptiles, is the basis of author's new classification of land vertebrates.

many species grew to an adult weight greater than 22 pounds, too large for true hibernation. These early therapsids must have had physiological adaptations that enabled them to feed in and move through the snows of the cold Gondwanaland winters. There were also some ectothermic holdovers from the Early Permian that survived into the Late Permian, notably the immense herbivorous caseid pelycosaurs and the big-headed, seed-eating captorhinids. As one might predict, large species of these two ectothermic families were confined to areas near the Late Permian Equator; big caseids and captorhinids are not found with the therapsids in cold Gondwanaland. In the Late Permian, then, there was a "modern" faunal zonation of large vertebrates, with endothermic therapsids and some big ectotherms in the tropics giving way to an all-endothermic therapsid fauna in the cold south.

In the earliest therapsid communities of southern Africa, superbly represented in collections built up by Lieuwe Boonstra of the South African Museum and by James Kitching of the University of the Witwatersrand, the predator-prey ratios are between 9 and 16 percent. That is much lower than in Early Permian finback communities. Equally low ratios are found for tropical therapsids from Russia even though the prey species there were totally different from those of Africa. The sudden decrease in predator-prey ratios from finbacks to early therapsids coincides exactly with the sudden change in bone histology from ectothermic to endothermic reported by Riqles, and with the sudden invasion of the southern cold-temperate zone by a rich therapsid fauna. The conclusion is unavoidable that even early therapsids were endotherms with high heat production.

It seems certain, moreover, that in the cold Gondwanaland winters, the therapsids would have required surface insulation. Hair is usually thought of as a late development that first appeared in the advanced therapsids, but it must have been present in the southern African endotherms of the early Late Permian. How did hair originate? Possibly the ancestors of therapsids had touch-sensitive hairs scattered over the body as adaptations for night foraging; natural selection could then have favored increased density of hair as the animals' heat production increased and they moved into colder climates.

The therapsid predator-prey ratios, although much lower than those of ectotherms, are still about three times higher than those of advanced mammals today. Such ratios indicate that the therapsids achieved endothermy with a moderately high heat production, far higher than in typical reptiles but still lower than in most modern mammals. Predator-prey ratios of early Cenozoic communities seem to be lower than those of therapsids, and so one might conclude that a further increase in metabolism occurred somewhere between the advanced therapsids of the Triassic and the mammal of the post-Cretaceous era. Therapsids may have operated at a lower body temperature than most living mammals do, and thus they may have saved energy with a lower thermostat setting. This suggestion is reinforced by the low body temperature of the most primitive living mammals: monotremes (such as the spiny anteater) and the insectivorous tenrecs of Madagascar; they maintain a temperature of about 86°F instead of the 94°F to 102°F of most modern mammals.

Thecodont Transition

The vigorous and successful therapsid dynasty ruled until the middle of the Triassic. Then their fortunes waned and a new group, which was later to include the dinosaurs, began to take over the roles of large predators and herbivores. These were the Archosauria, and the first wave of archosaurs were the thecodonts. The earliest thecodonts, small and medium-sized animals found in therapsid communities during the Permian-Triassic transition, had an ectothermic bone histology. In modern ecosystems, the role played by large freshwater predators seems to be one in which ectothermy is competitively superior to endothermy; the low metabolic rate of ectotherms may be a key advantage because it allows much longer dives. Two groups of thecodonts became large freshwater fish-eaters: the phytosaurs, which were confined to the Triassic, and the crocodilians, which remain successful today. Both groups have ectothermic bone. (The crocodilian endothermy was either inherited directly from the first thecodonts or derived secondarily from endothermic intermediate ancestors.) In most of the later, fully terrestrial advanced thecodonts, on the other hand, Riqles discovered a typical endothermic bone histology; the later thecodonts were apparently endothermic.

The predator-prey evidence for thecodonts is scanty. The ratios are hard to compute because big carnivorous cynodonts and even early dinosaurs usually shared the predatory role with thecodonts. One sample from China that has only one large predator genus, a big-headed erythrosuchid thecodont, does give a ratio of about 10 percent, which is in the endothermic range. The zonal evidence is clearer. World climate was moderating in the Triassic (the glaciers were gone), but a distinctive flora and some wood-growth rings suggest that southern Gondwanaland was not yet warm all year. What is significant in this regard is the distribution of phytosaurs, the big ectothermic fish-eating thecodonts. Their fossils are common in North America and Europe (in the Triassic tropics) and in India, which was warmed by the equatorial Tethys Ocean, but they have not been found in southern Gondwanaland, in southern Africa or in Argentina, even though a rich endothermic thecodont fauna did exist there.

Did some of the thecodonts have thermal insulation? Direct evidence came from the discoveries of A. Sharov of the (then) Academy of Sciences of the U.S.S.R. Sharov found a partial skeleton of a small thecodont and named it *Longisquama* for its long scales, strange parachute-like devices along the back that may have served to break the animal's fall when it leaped from trees. More important is the covering of long, overlapping, keeled scales that trapped an insulating layer of air next to its body. These scales lacked the complex anatomy of real feathers, but they are a perfect ancestral stage for the insulation of birds. Feathers are usually assumed to have appeared only late in the Jurassic with the first bird, *Archaeopteryx*. The likelihood that some thecodonts had insulation is supported, however, by another of Sharov's discoveries: a pterosaur, or flying reptile, whose fossils in Jurassic lakebeds still show the epidermal covering. This beast (appropriately named *Sordus pilosus*, the "hairy devil") had a dense growth of hair or hairlike feathers all over its body and limbs. Pterosaurs are descendants of Triassic thecodonts or, perhaps, of very primitive dinosaurs. The insulation in both *Sordus* and *Longisquama*, and the presence of big erythrosuchid thecodonts at the southern limits of Gondwanaland, strongly suggest that some endothermic thecodonts had acquired insulation by the Early Triassic.

The Dinosaurs

Dinosaurs, descendants of early thecodonts, appeared first in the Middle Triassic, and by the end of the period had replaced thecodonts and the remaining therapsids as the dominant terrestrial vertebrates. Zonal evidence for endothermy in dinosaurs is somewhat equivocal. The Jurassic was a time of climatic optimum, when the poleward temperature gradient was the gentlest that has prevailed from the Permian until the present day. In the succeeding Cretaceous period, latitudinal zoning of oceanic plankton and land plants seems, however, to have been a bit sharper. Rhinoceros-sized Cretaceous dinosaurs and big marine lizards are found in the rocks of the Canadian far north, within the Cretaceous Arctic Circle. Dale A. Russell of the National Museums of Canada points out that at these latitudes the sun would have been below the horizon for

Feathered dinosaur, *Syntarsus*, pursues a gliding lizard across the sand dunes of Rhodesia in the early Jurassic period some 180 million years ago. This small dinosaur (adult weight about 65 lbs.) and others were restored by Michael Raath of the Queen Victoria Museum in Rhodesia and the author on the basis of evidence that some thecodonts, ancestors of the dinosaurs, had insulation, and on the basis of close anatomical similarities between dinosaurs and early birds. Dinosaurs, it appears, were endothermic, and the smaller species required insulation. Feathers would have conserved metabolic heat in cold environments and reflected heat of the Sun in hot climates such as this.

months at a time. The environment of the dinosaurs would have been far more severe than the environment of the marine reptiles because of the lack of a wind-chill factor in the water and because of the ocean's temperature-buffering effect. Moreover, locomotion costs far less energy per mile in water than on land, so that the marine reptiles could have migrated away from the arctic winter. These considerations suggest, but do not prove, that arctic dinosaurs must have been able to cope with cold stress.

Dinosaur bone histology is less equivocal. All dinosaur species that have been investigated show fully endothermic bone, some with a blood-vessel density higher than that in living mammals. Since bone histology separates endotherms from ectotherms in the Permian and the Triassic, this evidence alone should be a strong argument for the endothermy of dinosaurs. Yet the predator-prey ratios are even more compelling. Dinosaur carnivore fossils are exceedingly rare. The predator-prey ratios for dinosaur communities in the Triassic, Jurassic, and Cretaceous are usually from 1 to 3 percent, far lower even than those of therapsids and fully as low as those in large samples of fossils from advanced mammal communities in the Cenozoic. I am persuaded that all the available quantitative evidence is in favor of high heat production and a large annual energy budget in dinosaurs.

Were dinosaurs insulated? Explicit evidence comes from a surprising source: *Archaeopteryx*. As an undergraduate, I was a member of a paleontological field party led by John H. Ostrom of Yale University. Near Bridger, Montana, Ostrom found a remarkably preserved little dinosaurian carnivore, *Deinonychus*, that shed a great deal of light on carnivorous dinosaurs in general. A few years later, while looking for pterosaur fossils in European museums, Ostrom came on a specimen of *Archaeopteryx* that had been mislabeled for years as a flying reptile, and he noticed extraordinary points of resemblance between *Archaeopteryx* and carnivorous dinosaurs. After a detailed anatomical analysis, Ostrom established beyond any reasonable doubt that the immediate ancestor of *Archaeopteryx* must have been a small dinosaur, perhaps one related to *Deinonychus*. Previously, it had been thought that the ancestor of *Archaeopteryx*, and thus of birds, was a thecodont rather far removed from dinosaurs themselves.

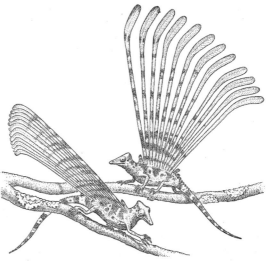

Longisquama, a small animal whose fossil was discovered in Middle Triassic lakebeds in Turkestan by the Russian paleontologist A. Sharov, was a thecodont. Its body was covered by long overlapping scales that were keeled, suggesting that they constituted a structural stage in the evolution of feathers. The long devices along the back were *V*-shaped in cross section; they may have served as parachutes and also threat devices, as shown here.

Archaeopteryx was quite thoroughly feathered, and yet it probably could not fly: the shoulder joints were identical to those of carnivorous dinosaurs and were adapted for grasping prey, not for the peculiar arc of movement needed for wing-flapping. The feathers were probably adaptations not for powered flight or gliding but primarily for insulation. *Archaeopteryx* is so nearly identical in all known features with small carnivorous dinosaurs that it is hard to believe feathers were not present in such dinosaurs. Birds inherited their high metabolic rate and most probably their feathered insulation from dinosaurs; powered flight probably did not evolve until the first birds, with flight-adapted shoulder joints appearing during the Cretaceous, long after *Archaeopteryx*.

It has been suggested a number of times that dinosaurs could have achieved a fairly constant body temperature in a warm environment by sheer bulk alone. Large alligators approach this condition in the swamps of the U.S. Gulf states. This proposed thermal mechanism would not give rise to endothermic bone histology or low predator-prey ratios, nor would it explain arctic dinosaurs or the success of many small dinosaur species with an adult weight of between 11 and 110 pounds.

Dinosaur Brains and Limbs

Large brain size and endothermy seem to be linked; most birds and mammals have a ratio of brain size to body size much larger than that of living reptiles and amphibians. The acquisition of endothermy is probably a prerequisite for the enlargement of the brain because the proper functioning of a complex central nervous system calls for the guarantee of a constant body temperature. It is not surprising that endothermy appeared before brain enlargement in the evolutionary line leading to mammals. Therapsids had small brains with reptilian organization; not until the Cenozoic did mammals attain the large brain size characteristic of most modern species. A large brain is certainly not necessary for endothermy, since the physiological feedback mechanisms responsible for thermoregulation are deep within the "old" region of the brain, not in the higher learning centers. Most large dinosaurs did have relatively small brains. Russell has shown, however, that some small and medium-sized carnivorous dinosaurs had brains as large as or larger than modern birds of the same body size.

Up to this point, I have concentrated on thermoregulatory heat production. Metabolism during exercise can also be read from fossils. Short bursts of intense exercise are powered by anaerobic metabolism within muscles, and the oxygen debt incurred is paid back afterward by the heart-lung system. Most modern birds and mammals have much higher levels of maximum aerobic metabolism than living reptiles, and can repay an oxygen debt much faster. Apparently, this difference does not keep small ectothermic animals from moving fast: the top running speeds of small lizards equal or exceed those of small mammals. The difficulty of repaying oxygen debt increases with increasing body size, however, and the living large reptiles (crocodilians, giant lizards, and turtles) have noticeably shorter limbs, less limb musculature, and lower top speeds than many large mammals, such as the big cats and the hoofed herbivores.

The Early Permian ectothermic dynasty was also strikingly short-limbed; evidently, the physiological capacity for high sprinting speeds in large animals had not yet evolved. Even the late therapsids, including the most advanced cynodonts, had very short limbs compared with the modern-looking running mammals that appeared early in the Cenozoic. Large dinosaurs, on the other hand, resembled modern running mammals, not therapsids, in locomotor anatomy and limb proportions. Modern, fast-running mammals utilize an anatomical trick that adds an extra limb segment to the forelimb stroke. The scapula, or shoulder blade, which is relatively immobile in most primitive vertebrates, is free to swing backward and forward and thus increase the stride length. Jane A. Peterson of Harvard has shown that living chameleonid lizards have also evolved scapular swinging, although its details are different from those in mammals. Quadrupedal dinosaurs evolved a chameleon-type scapula, and they must have had long strides and running speeds comparable to those of big savanna mammals today.

Dinosaurian ancestry of *Archaeopteryx*, and thus of birds, is indicated by its close anatomical relation to such small dinosaurs as *Microvenator* and *Deinonychus*; John H. Ostrom of Yale University demonstrated that they were virtually identical in all details of joint anatomy. The long forelimbs of *Archaeopteryx* were probably used for capturing prey, not flight.

When the dinosaurs fell at the end of the Cretaceous, they were not a senile, moribund group that had played out their evolutionary options. Rather they were vigorous, still diversifying into new orders and producing a variety of big-brained carnivores with the highest grade of intelligence yet present on land. What caused their fall? It was not competition, because mammals did not begin to diversify until after all the dinosaur groups (except birds!) had disappeared. Some geochemical and microfossil evidence suggests a moderate drop in ocean temperature at the transition from the Cretaceous to the Cenozoic, and so cold has been suggested as the reason. But the very groups that would have been most sensitive to cold, the large crocodilians, are found as far north as Saskatchewan and as far south as Argentina before and immediately after the end of the Cretaceous. A more likely reason is the draining of shallow seas on the continents and a lull in mountain-building activity in most parts of the world, which would have produced vast stretches of monotonous topography. Such geological events decrease the variety of habitats that are available to land animals, and thus increase competition. They can also cause the collapse of intricate, highly evolved ecosystems; the larger animals seem to be the more affected. At the end of the Permian, similar changes had been accompanied by catastrophic extinctions among therapsids and other land groups. Now, at the end of the Cretaceous, it was the dinosaurs that suffered a catastrophe; the mammals and birds, perhaps because they were so much smaller, found places for themselves in the changing landscape and survived.

The success of the dinosaurs, an enigma as long as they were considered cold-blooded, can now be seen as the predictable result of the superiority of their high heat production, high aerobic exercise metabolism, and insulation. They were endotherms. Yet the concept of dinosaurs as ectotherms is deeply entrenched in a century of paleontological literature. Being a reptile connotes being an ectotherm, and the dinosaurs have always been classified in the subclass Archosauria of the class Reptilia. The other land-vertebrate classes were the Mammalia and the Aves. Perhaps, then, it is time to reclassify.

Taxonomic Conclusion

What better dividing line than the invention of endothermy? There has been no more far-reaching adaptive breakthrough; and so the transition from ectothermy to endothermy can serve to separate the land vertebrates into higher taxonomic categories. For some time it has been suggested that the therapsids should be removed from the Reptilia and joined with the Mammalia. In light of the sudden increase in heat production and the probable presence of hair in early therapsids, I fully agree. The term Theropsida has been applied to mammals and their therapsid ancestors. Let us establish a new class Theropsida, with therapsids and true mammals as two subclasses.

How about the class Aves? All the quantitative data from bone histology and predator-prey ratios, as well as the dinosaurian nature of *Archaeopteryx*, show that all the essentials of avian biology—very high heat production, very high aerobic exercise metabolism, and feathery insulation—were present in the dinosaur ancestors of birds, thus I do not believe birds deserve to be put in a taxonomic class separate from dinosaurs. Peter Galton of the University of Bridgeport and I have suggested a more reasonable classification: putting the birds into the Dinosauria. Since bone histology suggests that most thecodonts were endothermic, the thecodonts could then be joined with the Dinosauria in a great endothermic class Archosauria, comparable to the Theropsida. The classification may seem radical at first, but it is actually a good deal neater bioenergetically than the traditional Reptilia, Aves and Mammalia. And for those of us who are fond of dinosaurs, the new classification has a particularly happy implication: the dinosaurs are not extinct; the colorful and successful diversity of the living birds is a continuing expression of basic dinosaur biology.

Chapter Seven
What Caused the Mass Extinction?

Introduction

All good things must come to an end. The great and marvelous dinosaurs ruled the land for more than 150 million years, increasing in diversity and sophistication along the way. They seemed set to continue to dominate terrestrial habitats—and even the sky—indefinitely. But then, suddenly, it came crashing down. Not a single dinosaur outside Aves lived into the Cenozoic. Not that this was necessarily a bad thing. Whether humans or a similarly intelligent technology-developing creature would have evolved in a dinosaurian world is open to question. And we still have the dinosaurs of the air to admire and even to keep as pets.

The big question is why dinosaurs experienced such an overwhelming failure, and over such a short span of time. If they had lost a long, gradual battle of competition with mammals, that would be understandable—and it does seem that birds outcompeted a dwindling diversity of pterosaurs during the Cretaceous, leaving only a few pterodactyloids left, to disappear at the end of the period—but mammals large enough to compete with dinosaurs did not appear until long after the latter were gone. One way to explain why dinosaurs went belly-up so fast is to look to intervention from the heavens, in the form of extraterrestrial explosions of one sort or another. It is interesting that our complex, life-packed solar system is in an unusually circular orbit that perpetually keeps us in the outer few percent of the galaxy's stars. Most stars and their planets orbit closer to the galactic center, where the stars are much denser, so they may be subjected too often to the damaging effects of nearby supernovae explosions to allow the evolution of complex life-forms like us. Even out here in the galactic boonies, the odd supernova goes off a tad close for comfort. When Dale Russell wrote his essay in 1982, he speculated that such a stellar bomb did in the dinosaurs. This interesting concept never gained majority support, but a more intimate visitor from space has become the favorite explanation of why dinosaurs went out with a big bang.

An Extraterrestrial Impact

by Walter Alvarez and Frank Asaro

About 65 million years ago, something killed half of all the life on the Earth. This sensational crime wiped out the dinosaurs, until then undisputed masters of the animal kingdom, and left the humble mammals to inherit their estate. Human beings, descended from those survivors, cannot avoid asking who or what committed the mass murder and what permitted our distant ancestors to survive.

For the past dozen years, researchers from around the world, in disciplines ranging from paleontology to astrophysics, have mustered their observational skills, experimental ingenuity, and theoretical imagination to answer these questions. Those of us involved in it have lived through long months of painstaking measurement, periods of bewilderment, flashes of insight, and episodes of great excitement when parts of the puzzle finally fell into place.

We now believe that we have solved the mystery. Some 65 million years ago, a giant asteroid or comet plunged out of the sky, striking the Earth at a velocity of more than 22,400 miles per hour. The enormous energy liberated by that impact touched off a nightmare of environmental disasters, including storms, tsunamis, cold and darkness, greenhouse warming, acid rains, and global fires. When quiet returned at last, half the flora and fauna had become extinct. The history of the Earth had taken a new and unexpected path.

Other suspects in the dinosaur murder mystery, such as sea-level changes, climatic shifts, and volcanic eruptions, have alibis that appear to rule them out. Some issues, however, are still unclear: Was it a single or multiple impact? Have such impacts occurred on a regular, periodic timetable? What is the role of such catastrophes in evolution?

The puzzle presented by a mass extinction is both like and unlike that of a more recent murder. There is evidence—chemical anomalies, mineral grains, and isotopic ratios, instead of blood or fingerprints or torn matchbooks—scattered throughout the world. No witnesses remain, however, and no chance exists of obtaining a confession. The passage of millions of years has destroyed or degraded most of the evidence in the case, leaving only the subtlest clues.

Indeed, it is difficult even to be sure which of the individual fossils that survive are those of victims killed by the impact. But paleontologists know there must have been victims because fossil-bearing sedimentary rocks show a great discontinuity 65 million years ago. Creatures such as dinosaurs and ammonites, abundant for tens of millions of years, suddenly disappeared forever. Many other groups of animals and plants were decimated.

This discontinuity defines the boundary between the Cretaceous period, during which dinosaurs reigned supreme, and the Tertiary, which saw the rise of the mammals. (It is known as the K/T boundary after *Kreide*, the German word for Cretaceous.)

When we began to study the K/T boundary, we wanted to find out just how long the extinction had taken to occur. Was it sudden—a few years or centuries—or was it a gradual event that took place over millions of years? Most geologists and paleontologists had always assumed that the extinction had been slow. (These fields have a long tradition of gradualism and are uncomfortable with invoking catastrophes.) Because dinosaur fossils are relatively rare, their age provides little detailed information on the duration of the extinction. It was possible to view the extinction of dinosaurs as gradual.

When paleontologists looked at the fossils of pollen or single-celled marine animals called foraminifera, however, they found the extinction to be very abrupt. In general, smaller organisms produce more abundant fossils and so yield a sharper temporal picture. The extinction also appeared more sudden as paleontologists studied closely the fossil record for medium-sized animals such as marine invertebrates. Among these are the ammonites (relatives of the modern chambered nautilus), which died out at the end of the Cretaceous period. The best record of their extinction is found in the coastal outcrops of the Bay of Biscay on the border between Spain and France.

In 1986, Peter L. Ward and his colleagues at the University of Washington made detailed studies of these outcrops at Zumaya in Spain. Ward found that the ammonites appeared to die out gradually—one species disappearing after another over an interval of about 560 feet, representing about 5 million years. But in 1988, Ward studied two nearby sections in France and found evidence that these ammonite species actually survived right up to the K/T boundary. The apparent gradual extinction at Zumaya was merely the artifact of an incomplete fossil record. If organisms whose fossils are well preserved died out abruptly, then it is likely that others that perished about the same time, such as dinosaurs, whose remains are more sparsely preserved, did so as well.

This establishes that the extinction was abrupt in geologic terms, but it does not establish how many years this extinction took, because it is a major accomplishment to date a rock to an accuracy of a million years. Intervals in the geologic records can be determined with precision only to within 10,000 years, a period longer than the entire span of human civilization. The duration of the mass extinction that marks the K/T boundary can be estimated more precisely than this. In the deep-water limestones at Gubbio in

Italy, a thin layer of clay separates Cretaceous and Tertiary sediments. The layer discovered by Isabella Premoli Silva of the University of Milan is typically about one-third inch thick. In the 1970s, one of us (Alvarez) was part of a group that found the clay falls within a 20-foot thickness of limestone deposited during the half-million-year period of reversed geomagnetic polarity designated 29R. On the face of it, this suggests that the clay layer, and the mass extinction it marks, represents a span of no more than about 1,000 years.

Jan Smit of the University of Amsterdam did a similar study of sediments at Caravaca in southern Spain, where the stratigraphic record is even more precise, and estimated the extinction lasted no more than 50 years. By geologic standards this is blindingly fast!

Our work on the K/T boundary began in the late 1970s when we and our Berkeley colleagues Luis W. Alvarez and Helen V. Michel tried to develop a more accurate way to determine how long the Gubbio K/T clay layer took to be deposited. Our efforts failed, but they did provide a crucial first clue to the identity of the mass killer. (That is what detectives and scientists need: a lot of hard work and an occasional lucky break.) The method depended on the rarity of iridium in the Earth's crust—about .03 part per billion as compared with 500 parts per billion, for example, in the primitive stony meteorites known as carbonaceous chondrites. Iridium is rare in the Earth's crust because most of the planet's allotment is alloyed with iron in the core.

We suspected that iridium would enter deep-sea sediments, such as those at Gubbio, predominantly through the continual rain of micrometeorites, sometimes called cosmic dust. This constant infall would provide a clock: the more iridium in a sedimentary layer, the longer it must have taken to lay down. Moreover, iridium could be measured at very low concentrations by means of neutron-activation analysis, a technique in which neutron bombardment converts the metal into a radioactive and hence detectable form.

One scenario we considered was that the K/T boundary clay layer formed over a period of about 10,000 years, when organisms that secrete calcareous shells died out, and so no calcium carbonate (which makes up most of the limestone) was deposited. Most layers at Gubbio contain about 95 percent calcium carbonate and 5 percent clay; the boundary layer contains 50 percent clay. If this scenario was correct, the ratio of iridium to clay would be the same in the boundary clay as in higher and lower layers. If clay deposition had slowed at the same time as calcium carbonate deposition, the ratio would be higher than that in adjacent rocks.

In June 1978, our first Gubbio iridium analyses were ready. Imagine our astonishment and confusion when we saw that the boundary clay and the immediately adjacent limestone contained far more iridium than any of our scenarios predicted—an amount comparable to that in all the rest of the rock deposited during the 500,000 years of interval 29R.

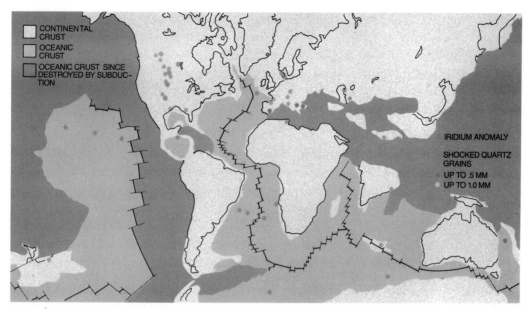

CONTINENTAL CRUST
OCEANIC CRUST
OCEANIC CRUST SINCE DESTROYED BY SUBDUCTION

IRIDIUM ANOMALY

SHOCKED QUARTZ GRAINS
UP TO .5 MM
UP TO 1.0 MM

Evidence of giant impact that caused the mass extinction at the end of the Cretaceous period is scattered around the globe. This map shows continents as they were at the time of the extinction 65 million years ago. Anomalous levels of iridium, which suggest a large impact, have been found in sediment from the Cretaceous-Tertiary (K/T) boundary, as have impact-generated quartz crystals.

Clearly, this concentration could not have come from the usual sprinkling of cosmic dust. For a year we debated possible sources, testing and rejecting one idea after another. Then, in 1979, we proposed the one solution that had survived our testing: a large comet or asteroid about 6 miles in diameter had struck the Earth and dumped an enormous quantity of iridium into the atmosphere. Since we first proposed the impact hypothesis, so much confirming evidence has come to light that most scientists working in the field are persuaded that a great impact occurred. More than 100 scientists in 21 laboratories in 13 countries have found anomalously high levels of iridium at the K/T boundary at about 95 sites throughout the world. The anomaly has been found in marine and nonmarine sediments, at outcrops on land, and in oceanic sediment cores. Further, we have analyzed enough other sediments to know that iridium anomalies are very rare. As far as we know, the one at the K/T boundary is unique.

The iridium anomaly is well explained by impact because the ratio of iridium to elements with similar chemical behavior, such as platinum, osmium, ruthenium, rhodium, and gold, is the same in the boundary layer as it is in meteorites. Miriam Kastner of the Scripps Institution of Oceanography, working with our group, has determined that the gold-iridium ratio in the carefully studied K/T boundary at Stevns Klint in Denmark agrees to within 5 percent with the ratio in the most primitive meteorites (type I carbonaceous chondrites).

Indeed, the ratios of all the platinum-group elements found in the K/T boundary give evidence of extraterrestrial origin. George Bekov of the Institute of Spectroscopy in Moscow and one of us (Asaro) have found that the relative abundances of ruthenium, rhodium, and iridium can distinguish stony meteorites from terrestrial samples. Analysis of K/T boundary samples from Stevns Klint, Turkmenia in the former Soviet Union, and elsewhere support the impact hypothesis.

So do ratios of isotopes. Jean-Marc Luck, then at the Institute of Physics of the Earth in Paris, and Karl K. Turekian of Yale University found that most of the osmium in K/T boundary samples from Denmark and New Mexico could not have come from a continental source because the abundance of osmium 187 is too low. The ratio of osmium 187 to osmium 186 is higher in continental rocks than in meteorites or in the Earth's mantle because those rocks are relatively enriched in rhenium, whose radioactive isotope, rhenium 187, decays to osmium 187. The osmium in K/T samples must be extraterrestrial or from the Earth's mantle.

Not only does the composition of rocks at the K/T boundary suggest impact, but so does their mineralogy. In 1981, Smit discovered another tell-tale clue: mineral spherules as large as a $\frac{1}{25}$ of an inch in diameter in the Caravaca K/T clay. (Alessandro Montanari of Berkeley confirmed their presence in Italy as well.) The spherules originated as droplets of basaltic rock, shockmelted by impact and rapidly cooled during ballistic flight outside the atmosphere, then chemically altered in the boundary clay. They are the basaltic equivalent of the more silica-rich glassy tektites and microtektites that are the known result of smaller impacts. The basaltic chemistry suggests that the impact took place on oceanic crust.

In addition to the spherules, shocked grains of quartz have been discovered by Bruce F. Bohor of the U.S. Geological Survey (USGS) in Denver and Donald M. Triplehorn of the University of Alaska. Painstaking studies by E. E. Foord, Peter J. Modreski and Glen A. Izett of the USGS show that the grains carry the multiple intersecting planar "lamellae"—bands of deformation—symptomatic of hypervelocity shock. Such grains are found only in known impact craters, at nuclear test sites, in materials subjected to extreme shock in the laboratory, and in the K/T boundary.

How would an impact disperse shocked and molten materials around the globe? A 6-mile diameter asteroid moving at more than 22,000 miles an hour would ram a huge hole in the atmosphere. When it hit the ground, its kinetic energy would be converted to heat in a nonnuclear explosion 10,000 times as strong as the total world arsenal of nuclear weapons. Some vaporized remains of the asteroid and rock from the ground near the impact point would then be ejected through the hole before the air had time to rush back in.

The fireball of incandescent gas created by the explosion would also propel material out of the atmosphere. The fireball of an atmospheric nuclear explosion expands until it reaches the same pressure as the surrounding atmosphere, then rises to an altitude where its density matches that of the surrounding air. At that point, usually around 6 miles high, the gas spreads laterally to form the head of the familiar mushroom cloud.

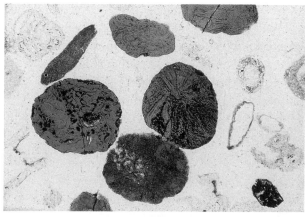

Basaltic spherules have been found embedded in clays at the K/T boundary at several locations. Now chemically altered, the spherules originated as molten droplets of ocean subfloor that were deposited globally by the K/T impact.

Computer models of explosions with energies of 1,000 megatons—about 20 times the energy of the largest nuclear bombs but only 1/100,000 the energy of the K/T impact—have shown that the fireball never reaches pressure equilibrium with the surrounding atmosphere. Instead, as the fireball expands to altitudes where the density of the atmosphere declines significantly, its rise accelerates and the gas leaves the atmosphere at velocities fast enough to escape the Earth's gravitational field. The fireball from an even greater asteroid impact would simply burst out the top of the atmosphere, carrying any entrained ejecta with it, sending the material into orbits that could carry it anywhere on the Earth.

The impact of a comet-sized body on the Earth, creating a crater 93 miles in diameter, would clearly kill everything within sight of the fireball. Researchers are refining their understanding of the means by which an impact would also trigger extinction worldwide. Mechanisms proposed include darkness, cold, fire, acid rain, and greenhouse heat.

Originally, we proposed that impact-generated dust caused global darkness that resulted in extinctions. According to computer simulations made in 1980 by Richard P. Turco of R&D Associates, O. Brian Toon of the National Aeronautics and Space Administration, and their colleagues, dust lofted into the atmosphere by the impact of an object would block so much light that for months you would literally be unable to see your hand in front of your face. Without sunlight, plant photosynthesis would stop. Food chains everywhere would collapse. The darkness would also produce extremely cold temperatures, a condition termed "impact winter." (After considering the effects of the impact, Turco, Toon, and their colleagues went on to study nuclear winter, a related phenomenon as capable of producing mass extinctions today as impact winter was 65 million years ago.)

In 1981, Cesare Emillíani of the University of Miami, Eric Krause of the University of Colorado, and Eugene M. Shoemaker of the USGS pointed out that an oceanic impact would loft not only rock dust but also water vapor into the atmosphere. The vapor, trapping the Earth's heat, would stay aloft much longer than the dust, and so the impact winter would be followed by greenhouse warming. More recently, John D. O'Keefe and Thomas J. Ahrens of the California Institute of Technology suggested that the impact might have occurred in a limestone area, releasing large volumes of carbon dioxide, another greenhouse gas. Many plants and animals that survived the extreme cold of impact winter could well have been killed by a subsequent period of extreme heat.

Meanwhile, John S. Lewis, G. Hampton Watkins, Hyman Hartman, and Ronald G. Prinn of the Massachusetts Institute of Technology have calculated that shock heating of the atmosphere during impact would raise temperatures high enough for the oxygen and nitrogen in the air to combine. The resulting nitrous oxide would eventually rain out of the air as nitric acid—an acid rain with a vengeance. This mechanism may well explain the widespread extinction of marine invertebrate plants and animals, whose calcium carbonate shells are soluble in acidic water.

Another killing mechanism came to light when Wendy Wolbach, Ian Gumore, and Edward Anders of the University of Chicago discovered large amounts of soot in the K/T boundary clay. If the clay had been laid down in a few years or less, the amount of soot in the boundary would indicate a sudden burning of vegetation equivalent to half of the world's current forests. Jay Meos of the University of Arizona and his colleagues calculated that infrared radiation from ejecta heated to incandescence while reentering the atmosphere could have ignited fires around the globe.

Detailed studies of the K/T boundary sediments may eventually provide evidence supporting a particular killing mechanism. For example, dissolution patterns in the Italian limestone show that bottom waters were acidic immediately after the extinction. And work we have done with William Lowrie of ETH-Zurich shows that those waters also changed briefly from their normal oxidizing state to a reducing condition, possibly because of the massive death of marine organisms.

A newly emerging point of view suggests, unlikely as it may seem, that the K/T extinction may have been caused by two or more nearly simultaneous impacts. Shoemaker and Piet Hut of the Institute for Advanced Study in Princeton, New Jersey, have identified a number of mechanisms that could yield multiple impacts, either on the same day or over the course of many years. Double or multiple craters have been found on the Earth, the moon, and other planets, suggesting that some asteroids may consist of two or more objects mutually orbiting one another. Alternatively, the Earth may have been struck by two or more large fragments of a comet nucleus in the process of breaking up.

The comet theory gains credibility from the discovery of apparently extraterrestrial materials near the K/T boundary; Meixun Zhao and Jeffrey L. Bada of the University of California at San Diego analyzed chalk layers just above and below the K/T boundary in Denmark. They found amino acids that are not used by life on the Earth but that do occur in carbonaceous chondrite meteorites. It seems unlikely that amino acids could survive the heat of a large impact, and they in fact do not appear in the K/T boundary itself.

Kevin Zahnle and David Grinspoon of NASA proposed that dust from a disintegrating comet entered the Earth's atmosphere over an extended period and carried these extraterrestrial amino acids with it. During that interval the impact of a large fragment of the comet would have caused the K/T extinction.

An apparently unrelated line of inquiry, based on statistical rather than chemical analyses, has yielded a hypothesis explaining how comets could hit the Earth periodically. In 1984, David M. Raup and John J. Sepkoski, Jr., of the University of Chicago published an analysis of the fossil record, which seemed to indicate that mass extinctions have occurred at 32-million-year intervals. Like most scientists working with the K/T boundary, we were very skeptical of their results. But astrophysicist Richard A. Muller of the University of California at Berkeley reexamined Raup and Sepkoski's data and convinced himself that the periodicity was real.

Muller, Marc Davis of Berkeley, and Hut hypothesized that a dim, unrecognized companion star orbiting the sun every 32 million years (which they provisionally dubbed Nemesis) might regularly disturb the orbits of comets on the outer fringe of the solar system. The disturbance would send a million-year storm of comets into the inner solar system, greatly increasing the chance of a large impact (or multiple impacts) on the Earth. Daniel Whitmire of the University of Southwestern Louisiana and Albert Jackson of Computer Sciences Corporation independently proposed the same hypothesis.

When Muller showed me the paper proposing Nemesis, I (Alvarez) was very skeptical. I told him I thought that it was "an ingenious solution to a nonproblem" because I was not convinced of Raup and Sepkoski's evidence for periodic mass extinctions. If the hypothesis were correct, I pointed out, terrestrial impact craters should show the same periodicity in their ages. Muller and I found, to his delight and to my surprise, that crater ages do show essentially the same periodicity as mass extinctions. Since then, I have felt that the hypothesis must at least be taken seriously.

It turns out, however, that it is very difficult to find a dim red star close to the sun when one has no idea where to look. Muller and Saul Perlmutter of Berkeley are now about halfway through a computerized telescopic search for a star with the characteristics of Nemesis; they expect to finish in a couple of years. [Ed. Note: Nothing has been found

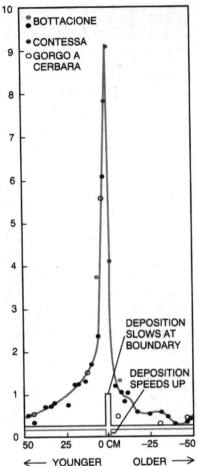

Iridium anomaly in limestone sediments is a clear sign of extraterrestrial impact. Iridium is rare in the Earth's crust; only an extraterrestrial body could have deposited the amount found in the K/T boundary. Other extinction scenarios have been suggested in which the rate of limestone deposition varied while cosmic iridium infall remained constant, but they cannot explain such high levels of iridium.

yet.] Meanwhile, new analysis of crater ages and extinction dates has raised questions about whether they actually are periodic. The small numbers of events and the sketchy information available make the question difficult to answer unequivocally.

Murder suspects typically must have means, motive, and opportunity. An impact certainly had the means to cause the Cretaceous extinction, and the evidence that an impact occurred at exactly the right time points to opportunity. The impact hypothesis provides, if not motive, then at least a mechanism behind the crime. How do other suspects in the killing of the dinosaurs fare? Some have an airtight alibi: they could not have killed all the different organisms that died at the K/T boundary. The venerable notion that mammals ate the dinosaurs' eggs, for example, does not explain the simultaneous extinction of marine foraminifera and ammonites.

Stefan Gartner of Texas A&M University once suggested that marine life was killed by a sudden huge flood of fresh water from the Arctic Ocean, which apparently was isolated from other oceans during the late Cretaceous and filled with fresh water. Yet this ingenious mechanism cannot account for the extinction of the dinosaurs or the loss of many species of land plants.

Other suspects might have had the ability to kill, but they have alibis based on timing. Some scientific detectives have tried to pin the blame for mass extinction on changes in climate or sea level, for example. Such changes, however, take much longer to occur than did the extinction; moreover, they do not seem to have coincided with the extinction, and they have occurred repeatedly throughout the Earth's history without accompanying extinctions.

Others consider volcanism a prime suspect. The strongest evidence implicating volcanoes is the Deccan Traps, an enormous outpouring of basaltic lava in India that occurred approximately 65 million years ago. Recent paleomagnetic work by Vincent E.

Courtillot and his colleagues in Paris confirms previous studies. They show that most of the Deccan Traps erupted during a single period of reversed geomagnetic polarity, with slight overlaps into the preceding and succeeding periods of normal polarity. The Paris team found that the interval in question is probably 29R, during which the K/T extinction occurred, although it might be the reversed-polarity interval immediately before or after 29R as well. Because the outpouring of the Deccan Traps began in one normal interval and ended in the next, the eruptions that gave rise to them must have taken place over at least half a million years ago. Most workers interested in mass extinction, therefore, have not considered volcanism a serious suspect in a killing that evidently took place over 1,000 years or less.

Some researchers have argued that, contrary to the fossil record, the K/T extinctions took place over many thousands of years and that volcanism can account for quartz grains, spherules, and the iridium anomaly. In 1983, William H. Zoller and his colleagues at the University of Maryland at College Park discovered high concentrations of iridium in aerosols from Kilauea volcano in Hawaii collected on filters over 30 miles away; however, the ratio between iridium and other rare elements in the volcanic aerosols does not match the ratio found at the K/T boundary. The ratio of gold to iridium in the Kilauea aerosols is more than 35 times that in the K/T boundary at Stevns Klint.

There has also been debate as to whether an explosive volcanic eruption might produce shocked quartz. It now seems agreed, however, that volcanic explosions can produce some deformation but that the distinctive multiple lamellae seen in the K/T boundary quartz can only be formed by impact shocks. In addition, John McHone of Arizona State University has found that they contain stishovite, a form of quartz produced only at pressures far greater than those of volcanic eruptions. And Mark H. Anders of Columbia University and Michael R. Owen of St. Lawrence University have used a technique known as cathode luminescence, in which an electric field causes quartz to glow, to determine the origin of the K/T grains. The colors produced by the grains are not volcanic; they argue instead for impact on an ordinary sedimentary sandstone. Moreover, basaltic spherules in the K/T boundary argue against explosive volcanism in any case; spherules might be generated by quieter forms of volcanism, but then they could not be transported worldwide.

The apparent global distribution of the iridium anomaly, shocked quartz, and basaltic spherules is strong evidence exonerating volcanism and pointing to impact. Eruptions take place at the bottom of the atmosphere; they send material into the high stratosphere at best. Spherules and quartz grains, if they came from an eruption, would quickly be slowed by atmospheric drag and fall to the ground. Nevertheless, the enormous eruptions that created the Deccan Traps did occur during a period spanning the K/T extinction. Further, they represent the greatest outpouring of lava on land in the past quarter of a billion years (although greater volumes flow continually out of mid-ocean ridges). No investigator can afford to ignore that kind of coincidence.

It seems possible that impact triggered the Deccan Traps volcanism. A few minutes after a large body hit the Earth the initial crater would be 25 miles deep, and the release of pressure might cause the hot rock of the underlying mantle to melt. Authorities on the origin of volcanic provinces, however, find it very difficult to explain in detail how an impact could trigger large-scale basaltic volcanism.

In the past few years, the debate between supporters of each scenario has become polarized: impact proponents have tended to ignore the Deccan Traps as irrelevant, while volcano backers have tried to explain away evidence for impact by suggesting that it is also compatible with volcanism. Our sense is that the argument is a Hegelian one, with an impact thesis and a volcanic antithesis in search of a synthesis whose outlines are as yet unclear.

Even in its present incompletely solved form, the mystery of the K/T mass killing carries a number of lessons. The late eighteenth and early nineteenth centuries, when the study of the Earth was first becoming a science, was a period marked by a long battle between catastrophists, who thought that sudden great events were crucial to the evolution of the planet, and uniformitarians, who explained all history in terms of gradual change. Stephen Jay Gould of Harvard University has shown how the uniformitarians so thoroughly won this battle that generations of geology students have been taught catastrophism is unscientific. The universe, however, is a violent place, as astronomy has shown, and it is now becoming clear that the Earth has also had its violent episodes.

Evidence that a giant impact was responsible for the extinctions at the end of the Cretaceous has finally rendered the catastrophic viewpoint respectable. Future geologists, with the intellectual freedom to think in both uniformitarian and catastrophic terms, have a better chance of truly understanding the processes and history of the planet than did their predecessors.

Catastrophes have an important role to play in evolutionary thinking as well. If a chance impact 65 million years ago wiped out half the life on the Earth, then survival of the fittest is not the only factor that drives evolution. Species must not only be well adapted, they must also be lucky. If chance disaster occasionally wipes out whole arrays of well-adapted organisms, then the history of life is not preordained. There is no inevitable progress leading inexorably to intelligent life—to human beings. Indeed, Norman Sleep of Stanford University and his colleagues have suggested that in the very early history of the Earth, when impacts were more frequent, incipient life may have been extinguished more than once.

Impact catastrophes may also prevent evolution from bogging down. The fossil record indicates that, in normal times, each species becomes increasingly well adapted to its particular ecological niche. Thus, it becomes ever more difficult for another species to evolve into that niche. As a result, the rate of evolution slows. Wholesale removal of species by impact, however, provides a great opportunity for the survivors to evolve into newly vacant niches. (We have heard graduate students compare this situation with the excellent job prospects they would face if half of all tenured professors were suddenly fired.) Indeed, the fossil record shows that the rate of evolution accelerated immediately after the end of the Cretaceous.

Among the happy survivors of the K/T extinction were the early Tertiary mammals, our ancestors. When dinosaurs dominated the Earth, mammals seem always to have been small and insignificant. Warm-blooded metabolism, small size, and a large number or other traits may have suited them to endure the harsh conditions imposed by impact—or they may just have been lucky. And with the removal of the huge reptiles from the scene, mammals began an explosive phase of evolution that eventually produced human intelligence. As detectives attempting to unravel this 65–million-year-old mystery, we find ourselves pausing from time to time to reflect that we owe our very existence as thinking beings to the impact that destroyed the dinosaurs.

A Volcanic Eruption

by Vincent E. Courtillot

The mysterious mass extinction that took place 65 million years ago has been attributed to either the impact of a large asteroid or a massive volcanic eruption. Both hypotheses presume that clouds of dust and chemical changes in the atmosphere and oceans created an ecological domino effect that eradicated large numbers of animal and plant families. The geologic record generally is consistent with either scenario; the central issue has been how rapid the event was. New evidence implies that the mass extinction occurred over tens or even hundreds of thousands of years. Such a duration closely corresponds to an episode of violent volcanic eruptions in India that occurred at the time of the mass extinction. Moreover, other extinction events also appear to be roughly simultaneous with periods of major volcanic activity.

The conventional divisions of geologic history reflect times of significant geologic and biological change. The mass extinction 65 million years ago defined the end of the Mesozoic era, when reptiles enjoyed great evolutionary success, and the beginning of the Cenozoic era, when mammals became extremely prevalent. Because the last period of the Mesozoic is the Cretaceous, and the first period of the Cenozoic the Tertiary, the time of the most recent mass extinction is called the Cretaceous-Tertiary, or K/T boundary. At this boundary, the dinosaurs met their demise, and, even more remarkable, 90 percent of all genera of protozoans and algae disappeared. John J. Sepkoski, Jr., and David M. Raup of the University of Chicago conclude that from 60 to 75 percent of all species vanished then. Equally important, many species, among them the ancestors of human beings, survived.

In 1980, Luis W. and Walter Alvarez (father and son) of the University of California at Berkeley, along with their colleagues Frank Asaro and Helen V. Michel, discovered unusually high concentrations of the metal iridium—from 10 to 100 times the normal levels—in rocks dating from the K/T boundary in Italy, Denmark, and New Zealand. Iridium is rare in the Earth's crust but can be relatively abundant in other parts of the solar system. The Berkeley group, therefore, concluded that the iridium came from outer space, and thus the asteroid hypothesis was born.

A large asteroid impact would have cloaked the Earth with a cloud of dust, resulting in darkness, suppression of photosynthesis, the collapse of food chains and, ultimately, mass extinction. The iridium is contained in a thin layer of clay whose chemical composition differs from that of the layers both above and below the boundary. Alvarez's group interpreted the clay as being the altered remains of the dust thrown up by the impact. In this view, the boundary layer was laid down in less than one year, a flickering instant in geologic time. Other unusual findings at the K/T boundary, most notably quartz crystals that appear to have been subjected to extremely powerful physical shocks, also could be explained by an asteroid impact.

An alternative to the asteroid hypothesis had already been brewing for some time. As early as 1972, Peter R. Vogt of the Naval Research Laboratory in Washington, D.C., pointed out that extensive volcanism had taken place at roughly the time of the K/T boundary, principally in India. The volcanism produced extensive lava flows, known as the Deccan Traps (deccan means southern in Sanskrit, and trap means staircase in Dutch). Vogt suggested that the traps might be connected to the many changes that took place at the end of the Cretaceous period.

In the mid-1970s, Dewey M. McLean of the Virginia Polytechnic Institute proposed that volcanoes could produce mass extinctions by injecting vast amounts of carbon dioxide into the atmosphere that would trigger abrupt climate changes and alter ocean chemistry. Charles B. Officer and Charles L. Drake of Dartmouth College analyzed sediments from K/T boundary sections and concluded that the iridium enrichment and other chemical anomalies at the boundary were not deposited instantaneously but rather over a period of 10,000 to 100,000 years. They also argued that the anomalies were more consistent with a volcanic rather than meteoritic origin.

The amount of time represented by the clay layer at the K/T boundary emerged as a major point of contention. Dating a 100-million-year-old rock with a precision of one part in 1,000 (that is, to within 100,000 years) is not yet possible. Yet much of the debate focuses on whether the boundary clay was deposited in less than one year (as would be expected from an impact) or in 10,000 (from an extended period of volcanism).

The sheer size of the Deccan Traps suggests that their formation must have been an important event in the Earth's history. Individual lava flows extend well over 3,900 square miles and have a volume exceeding 2,400 cubic miles. The thickness of the flows averages from 33–164 feet and sometimes reaches almost 500 feet. In western India, the accumulation of lava flows is almost 8,000 feet thick (more than a quarter the height of Mount Everest). The flows may have originally covered more than 772,000 square miles, and the total volume may have exceeded 48,000 cubic miles.

An important, unresolved question was whether the date and duration of Deccan volcanism are compatible with the age and thickness of the K/T boundary. Until recently, the lava samples from the Deccan Traps were thought to range in age from 80 to 30 million years (estimated by measuring the decay of the radioactive isotope potassium-40 in rocks). Whether this range was real or reflected an error in measurement was unknown. So in 1985, I joined forces with a number of colleagues to try to clarify the picture.

One important clue emerged from the fact that the Deccan rocks are basalts, volcanic rocks rich in magnesium, titanium, and iron, that are rather strongly magnetic. When basaltic lava cools, the magnetization of tiny crystals of iron-titanium oxides in the rock becomes frozen, aligned with the Earth's magnetic field. The polarity of the field occasionally reverses, so that the magnetic north pole becomes south, and vice versa. These brief reversals—about 10,000 years long—occur in random fashion at a rate that has varied from about one reversal every million years at the end of the Cretaceous to roughly four every million years in recent times.

Jean Besse and Didier Vandamme at the Institute of Physics of the Earth in Paris and I found that more than 80 percent of the rock samples from the Deccan Traps had the same reversed polarity. Had the volcanism truly continued from 80 to 30 million years ago, we would have expected to find approximately equal numbers of normal- and reverse-magnetized samples, because tens of reversals took place during that 50-million-year stretch. In fact, the thickest (3,280-feet-thick) exposed sections of the traps record only one or two reversals. We, therefore, concluded in 1986 that Deccan volcanism began during an interval of normal magnetic activity, climaxed in the next reversed interval, then waned in a final normal interval. Judging from the usual frequency of reversals, our results implied that the volcanism could not have lasted much more than 1 million years.

If so, the spread of ages found by potassium-40 dating must have been wrong. My colleagues Henri Maluski of the University of Montpellier, Gilbert Féraud of the University of Nice, and other researchers used a newer, more reliable technique—argon-argon dating—to determine how much potassium-40 had decayed during the lifetime of the rock samples. Their results confirmed that the Deccan flows were laid down over a relatively brief period. Age estimates for the Deccan lavas now cluster between 64 and 68 million years, and much of the remaining scatter in ages may result from alteration of the samples or differing laboratory standards.

Although accurate dating of sedimentary rock is difficult, recent findings by Ashok Sahni of the University of Chandigarh, J. J. Jaeger of the University of Montpellier, and their colleagues further narrow estimates of the age of the Deccan Traps. Sediments immediately below the Deccan flows contain dinosaur fossil fragments that seem to date from the Maastrichtian stage, the last 8 million years of the Cretaceous. Dinosaur and mammalian teeth and dinosaur egg fragments that appear to be of Maastrichtian age have also been found in layers of sediment between the flows. This implies that Deccan volcanism began during the very last stage of the Cretaceous.

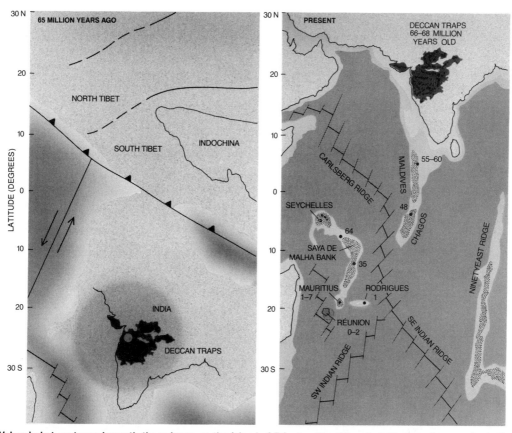

Volcanic hot spot now beneath the volcano on the island of Réunion (the Piton de la Fournase) was under India when the Deccan traps erupted, according to research by the author. Ages of the seamounts south of India increase steadily from Réunion to the Deccan Traps where the line of volcanic activity made its first appearance.

More precise data come from oil-exploration wells on the east coast of India, which crossed three thin trap flows, each separated by a layer of sedimentary rock. The lowest level of lava rests on sedimentary layers that contain fossils of a plankton called *Abatomphalus mayaroensis*, which thrived during the last 1 million years of the Cretaceous and became extinct shortly thereafter. The sedimentary rock layers between the lava flows also contain fossils from the same time, but the layers above the flows do not. *A. mayaroensis* fossils appear in strata with normal magnetic polarity that lie below (before) the K/T boundary, and disappear at the boundary itself, which is located in the next magnetically reversed set of strata.

The most reasonable conclusion from the various evidence is that Deccan volcanism began during the last normal magnetic interval of the Cretaceous, climaxed during the following reversed interval (at or very near the Cretaceous-Tertiary boundary) and ended in the first normal magnetic interval of the Cenozoic era. Magnetic and fossil studies together reduce the estimated duration of Deccan volcanism to about 500,000 years, the best time resolution that can be obtained using present techniques. The fact that Deccan volcanism—one of the largest and fastest episodes of lava flow of the past 250 million years—coincided with the K/T boundary to within the best time accuracy now attainable made it hard for us to escape the conclusion that a link existed between the Deccan Traps and the mass extinction.

Having established that the Deccan Traps erupted roughly simultaneously with the extinction at the end of the Cretaceous period, we next sought to determine whether a volcanic eruption could explain the observed features of the K/T boundary layers. In general, either a huge volcanic eruption or an asteroid impact could plausibly have produced these features.

The unusual iridium-rich deposit that appears to have been laid down simultaneously around the Earth need not have come from outer space. William H. Zoller, Ilhan Olmez, and their colleagues at the University of Maryland at College Park discovered unusual iridium enhancements in particles emitted by the Kilauea volcano in Hawaii. J. P. Toutain and G. Meyer of the Institute of Physics of the Earth found iridium in particles emitted by another volcano, the Piton de la Fournaise on the island of Réunion, which (as discussed below) is related to the Deccan volcanism. Iridium-rich volcanic dust has been found embedded in the Antarctic ice sheet, thousands of miles from the source volcanoes.

The composition of the clay at the boundary layer differs from that of the clays above and below the layer. The usual mineral in clay, illite, is replaced by smectite, which can be created when basaltic rock is altered. Recent studies of the mineralogy of the K/T boundary clay at Stevns Klint in Denmark led W. Crawford Elliott of Case Western Reserve University and his coworkers and Birger Schmitz of the University of Göteborg to conclude that the clay consists of a distinctive kind of smectite that in fact is altered volcanic ash.

The K/T boundary clay can be simulated by mixing 10 parts of material from the Earth's crust with one part of material from common stony meteorites. The Earth's mantle (the layer below the crust), however, has a composition similar to that of stony meteorites, and so could generate the same chemical anomalies. Karl K. Turekian of Yale University and Jean-Marc Luck, then at the Institute of Physics of the Earth, found that the relative abundance of the elements rhenium and osmium in the clay resembles the ratio in both meteorites and in the Earth's mantle.

Peculiar physical features in material from the K/T boundary also can be explained by either hypothesis. Boundary layers contain large numbers of tiny spherules, some of which consist of clay minerals that appear to be altered remains of molten basaltic droplets; but it is impossible to say whether they originated as volcanic ejecta or from oceanic crust melted by an asteroid impact. Matters are somewhat confused by the fact that at least some of the spheres turned out to be round fossil algae or even recent insect eggs that contaminated the material.

The discovery of shocked deformed grains of quartz crystal in K/T boundary layers, first made by Bruce F. Bohor and Glen A. Izett of the U.S. Geological Survey in Denver, is often considered the strongest evidence in favor of the impact hypothesis. Such shocked grains had been found previously only from known impact craters (such as Meteor Crater in Arizona) or from sites of underground nuclear explosions. They are produced by dynamic shock stress at more than 100,000 times atmospheric pressure; but shocked structures can be produced at much lower pressures if the rock is heated before the shock occurs, as would be the case in a volcanic eruption.

As magma rises to the Earth's surface, it decompresses and releases dissolved gases. At the same time, the magma often cools and thickens. If it cools particularly quickly, it becomes so stiff that the gases cannot escape. Pressure therefore builds up, possibly leading to an explosion and powerful shock waves. Such stresses might be sufficient to shock quartz crystals if the temperatures and duration were great enough.

Magma that is rich in silicate material is viscous and especially prone to provoke explosive eruptions; examples of silicic volcanism include Vesuvius and Mount St. Helens. In 1986, Neville L. Carter of Texas A&M University and his associates discovered evidence of shock features similar to those at the K/T boundary in rocks from some geologically recent silicic volcanic explosions, such as the large Toba, Sumatra, eruption of 75,000 years ago. Using transmission electron microscopy, Jean-Claude Doukhan of the University of Lille recently found that shock features produced by laboratory impact, meteorite impact, and those observed in samples from the K/T boundary are all different from one another in some respects and that the similarity between laboratory and meteorite features has been overstated. Shock features from K/T samples are decorated with microscopic bubbles that are not observed in samples from meteorite impacts and that seem to indicate a higher formation temperature, compatible with a volcanic origin.

Explosive silicic volcanism commonly precedes periods of relatively quiet, Deccan-type (flood basaltic) volcanism, during which basaltic lava flows freely and copiously. Of the volume of lava from known Deccan-type flows 10 to 15 percent erupts in episodes of explosive silicic volcanism. A rising plume of hot magma would melt its way through the continental crust, producing the viscous silicic (acidic) magmas that lead to explosive volcanism.

The unusual chemical and physical features in the K/T boundary layers are present worldwide. An asteroid impact could have propelled material into the stratosphere, where it would have been transported around the globe. On the other hand, Richard B. Stothers and his coworkers at the National Aeronautics and Space Administration's Goddard Space Flight Center in Greenbelt, Maryland, modeled the manner in which fountains of lava, such as those from Kilauea in Hawaii, expel dust and ejecta. When scaled up to the dimensions of the Deccan volcanism, their models predict that large amounts of material should also be lofted into the stratosphere. Atmospheric circulation would distribute material rather evenly between the two hemispheres, no matter where it was originally emitted.

The appalling consequences of an asteroid impact and a massive volcanism would be quite similar. The first effect would have been darkness resulting from large amounts of dust (either impact ejecta or volcanic ash) into the atmosphere. The darkness would have halted photosynthesis, causing food chains to collapse. Such environmental trauma appears to be reflected in the fossil record. Freshwater creatures were much less affected than land- or sea-based ones, perhaps because freshwater animals did not feed on vascular plants (as do many land-dwelling animals) or on photosynthetic plankton (an important food source for marine vertebrates that was devastated at the end of the Cretaceous).

Life would also have been confronted by large-scale toxic acid rain. The heat of a large impact would have triggered chemical reactions in the atmosphere that would in turn produce nitric acid. Alternatively, volcanic eruptions would have emitted sulfur that would form sulfuric acid in the air. The environmental effects of sulfur-rich volcanism can be significant even in the case of fairly moderate eruptions. The 1783 eruption at Laki, Iceland, killed 75 percent of all livestock and eventually 24 percent of the country's population, even though it released only 2-plus cubic miles of basaltic lava. The event was followed by strange dry fogs and an unusually cold winter in the Northern Hemisphere.

Using the Kilauea eruption as a model, Terrence M. Gerlach of Sandia National Laboratory in Albuquerque, New Mexico, estimated that the Deccan Traps injected up to 30 trillion tons of carbon dioxide, 6 trillion tons of sulfur and 60 billion tons of halogens (reactive elements such as chlorine and fluorine) into the lower atmosphere over a few hundred years. The emissions from the Laki eruption seem to have been far greater than would be expected from simply scaling up the figures for Kilauea, so the estimates may represent a lower limit. Airborne sulfur and dust from a 240-cubic-mile lava flow could decrease average global temperatures by 5–9°F.

Other factors could contribute to an opposite effect, however. Marc Javoy and Gil Michard, both of the Institute of Physics of the Earth and the University of Paris, propose that sulfur dioxide from Deccan volcanoes turned the ocean surface acidic, killing the algae that normally extract carbon dioxide from the atmosphere and then carry it to the ocean bottom when they die. Acidic ocean waters also would have dissolved carbonate sediments at the bottom, releasing trapped carbon dioxide. Altogether, atmospheric carbon dioxide levels would shoot up to about eight times the present concentration, producing a rise in temperature of 9°F. The interaction between cooling from dust and warming from carbon dioxide (which may occur on widely different time scales) is unclear, but the resulting climate gyrations probably would have been especially traumatic for the global ecosystem. Both the asteroid and volcanic hypotheses predict overlapping cooling and warming effects.

So far, the evidence discussed has been equally consistent with both hypotheses. But many details suggest that the mass extinction and odd physical processes that occurred at

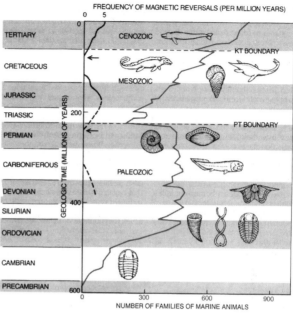

Marine animal diversity is plotted alongside the rate of reversals of the Earth's magnetic field (data prior to 165 million years ago are approximate). Two long periods with no reversals stand out: one before the Permian-Triassic extinction of 250 million years, the other before the Cretaceous-Tertiary extinction of 65 million years ago. The correlations suggest a causal relationship between the behavior of the earth's core, where the magnetic field is generated, and mass extinctions.

the end of the Cretaceous took place over hundreds of thousands of years. This period is comparable to the duration of Deccan volcanism but incompatible with a sudden asteroid impact.

A number of paleontologists have pointed out that the extinction at the end of the Cretaceous was not a single, instantaneous event. Extinction rates appear to have started to increase up to a million years before the K/T boundary. Even near the boundary, the pattern is not uniform: for instance, planktonic foraminifera and nanoplankton (microscopic calcareous algae) species exhibit different patterns of extinction and recovery. This ragged sequence is known as stepwise mass extinction.

One of the most thorough recent studies of the pattern of extinctions was conducted by Gerta Keller of Princeton University. When she analyzed the well-preserved sections of the K/T boundary in Tunisia and Texas, Keller found evidence for a first phase of extinction (also seen in the macrofossil record) that began 300,000 years before the K/T iridium event and for another extinction event that took place 50,000 years after the boundary. Keller attributes the first event to falling sea levels and global cooling.

Other evidence confirms that the Earth experienced not one but many disruptions at the end of the Cretaceous. Abrupt change occurred, for example, in the abundance of carbon-13 and oxygen-18 (respectively, light and heavy versions of these elements, whose concentrations vary according to the ocean temperature and acidity and to the number of living creatures present). Extinctions and carbon-13 fluctuations observed in strata in Spain occur in magnetic intervals that fit the same normal-reversed-normal polarity pattern found in the Deccan Traps.

Even the iridium appears to display a number of fine fluctuations near the K/T boundary. Robert Rocchia and his colleagues at the Atomic Energy Commission and National Center for Scientific Research in Gif-sur-Yvette and Saclay, France, found secondary iridium peaks above and below the primary iridium layer (corresponding to time intervals of about 10,000 years) in K/T boundary clay in Spain and Denmark. Rocchia, I, and our colleagues found that the layer of iridium enrichment in Gubbio, Italy, seems spread over about 500,000 years. The much-discussed shocked quartz crystals exhibit a similar pattern of distribution. Officer and Carter discovered that shocked minerals extend through 13-plus feet of the Gubbio section, again corresponding to a time span of about 500,000 years.

James C. Zachos of the University of Rhode Island and his coworkers measured the chemical composition of microscopic fossils from the North Pacific seafloor and found that the productivity of open-sea marine life was suppressed at the time of the K/T boundary and for about 500,000 years thereafter. They also concluded that significant environmental changes, including cooling, began at least 200,000 years before the boundary.

Some proponents of the impact theory, most prominently Piet Hut of the Institute for Advanced Study in Princeton, New Jersey, and his colleagues, quickly substituted a series of comet impacts for the single asteroid impact to explain these findings. The search for an all-encompassing answer also led to the suggestion that the Deccan Traps might mark the site of the asteroid impact. But there are many difficulties with that idea: no traces of an impact have been found in India. Robert S. White of the University of Cambridge has shown that large impacts cannot trigger massive volcanism because the section of the mantle just below the lithosphere (the relatively rigid crust and upper mantle) does not normally contain large reserves of molten rock. Moreover, Deccan volcanism started during a normal geomagnetic interval, a few hundred thousand years before the reversed magnetic interval containing the K/T iridium anomaly and the clay layer.

During the Cretaceous period, volcanism increased, the sea level rose and fell drastically, and the global mantle shifted significantly. The Cretaceous period and the one that preceded it, the Jurassic, were also times of major continental breakups. Between 120 and 85 million years ago, the Earth's magnetic field did not undergo a single magnetic reversal; but 15 to 20 million years before the K/T boundary, the field started reversing again. Reversal frequency, which indicates activity in the Earth's core and at the core-mantle boundary, has increased regularly since then to about once every 250,000 years at present.

All these features can be related to an episode of energetic mantle convection that began tens of millions of years before the K/T boundary. To me, the existence of overlapping short- and long-term geodynamic, geologic, and paleontological anomalies points to a common internal cause.

What might that cause be? A likely answer comes from the theory of mantle hotspots, developed most prominently by W. Jason Morgan of Princeton University and others. Peter L. Olson and Harvey Singer of Johns Hopkins University developed a model that may explain these regions of persistent volcanic activity. A plume of hot, low-density, and low-viscosity material rises from the lowermost parts of the mantle, forming a quasi-spherical head as it pushes its way through cooler, thicker mantle. The head keeps growing as long as it is fed by a conduit of molten rock rising from below.

White and Dan P. McKenzie, also of Cambridge, along with Mark Richards and Robert A. Duncan of Oregon State University and I, think that as a hot mantle plume rises, the crust above the plume lifts and stretches, leading to continental rifting (see "Volcanism at Rifts," by Robert S. White and Dan P. McKenzie, *Scientific American*, July 1989.) The plume material decompresses as it reaches the surface and so melts rapidly (in less than 1 million years). The head of the plume would elevate a large area of crust, so that when the magma finally broke through to the surface, it would run rapidly downhill, producing extensive flows.

The Deccan eruptions could have followed the arrival of such a head at the base of the lithosphere. Volcanism from a hot plume would be rapid and highly episodic. Individual flows would be extruded in days or weeks; the next flow would follow years to thousands of years later. The far-reaching ecological consequences of each flow could explain the stepwise mass extinctions.

The giant mantle plume that produced the Deccan Traps should have left structural and dynamic relics. In 1987, the Ocean Drilling Program, led by Duncan, explored and dated an undersea chain of volcanoes that extends from southwest India, near the Deccan Traps, to Réunion, the active volcano east of Madagascar. Réunion is a hotspot volcano—one powered by a deep, rising flow of hot magma from the mantle—that burned its way through the Indian and African continents as they drifted over it. The ages of the Réunion seamounts increase steadily from 0 to 2 million years around Réunion itself to 55 to 60 million years just south of the Deccan Traps.

Richards, Duncan, and I believe that the Réunion hotspot may represent the tail of hot magma that would be expected to follow in the wake of the plume that produced the traps. Besse, Vandamme, and I verified that the mantle hotspot now beneath Réunion was located precisely under the Deccan Traps at the end of the Cretaceous. There is no trace of the hotspot from before the K/T boundary; the episode of violent Deccan volcanism appears to mark the appearance of the hotspot at the surface of the Earth.

The internal geologic activity associated with a rising mantle plume fits the behavior of the Earth's magnetic field at the time of the K/T boundary. Slow convection of the molten iron in the Earth's outer core—6-plus miles per year—is thought to produce the Earth's magnetic field. Instabilities at the boundary between the core and the mantle above it may cause magnetic reversals.

Heat escaping from the core raises the temperature and so lowers the density of material in the deepest layer of the mantle (called the D"), which grows thicker until it becomes unstable and forms rising plumes of magma. Long durations with few or no magnetic reversals, such as the span from 120 to 85 million years ago, indicate a lack of outer core activity and the growth of the D" layer.

About 80 million years ago the layer broke up, sending enormous hot magma plumes upward. At this point, flow of heat from the core to the mantle would have increased, and magnetic reversals would have resumed. At typical mantle velocities of about 3¼ feet a year, the plumes would have traveled a few million years before reaching the surface, where the sudden decompression of the plumes would have led to explosive volcanism followed by large lava flows. Smaller, secondary plumes would not have reached the surface but could have accelerated mantle convection, seafloor spreading, sea-level changes, and other geologic disruptions that took place during the Cretaceous.

This kind of geologic upheaval may be a natural consequence of the fact that the Earth is an active, complex heat engine, composed of layers that have vastly different physical and chemical properties. Smooth, well-regulated mantle convection and brutal, plume-like instabilities are perhaps just two extremes of the ways in which the Earth's internal heat escapes to the outside.

If this is indeed the way the Earth functions, similar catastrophes should have taken place. Most major, relatively recent extinction events (those since the Mesozoic era began 250 million years ago) seem to correlate in time with a large flood basalt eruption. Interestingly, the longest known period during which the Earth's magnetic field did not reverse also ended with the largest mass extinction, the one that marked the dawn of the Mesozoic era. More than 95 percent of the marine species disappeared at that time. The 250–million-year-old Siberian Traps are a prime candidate for having caused this extinction.

Both the asteroid impact and volcanic hypotheses imply that short-term catastrophes are of great importance in shaping the evolution of life. This view would seem to contradict the concept of uniformitarianism, a guiding principle of geology that holds that the present state of the world can be explained by invoking currently occurring geologic processes over long intervals. On a qualitative level, volcanic eruptions and meteorite impacts happen all the time; they are not unusual. On a quantitative level, however, the event witnessed by the dinosaurs is unlike any other of at least the past 250 million years.

Magnetic reversals in the Earth's core and eruptions of large plumes in the mantle may be manifestations of the fact that the Earth is a chaotic system. Variations in the frequency of magnetic reversals and breakup of continents over the past few hundred million years hint that the system may be quasi-periodic: catastrophic volcanic episodes seem to occur at intervals of 200 million years, with lesser events spaced some 30 million years apart.

It is tempting to speculate that the dawn of the Paleozoic era 570 million years ago, when multicellular life first appeared, might have coincided with one such episode. Large extinctions abruptly open broad swaths of ecological space that permit new organisms to develop. Events that at first seem to have been disasters may have been agents essential in the evolution of complex life.

The Mass Extinctions of the Late Mesozoic

by Dale A. Russell

One of the most striking events in the record of life on our planet is the simultaneous disappearance at the end of the Mesozoic era, some 65 million years ago, of many kinds of reptiles, certain kinds of marine invertebrates, and certain kinds of primitive plants. For generations scholars have sought unsuccessfully to explain this event. New evidence, however, has led to a novel hypothesis: the disappearances were the result of a catastrophic disruption of the biosphere by an extraterrestrial agency.

Catastrophism is not a new doctrine in efforts to account for episodes in the history of the Earth, but it has not been a particularly popular one. Early in the nineteenth century, when geology was in its infancy, the French anatomist Georges Cuvier suggested that the past had been marked by a series of environmental "revolutions," or catastrophes. In his view, such disruptions would account for three animal disappearances: that of the mammoths at the end of the ice age, that of the many primitive mammals fossilized in rocks lying deeper than the ice-age gravels, and that of the giant reptiles fossilized in chalkbeds lying deeper still. In the decades that followed, however, the work of such pioneer geologists as Charles Lyell made it apparent that the processes of change in Earth history were of far greater duration than Cuvier had believed. Catastrophism fell from favor, to be replaced by the doctrine of gradualism. For more than a century now, paleontologists have generally agreed that whatever may have caused the disappearances at the end of the Mesozoic era, it could not have been a worldwide catastrophe.

The principal casualties among the reptiles were the dinosaurs. As an example, late in the Cretaceous period, the closing chapter of the Mesozoic, at least 15 separate families of dinosaurs, possibly representing between 50 and 70 distinct species, inhabited North America. In the rocks that were formed immediately after the Cretaceous, not one dinosaur skeleton has been found. That is why the end of the Mesozoic is generally characterized as the time when the dinosaurs became extinct. The dinosaurs were not, however, the only organisms to disappear. Among the 33 other families of reptiles that inhabited North America late in the Cretaceous were the following losses: all four of the families of marine turtles (although three of the four survived elsewhere); one of the three families of crocodilians, the Goniopholidae; two pterosaur (flying reptile) families, the Ornithocheiridae and Azhdarchidae; two ichthyosaur (marine reptile) families; and all three of the plesiosaur (also marine reptile) families, the Elasmosauridae, the Polycotylidae, and the Cimoliasauridae; and, finally, two of the eight families of lizards, the Polyglyphanodontidae (primitive skinklike land forms) and the Mosasauridae (large marine forms).

What happened? Was there a gradual or a catastrophic extinction? My own interest is primarily in the larger reptiles of the Mesozoic in North America, and so the examination of these questions I undertake here will focus mainly on the disappearance of those animals. Among the many hypotheses put forward to account for their disappearance are disruptions of the food chain, both at sea and on land; a general alteration of the environment as the sea level began to drop at the end of the Mesozoic; a sharp rise in temperature; a fall in temperature caused by volcanic dust in the atmosphere; and so on. None of these phenomena, however, would seem, by itself, to be a convincing cause of the reptilian extinctions.

In 1979, paleontologists interested in the problem were presented with a new possibility. A group of workers at the University of California at Berkeley—the geologist Walter Alvarez, his father, the physicist Luis W. Alvarez, and two physical chemists, Frank Asaro and Helen V. Michel—announced the discovery of abnormally large traces of the heavy element iridium in a marine formation near Gubbio in the Apennine mountains of Italy. The iridium was concentrated in a layer of clay, one-half to three-quarters of an inch thick, that separates marine limestone of late Cretaceous age from an overlying marine limestone of early Paleocene age. The limestone below the clay contains fossil marine organisms typical of the latest part of the Cretaceous. No organisms are preserved in the clay. In the limestone above the clay, the Cretaceous organisms are absent; they have been replaced by other organisms typical of the Paleocene.

Iridium is one of several elements geologists call siderophiles, "iron lovers." It is rarely present in the rocks of the Earth's crust but is comparatively abundant in meteorites. The steady rain of micrometeorites on the surface of the Earth (more than 70 percent of which fall into the oceans) results in modest concentrations of iridium and other siderophilic elements in the sediments that accumulate in the ocean basins.

In 1977, Walter Alvarez was working with an international group of scholars, including the paleontologist Isabella Premoli Silva of the University of Milan, who were examining the marine strata near Gubbio that include the layer of clay. Because the infall of micrometeoritic material is thought to be more or less constant, Luis Alvarez suggested that by measuring the amount of iridium in the clay it would be possible to calculate how much time had passed during the deposition of the layer. When Asaro and Michel did so, they discovered to their surprise that the iridium in the clay layer was 30 times more abundant than it was in clays from adjacent limestone strata.

Layer of clay in this photograph is about 3/4 of an inch thick. It separates two beds of marine limestone exposed near Gubbio in the Apennine mountains of Italy. The white limestone below the clay is late Mesozoic in age and the grayish limestone above it is early Cenozoic. Analysis of the clay showed it to be 30 times richer in the heavy element iridium than the clays from adjacent marine strata. This had led the geologist Walter Alvarez, his father, physicist Luis W. Alvarez, and two chemists, Frank Asaro and Helen V. Michel, all of the University of California at Berkeley, to the hypothesis that the surplus of iridium came from an extraterrestrial object, perhaps an asteroid-sized meteorite, that crashed into the Earth at the end of the Mesozoic. The Alvarez group further hypothesized that the collision was the cause of the many extinctions of marine and terrestrial organisms at that time. The coin is the size of a 25-cent piece.

If this excess of iridium had somehow been derived from terrestrial sources, the clay should have shown comparable enhancements in the other elements normally associated with the minerals that form clay. The Berkeley group's analysis disclosed a different pattern of enhancement, closer to that of the relative abundances of the elements found in meteorites. Could the surplus iridium have come from the oceanic reservoir of elements derived from micrometeorites, suddenly precipitated by some chemical event? Evidently not; neither above nor below the clay stratum was there any evidence that the normal rate of siderophile accumulation had fallen off as it should have if a precipitation had occurred. In this connection, Charles J. Orth of the Los Alamos Scientific Laboratory and his collaborators have found a similar surplus of iridium at the top of Cretaceous

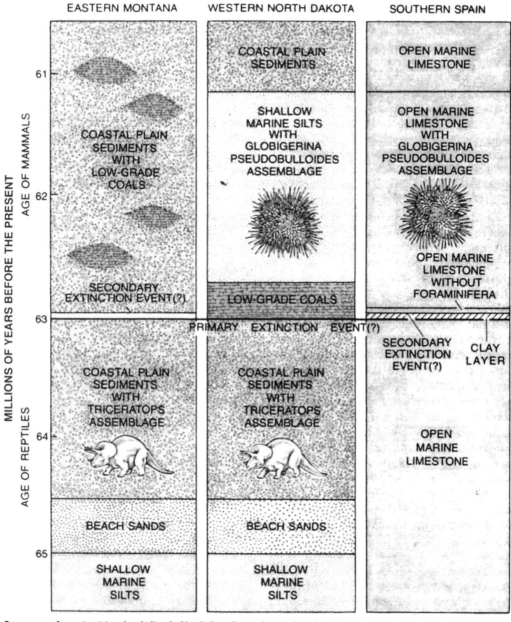

Sequence of events at two land sites in North America and a marine site in Spain at the close of the Mesozoic and the opening of the Cenozoic is presented in this table. The presence of the same replacement foraminiferan, *Globigerina pseudobulloides*, in North Dakota and Spain in the Cenozoic suggests a coincidence of marine and terrestrial extinctions.

The limestones Smit and his colleagues have been studying in Spain record a remarkable series of events. They were deposited on the floor of an open tropical sea that intruded into southern Spain in late Mesozoic times. The formations are composed almost entirely of the calcium carbonate shells and platelets of tiny foraminifera: free-floating protozoan members of the sea's zooplankton. Here, for more than 10 million years, planktonic productivity remained high, and there was no significant change in the character of the organic debris deposited on the sea floor.

Then within a layer of rock less than a quarter-inch thick (representing less than 200 years of deposition) nearly 90 percent of the foraminifera species found lower in the formation simply vanish. Those protozoans that survived attained only a tenth the size of their predecessors. As the rain of shells and platelets nearly ceased, so did the burrowing activity of bottom-dwelling invertebrate animals. A blanket of laminated red and green clays accumulated on the sea floor, reaching a thickness of about 4 inches. Conditions apparently remained stable for perhaps 20,000 years; by then, all but one species of the surviving foraminifera had dwindled to extinction.

Thereafter, life began to proliferate once more. The deposition of sediments resumed, and the ocean floor was once again plowed by bottom-dwelling invertebrates. A new assemblage of foraminifera arose and was soon followed by another, characterized by the presence of, among others, the species *Globigerina pseudobulloides*. The resurgent protozoans inhabited the ancient Spanish sea for the next 2 million years.

Half a world away, in what is now North Dakota, a great interior sea spread westward at this same time, flooding a delta area where the remains of *Triceratops* had become fossilized and had been covered by coal-bearing strata. The marine siltstones that were laid down on top of the coal hold the shells of foraminifera species belonging to the same *G. pseudobulloides* assemblage that appeared in the Spanish sea after the great foraminiferan extinction. Given the uncertainties in estimating elapsed time from the thickness of sedimentary deposits, it seems possible that the story told by the sediments here, at Fort Peck Reservoir and in southern Spain, is the same. If that is the case, the extinction of the dinosaurs on land and of the foraminifera at sea would have coincided.

The foraminifera were not the only marine organisms to die out at the end of the Mesozoic. As I noted above, so did a number of marine reptiles. So did various mollusks: the coiled-shell cephalopods known as ammonites, the squidlike cephalopods known as belemnites, and the peculiar coral-like bivalves known as rudists. Most of the major families of marine animals survived, but they lost many genera and species.

The fossil record at this crucial boundary is not as well understood with respect to larger marine animals as it is with respect to the microfauna such as the foraminifera. The reason is that the larger animals are numerous and diverse and the number of paleontologists is finite. As one example, even in such relatively well-studied formations as the chalks of Denmark, the survival rate among such important animal groups as sponges, lampshells, marine snails, and crustaceans remains uncalculated. As another example, the record of animal life in the tropical regions of the globe at this time is still poorly known. In view of the paucity of data, it is no wonder that the issue of gradualism versus catastrophism is so vigorously debated.

A crude tabulation is helpful in suggesting the magnitude of the extinctions. Compare the number of animal genera in the fossil record some 10 million years before the end of the Cretaceous with the number of genera in the record in a comparable period after the crisis. It is obvious how insecure these numerical values are. Nevertheless, the numbers appear to reflect a 50 percent decline in generic diversity worldwide. When one repeats this numbers game, counting the number of species recorded for certain plant and animal genera before and after the crisis, the result is similar. In a sample that includes mammals as representative land animals, chitinous marine algae as representative plants, and sand dollars, starfishes, and oysters as representative marine animals, the decline in species during the extinction interval is from about three species per genus to one and a half. Therefore, it seems reasonable to estimate that the biological crisis associated with the extinction of the dinosaurs also caused 75 percent of the previously existing plant and animal species to disappear. Indeed, this estimate is probably somewhat conservative.

The record of extinctions shows certain anomalies. For example, no land animal weighing more than about 55 pounds survived, and many of those that disappeared were considerably smaller. Again, the terrestrial plants of the northern regions of the Temperate Zone suffered more losses than those farther south. Yet the plants and animals of freshwater communities were scarcely affected. Much the same was probably true of deepwater marine mollusks, in the opinion of Arthur H. Clarke of Ecosearch, Inc., Mattapoisett, Massachusetts. Shallow-water marine life, however, particularly the fauna of tropical reefs, was much more profoundly altered.

Even animals that shared the same environment were not identically affected. As Eric Buffetaut of the University of Paris has pointed out, crocodiles that occupied shallow marine waters survived the extinctions whereas the mosasaurs that occupied the same habitat did not. Whatever the agents of biological stress were, disturbances in food chains included, the ability of the biosphere to resist them was evidently varied.

What is the significance of the apparently dual nature of animal and plant extinctions at the end of the Mesozoic? Were the extinctions truly separate events, with the land animals dying out first and the plants second? If they were, was the second extinction the result of stresses as severe as those that caused the first, or was it simply a quasi-successional phenomenon of biology? Whatever the answer to these questions is, humankind may have been the long-term beneficiary of the evident catastrophe. As the Mesozoic drew to a close, certain small carnivorous dinosaurs had achieved the ratio of brain weight to body weight that is characteristic of early mammals. If these presumably more intelligent reptiles had survived, their descendants might conceivably have continued to suppress the rise of the mammals, thereby preempting our own position as the brainiest animals on the planet.

The Yucatan Impact and Related Matters

by Gregory S. Paul

Within a year after the 1990 essay by Courtillot and Alvarez and Asaro appeared in the pages of *Scientific American* came a major—and rather unexpected—discovery: the identification of the enormous Chicxulab impact crater underlying the Yucatan peninsula of southern Mexico. The crater, buried under Cenozoic sediments—half under water and half on land—had been tentatively discovered a number of years before but was forgotten until researchers realized the connection between the circular geological structure and the end of the Mesozoic. At 60 to 100 miles across (depending upon competing estimates), Chicxulab is among the larger craters known in the solar system. Its formation at a location that is readily accessible is a stroke of scientific luck. It was definitely made at the K/T boundary; indeed, its effects contributed to forming the boundary. In the surrounding region, the super-waves created as the meteorite hit the then-shallow waters appear to have created extensive deposits that had long perplexed geologists. Made by a comet or asteroid about 6 to 8 miles across (the size of Mt. Everest), Chicxulab has preempted all other craters as the probable "dinosaur killer." The extraterrestrial impact hypothesis is now far and away the leading candidate for the K/T extinction. Even so, questions still surround what happened and why.

It is interesting that, despite the hard work of geologists, impacts have not been correlated with most other mass extinctions, including the Permian extinction, which, if anything, was more extensive than the later Cretaceous event. On the other hand, a number of large land animals survived the P/T (Permian/Triassic) extinction, unlike the K/T (Cretaceous/Tertiary) disaster. Various theories that comet swarms—initiated by a "Nemesis" star or other cyclical extraterrestrial patterns—have caused mass extinctions on a 30 or so million year schedule have not been supported by new evidence. Nor has the notion that a short series of impacts killed off the dinosaurs step by step. Therefore, the terrestrial mechanisms that probably lie behind the great majority of mass extinctions remain poorly understood. As it is, we have only the one major K/T event to consider in terms of extraterrestrial intervention.

At this time, four super-impact events are known from the Mesozoic before the K/T Chicxulab event, and one is known from the Cenozoic. Of these, it is possible, but not yet certain, that three of the impact events—a possible multiple impact before the end of the Triassic, another at or near the J/K (Jurassic/Cretaceous) boundary, and a linked pair of craters from the middle Cenozoic (one of which formed the Chesapeake Bay)—were about as energetic as the K/T explosion. The Mesozoic-through-Cenozoic crater survey is also incomplete. Most of the craters should have excavated the deep ocean floor, where they lie undetected, or more probably were destroyed by tectonic subduction. The last 250 million years' worth of sediments have not been carefully surveyed in search of reentry debris associated with big impacts—the K/T layer was found only because people were intensely interested in that particular zone. It is, therefore, possible that we do not know about Mesozoic and Cenozoic impacts approaching or even surpassing the scale of Chicxulab. At the same time, a recent survey of the K/T crater indicates it may not have been as extremely large as some thought.

In general, the hypothesis has been that the largest extinction of the Mesozoic was so bad because it was associated with the biggest impact of the age. An obvious problem with this premise is that the K/T impact may not have been unique after all. The other problem is that the other super-impacts did not produce correspondingly extreme extinctions of dinosaurs or other large creatures. The Late Triassic Manicouagan impact (Quebec) may or may not have been associated with some minor extinctions, but there is no evidence that dinosaurs suffered any lasting effects. The Late Jurassic, Morokweng impact (South Africa)—which possibly exceeded the power of Chicxulab—may have been associated with dinosaur and pterosaur extinctions at the J/K boundary, but these were at most modest. The well-dated middle Cenozoic Popigai-Chesapeake impact (Siberia/United States) does not closely coincide with a major extinction. This leads to an obvious question: If dinosaurs and other large land animals survived a number of super-impacts with few or no losses, why did they fail so totally when yet another piece of space debris hit planet Earth?

An important assumption about most K/T extinction scenarios is that dinosaurs were relatively easy to kill off, mainly because they were big. To understand why, we must take a look at how animals reproduce. Organisms can be sorted into two basic reproductive types: K-strategists, and r-strategists. The latter are those that produce large numbers of young, which experience high rates of mortality. Classic r-strategists are many insects and most small mammals. These are "weed" species, in that their high rates of reproduction allow them to achieve very high rates of population growth and dispersal when conditions are favorable enough to let a large percentage of juveniles survive. Because of this reproductive potential, r-strategists can quickly recover from population losses. The r-strategists are, therefore, very hard to kill off, as anyone who has targeted cockroaches and mice for extermination knows too well.

K-strategists reproduce slowly and try to keep juvenile mortality to a minimum, often via intense parental care. Large mammals are classic K-strategists. Because they cannot churn out numerous young, maximum population growth rates are rather low even under the best of circumstances. Also, big animals are always relatively few in number because each individual eats so much. Another problem for K-strategist mammals is that the young cannot survive without parental care, especially during the nursing phase. This means that a lot of adults must care for a limited number of young.

Dinosaurs were r-strategists par excellence. As far as is known, dinosaurs of all types and sizes laid large numbers of eggs, a dozen or more per season. Therefore, rates of population recovery should have been very high. Although some may have fed their young, especially when they were small nestlings, dinosaurian parental care as a whole was not as intense as in K-strategist mammals and birds. This implies that just a few hundred adults of any particular dinosaur species needed to survive to reestablish their population over a short period. What is extraordinary is that this was true of giant dinosaurs as well as small ones. Ergo, even the biggest dinosaurs were weed species, whose survival and recovery potential was probably far superior to that of giant mammals.

The r-strategy reproduction of dinosaurs helps explain why they were so successful for so long. It is notable that few major dinosaur groups went entirely extinct before the end of the Mesozoic—exceptions being prosauropods and stegosaurs. Otherwise, dinosaur history was a story of accumulative increase in diversity, with older groups continuing to live alongside the new. At no time was there a major "size squeeze," in which most or all of the large dinosaurs went extinct at the same time, to be replaced by an entirely new set of large forms reevolved from small-bodied stock. Sauropods were persistently enormous and diverse for 130 million years. In contrast, K-strategist mammalian giants have not been so successful to date. Uintatheres, arsinotheres, titanotheres, indricotheres, and megatheres have all come and gone within brief spans. Even proboscideans (elephants and extinct related forms) have been extant for only 40 million years.

Another common tacit assumption about dinosaurs is that they were more vulnerable to climatic disruption than mammals. This is a holdover from the traditional view of dinosaurs as reptiles. The presence of dinosaurs in polar regions where reptiles were sometimes absent is especially important in this regard, because it implies that the archosaurs' ability to cope with a postimpact winter should have been better than often assumed. Nor is there reason to believe that the thermoregulation and energetics of the dinosaurs of the end of the Mesozoic were grossly inferior to those of the mammals and birds that survived. Because their energy intake was probably somewhat less than birds, the vulnerability of terrestrial dinosaurs to environmental pollutants should have been less. The large brains and sophisticated sensory systems of advanced theropod dinosaurs offered them the mental agility to adjust to new and adverse conditions. Birds did enjoy an advantage over land-bound dinosaurs. Their ability to fly allowed them to move away from bad local and even regional conditions in search of less odious venues.

There never where dinosaurs like these! But there might have been if just a few of the nonavian examples had not died out 65 million years ago. Perhaps horned dinosaurs would have evolved from surviving protoceratopsids to thunder across the American plains, accompanied by hadrosaurs with the square-tipped bills ideally suited for

cropping the new grasses. All to be hunted by long-legged tyrannosaurs that had lost their useless arms. Horned rodents peer from their burrows, a small ornithopod tries to stay out of the its big relative's way, and geese head north for the summer.

The rapid reproduction and/or sophisticated thermoregulatory abilities of dinosaurs may have been an important reason that they survived a number of Mesozoic super-impacts in good order. Which returns us to events at the K/T boundary. The immediate global result of a super-impact explosion on the scale of Chicxulab is the projecting of a debris cloud at suborbital velocities around the entire planet within 40 minutes. As the debris reenters en masse, it produces an incandescent, high-altitude pyrosphere that heats up the surface as hot as a kitchen oven for some minutes. This not only has adverse effects upon exposed animals, but it initiates mass forest fires. The atmosphere is massively polluted, a thousand times worse than the harshest modern smog. Sunlight is blocked out for many months, shutting down plant growth and causing a global winter that brings snow to the Equator. Acid rain—especially severe when the impact releases materials locked up in a sulfur-rich carbonate shelf—is so intense that it is corrosive; airborne toxic metals are lethal to nonburrowing animals. Water is also polluted, hence the marine food chain collapses. As the skies clear, high levels of carbon dioxide—again the result of disruption of a carbonate shelf—cause a greenhouse effect that drives global temperatures far above even the Mesozoic norm. Major droughts ensue. Where the K/T impact was most different from other Mesozoic meteoritic explosions was where it occurred. The crater was dug into a sulfur-rich carbonate shelf, a statistically rare event. The extremely high level of atmospheric acidification and the carbon dioxide boost that should have resulted may have distinguished the Chicxulab event from impacts of similar power.

Such conditions should crush animal life. And that is the problem with the scenario. The projected conditions are too severe; things could not have been *so* bad. We know this because had such conditions been prevalent everywhere, virtually every tetrapod would have been wiped out. Yet viable populations of reptiles, some of them large-bodied, as well as amphibians, mammals, and birds *did* survive, all around the planet. The survival of amphibians and birds is especially significant. The former are exceptionally sensitive to environmental toxins because their thin skins easily absorb whatever they contact. As for birds, their high metabolic rates have two effects. First, they must constantly breathe large volumes of air and eat lots of food, so their intake of any environmental toxins is rapid and high. Also, they starve quickly when denied food. Birds and amphibians are, therefore, considered key indicators of environmental degradation, whether in mines or the biosphere as a whole. That thin-skinned amphibians and hyper-energetic birds survived the K/T impact shows that the environmental toxin and acid load was not consistently intolerable, and that there was food to be found.

There is additional evidence that the K/T crisis was not as awful as some estimate. In North America, a K/T "fern spike" indicates that almost all of the shrubs and trees were wiped out, to be replaced for a period by colonizing ferns. This may have been the result of the continent being down-range of the majority of blast—a consequence of the

Chicxulab meteorite impacting at a shallow angle from the south. Half of the world's forests may have burned. But the glass-is-half-full view observes that half of the world's flora did not burn. Especially in the Southern Hemisphere, where relatively little of the blast was directed and where there is no evidence of significant floral extinctions at the time. The simple presence of heavy cloud cover would have provided an effective local thermal shield against the short-lived pyrosphere, and subsequent rains would have put out many of the fires.

The combined evidence shows that the postimpact environment was not so harsh as to be unsurvivable, and that refuges in which to survive were available for large numbers of tetrapods. Those dinosaurs that happened to be shielded by heavy cloud cover should have survived the initial pyrosphere. Postimpact pollution levels unable to destroy all hypersensitive amphibians and birds should have harmed dinosaurs even less. Sophisticated thermometabolics were available to cope with unusual climatic fluctuations. Enough flora apparently survived to support viable fauna populations of dinosaurs. Even if most or all r-strategist dinosaur species were nearly decimated, only a few hundred individuals of a given species needed to survive in order to lay the foundations for rapid recovery. It is understandable that some or even most dinosaur species succumbed to the aftereffects of the Chixculub impact, especially in the Northern Hemisphere where the habitat degradation was most severe. What is not yet explicable is why every single species of terrestrial dinosaur in every part of the world failed to survive when Southern Hemisphere forests survived largely intact. The small birdlike dromaeosaurs and troodonts, big-brained and anatomically sophisticated, could have hunted the small mammals and lizards that survived the catastrophe. Had just a few dinosaurs managed to hang on into the early Cenozoic, they could have been the seeds for a new radiation of terrestrial dinosaurs.

What are the alternatives? Super-vulcanism is superior to an impact as an extinction agent in that its effects are extended over time, causing an attrition effect. On the other hand, repeated extinction events are similar to repeated applications of pesticides: victims tend to develop resistance. In this case, those species that are best able to survive the first event do so, and are not likely to succumb to the next—others occurred without having a marked effect on dinosaurs.

Disease as an explanation for mass extinction of dozens of species suffers from the same problem as repeated eruptions or impacts—the classic Darwinian phenomenon of resistance. It is quite difficult to kill off even all of a single species with disease; the resistant individuals that almost invariably survive are well positioned to make a comeback. Over the last half-millennium, mortality rates among various human (e.g., Amerindian) and animal (wildebeest-rinderpest epidemic) populations have often exceeded 90 percent, but no species has yet gone extinct, and full recoveries have often occurred. Killing off even a fraction of the dozens or hundreds of dinosaur species via this mode may well be impossible. Besides, birds had been flying across and between the continents and spreading disease among themselves and between other tetrapods for tens of millions of years without disastrous results.

Supposedly, diseases spread like wildfire at the end of the Cretaceous because a global drop in sea level allowed mixing of previously separated faunas. The Mesozoic, especially the Cretaceous, was an era of unusually high sea levels, and the K/T regression was a strong one by the standards of the time. The severity of the drop has caused it to be proposed as the primary cause of the extinction. The problem here is that increasing the total area of land for dinosaurs and birds to live on would probably have *helped* their fortunes, not hurt them. Some local populations might have been adversely impacted, but arguments that terrestrial dinosaurs collapsed because of continental expansion are too convoluted and based on too few analyses of a few lowland populations to be convincing.

A reproduction-extinction link starts with the observation that some reptiles, including crocodilians, have temperature-dependent sex determination: the sex of the embryo is determined by the temperature at which a particular egg is incubated. Some researchers have concluded that dinosaur sexes probably were temperature-sensitive. They further suggested that fluctuating temperatures at the end of the Cretaceous skewed the sex ratios so badly that these dinosaurs went extinct. The problems with this hypothesis are legion. In many reptiles and birds, sex is genetically determined, and it is entirely possible that the same was true of some or all dinosaurs. But even if dinosaur sex ratios were temperature-dependent enough to be disrupted, it is hard to see how this problem would suddenly wipe out every single dinosaur species after they had been spawning successfully for 150 million years—especially since crocodilians and turtles with temperature-sensitive sex determination survived any temperature fluctuations at the K/T crisis!

This brings us to the matter of climatic change. Climatic change is the classic dinosaur killer, invoked by many a paleontologist since the 1800s. If this notion is so popular, why has it never been accepted as the premiere killing agent? The climate was changing throughout the Mesozoic—one example is the sudden and sharp drop and subsequent rebound in temperature that appears to have occurred well before the end of the Cretaceous, at a time when dinosaurs were increasing in diversity—and the weather change at the end of the era was by no means extreme. There was no onset of an ice age, or long-term super-heating that left even the poles hot in the winter. For that matter, Mesozoic climates were not quite as universally warm and balmy as is usually thought. Winters were probably quite chilly in continental interiors, and there may have even been modest continental glaciation at the south pole. Dinosaurs and birds had long been living and reproducing in climates ranging from polar to tropical, wet forests to desert. As explained earlier in the book, they appear to have had well-developed thermoregulatory

Things were different after the dinosaurs disappeared. This scene shows New Mexico about 7 million years after the great extinction, when mammals were starting to become fairly large. A primitive carnivore, *Ancalagon* emerges from its burrow as its partner dines on a small crocodilian. Hardwoods formed dense forests for the first time, and the small predator *Criacus* climbs a trunk while the insectivore *Deltotherium* scampers along a branch.

systems. Dinosaurs had the option of moving if changing climate in a particular location became a problem. Climatic change appears ill suited to explain the entire collapse of the Dinosauria.

Were dramatic floral changes responsible for the K/T debacle? In the Late Cretaceous, flowering angiosperms displaced conifers, cycad relatives, and ferns as the dominant land flora, but the change had been well underway for tens of millions of years before the end of the Mesozoic. If anything, the new plants were better food sources than the old. They tended to reproduce more rapidly, grow faster, and produce more palatable leaves, larger seeds, and more nutritious fruits. A whole array of dinosaurs and birds evolved along with the new flora, and birds would continue to thrive in the new forests and grasslands of the Cenozoic.

Finally, there are the more exotic, and perhaps crucial, implications of information-processing theory, complexity theory, and chaos theory. Computer simulations of evolutionary trends and processes suggest that chaos-driven instability causes complex species communities to periodically experience self-initiated mass extinctions. It has been shown that a nonlinear response to environmental perturbation, which by itself is insufficient to directly cause a mass extinction, can initiate a runaway effect that does result in the latter. It is the ultimate combination of Murphy's Law and the snowball effect, in which things quickly go from bad to worse to catastrophic.

Conclusions

Dinosaurs were such a large, diverse, and reproductively potent group that their total extinction at a time when numerous other tetrapods survived remains amazing. To date, the extraterrestrial impact hypothesis is at best incomplete, in that a viable mechanism by which the aftereffects of the impact could destroy the entire Dinosauria—except one branch of Aves—has not yet been demonstrated. Lacking such a mechanism for total extermination, and without confirmation that the Chixculub event was uniquely powerful, the impact hypothesis cannot be considered wholly verified, although it remains superior to the alternatives. It is possible that it was the combination of events at the time that conspired to do the job. Super-vulcanism and increased disease vectors may have reduced the numbers of dinosaurs. Then an impact that would not have killed off an entire healthy dinosaur population caused the population to crash to minimal levels, leaving a battered remnant that teetered on the edge of extinction or survival, until chaotic instability wiped out the last breeding individuals.

Chapter Eight: The Paleofuture

Hi-Tech Trends and Robotic Brains

by Gregory S. Paul

(Editor's Note: As this essay speculating on the future of paleontology is my own, I've chosen to forego the usual introductory comments.)

As one century transitions into another, it is a good time to take a forward peek at what may await in the world of dinosaur paleontology. People have been discovering and analyzing dinosaurs in a scientific manner for some 180 years. The results have been nothing less than spectacular in popular and scientific terms. Revealed to enthusiast and researcher alike has been a fantastic and long-lived global bestiary of a great multitude of strange and exotic lizard-birds dramatically different from the mammal-dominated megafauna of which we are a part.

For many years, analysis was based mainly on visual examination of the external features of the bones, the few available skin impressions and trackways, with limited attempts to look into the interior structures of bones at macroscopic and microscopic levels. Such visual work will always be important, but in recent decades, high-tech methodology has assumed increasing importance in the field. Eggs, bones and entire skeletons are probed with computerized 3-D X rays and sampled for differing isotopes. Computer simulations reconstruct dinosaur skeletons in virtual worlds where they can be articulated and manipulated in ways not practical with the real bones. Dinosaur feathers are put under scanning electron microscopes, and their beta keratin contents tested. These new means of examining dinosaur remains have led to a much deeper understanding of the extinct archosaurs. The first tentative steps to use ground-penetrating radar to probe below the surface limits of human eyes have been taken.

Yet, for all that has been done so far, we have just scratched the surface. Some dinosaur faunas have been well sampled after a century or more of scouring by paleontologists, but even such well-known dinosaur faunas will continue to produce surprises, as witness the discovery of *Baryonyx* in the old Wealden beds outside London, and the new giant theropod in Germany. Other dinosaur-bearing beds in western countries have only been lightly sampled so far. In the western United States and Canada, where dinosaurologists have been roaming about for so long, there remain little formations that are only now being revealed. Lots of new dinosaurs are to be found there.

But the real potential is in the nondeveloped world. South of the United States' border, the great badlands of Mexico are only starting to be explored. In southern South America, much good work has already been done, but there, too, barren stretches of virgin Mesozoic beds beckon the twenty-first-century paleontologist.

As abundant as New World dinosaurs will continue to be, the greatest future source is the Eastern Hemisphere with its enormous continents. Think of endless Siberia, which promises to produce polar dinosaurs to match those of Alaska and the Yukon. As China becomes an integral part of the global economy, regions north and south are proving to be chock-full of dinosaurs never seen before by modern science; and even in heavily vegetated southeastern Asia, there is much to find. India will continue to be dinosaur-rich, as will southern Africa. But the mother lode promises to be the arid lands that stretch from northwestern Africa, across the vast Sahara, through the Middle East, up into the Muslim lands newly separated from the former Soviet Union, into northwestern China and Mongolia: an area of exposed sediments greater than all of North America, and much of it is still out of reach to scientists due to remoteness and politics. What fabulous finds await us there?

To date, well over 300 valid dinosaur genera have been named. As impressive as that is, it is probable that the number of nonavian dinosaur genera that lived in the Mesozoic totaled in the low thousands. Many of them were not preserved in the fossil record—especially those that lived in the erosive highlands—but it is clear that the great majority of accessible dinosaurs await discovery. Many will not be all that different from what we already know—new tyrannosaurs, prosauropods, and hadrosaurs—but others will be dramatically divergent from what we have so far seen, and some will appear deeply strange to our late Cenozoic eyes. We already have parts of some of these oddities. What, for example, did the rest of Mongolian *Deinocheirus* look like, known so far only from its colossal ornithomimid-like arms?

One thing we can be sure of is that more feathered dinosaurs will show up, many from the Yixian, others from similar quiet water sediments capable of preserving such delicate structures. The dinosaur eggs that once seemed rare have proved abundant, so we can look forward to new embryos and great nesting sites. Probably to be found are dinosaur skeletons locked in mortal combat, and trackways showing the last crucial minutes of a Mesozoic hunt. Dinosaur bones found on the North Slope of Alaska are nearly fresh because they have always been kept cool. Perhaps in chilly northern sediments impregnated with oil lie the entire, pickled bodies of dinosaurs with the traces of their last meal in their guts. We can always hope.

The kinds of dinosaurs that await future discovery are not all that will be new; so will be the technologies used to find, study, and exhibit them. Today's robotic dinosaurs are semi-static, in that they cannot walk and their motions are limited. At MIT, an advanced research and development project to produce the first, practical bipedal-walking robots includes "Troody," a small machine based on the predatory dinosaur *Troodon*. As robotics become more sophisticated, subtle, and increasingly lifelike, we can look forward to fully mobile dinorobots appearing on the scene.

Will dinosaur DNA be used to bring the beasts themselves back to life? DNA is simply a quad-digital computer, and our ability to manipulate it is growing rapidly, and in following decades may become extreme. Whole new organisms small and large will be created. In such a super-biotech environment, re-creating actual dinosaurs should, in principal, become possible. The problem is where to get the genetic information. DNA is not a highly stable molecule in the long term, and it is questionable how much, if any, dinosaur DNA has survived 65-plus million years' worth of burial. Even amber may not be an adequate preservative. Bringing them back alive is a dubious proposition, but one should never rule anything out. A lot of dinosaur DNA is available in birds, and what may be feasible is "back-engineering" a near-dinosaur by altering the genetic code of a bird, preferably a ratite. In the run-of-the-mill bird, some nonavian coding is still present but suppressed. By figuring out what is avian and what is not, and activiting the latter while deactivating the former, it may prove possible to create a facsimile of a small predatory dinosaur.

Until that day, we are stuck with fossils. Even at the beginning of the twenty-first century, many of the techniques used to find and dig up dinosaurs remain surprisingly little changed from the last turn of the century. True, jets have largely replaced steam locomotives as the primary means of traveling to a site—but not necessarily in China or India—and other vehicles have displaced horse-drawn carts and camels for getting around the badlands. In more civilized lands, field crews can pile into an SUV and head into town for root-beer floats, pizza, and showers after a hot, grimy day's work. Aerial and even satellite images can be used to help locate promising sites; global positioning receivers are regularly used to precisely locate sites; and, on occasion, ground-penetrating radar is brought in to find old bones.

But finding dinosaurs still largely consists of human beings walking along, eyeballing sediments in search of the odd scraps of bone that can be traced upslope to where an already damaged skeleton is eroding out of the ground. Recovering the specimen then consists of the same human beings sitting on their behinds, swatting at the bugs, baking in the heat, wiping the dust from their eyes, and using various hammers, picks, and brushes to uncover a fossil whose actual form they can discern only as they uncover it. Aside from the occasional use of jackhammers, this is not too different from how old Barnum Brown got *Tyrannosaurus* out of the ground. When the bones are exposed, they are splinted, papered, and plastered, using a basic method developed in the late 1800s. Then grunt power is used to haul the specimen out of the quarry—okay, sometimes a helicopter is used to lift out especially heavy or isolated pieces.

In the next couple of decades, we can expect modest increases in the use of high-tech methods of finding and researching dinosaurs. Certainly, computers will have an increasing role in imaging dinosaur remains, and then processing the data. CT scanning will continue to replace mechanical removal of sediment for examining the interiors of skulls. Our ability to use sophisticated biochemical analysis will surely improve, and will so rapidly. But the fundamental methods of fieldwork will remain largely the same.

After that, matters may see a fundamental change. Biological evolution is not intelligently guided, so is it very inefficient and slow. It took many millions of years to upgrade brains from the relatively simple organs seen in fishes to the sophisticated system we humans have between our ears. Even the recent hominid expansion of brain size, as dramatic as it was by geological standards, was a multimillion-year affair. Technology is evolution by other means, and because it *is* intelligently guided, it works literally millions of times faster than bioevolution. In just one century the speed of calculating machines has expanded an astounding trillionfold! Even so, the low-cost computers most people can afford today can do only a few hundred million calculations each second, about the same as an insect can manage. The latest room-sized supercomputers top out at a trillion or few calculations per second.

In comparison, the human brain is a compact organ that uses a few watts worth of glucose to process the equivalent of a thousand trillion calculations per second, or a petaflop. But don't be too impressed with your mind machine; for the last 100,000 years, the slow-evolving human brain has been in an evolutionary rut, showing no significant increase in performance. The speed of human-devised computers is doubling every 18 months or so, and they are also becoming increasingly complex and brainlike. Some of the latest commercial software was developed via Darwinian, neural networking programs quite different from the line-by-line programs we have grown used to. When one information processing system is changing, and another is upgrading on a subyearly basis, the trend is clear. At some point, the speed and sophistication of the new artificial information processors will exceed that of the human brain. The question is when.

IBM has announced that it plans to construct a petaflop supercomputer by 2005, well ahead of projections based on the 18-month doubling rate. This massive machine will be a mere numbers-cruncher, but it suggests that the great computer speed-up is about to leap to a new, higher level. It is quite plausible that in 20 or so years it will be possible to go to a store and acquire at modest cost a computer that can process as much information as fast as its human owner. And it won't stop there: the power of the machines will quickly soar far beyond that of mere mortals. What will people do with such incredible devices, more cladistics? Probably things more interesting.

A blind man was recently fitted with a camera, connected via wires directly to his optic lobe, for the first time allowing such a person to navigate a room. The system was crude, but soon high-resolution vision will be available to those who cannot see. To those who can see, artificial visual upgrades will be available, including infrared and ultraviolet sensitivity and night vision, as well as telescopic and microscopic performance. Imagine the dinosaurs a person will be able to find and examine with such new eyes.

But that is just for starters. At the University of Southern California, an advanced research and development program is underway to produce biochips to replace defective brain parts; human trials are expected to begin in 10 years. Of course, the chips will get better and better, while the brain will not. Soon the chips will be better than even healthy brain parts, and people will start getting silicon implants that boost their mental capacity. Keven Warwick, a professor of cybernetics at the University of Reading, has announced plans to connect his brain directly to computers in the upcoming years. It is the barest of beginnings, equivalent to the Wrights' first struggling attempts to power their new machines into the air.

In two or three decades, researchers will be connecting their brains to small computers whose power and sophistication will rival and even exceed their original thinking organs. They will use these hypercomputers to figure out how the brain generates a conscious mind and then use the knowledge to construct new superbrains that are as conscious as we are, but far more potent. There will be no limit to the intellectual power of such minds, but one hopes they will be kind enough to provide the means for average humans to join the new cybercivilization via mind transfer. (Otherwise, we're cooked.)

Before you reject such radical views of the near future, consider that if dinosaurs circa 65 million years ago could have thought about it, no doubt they would have dismissed projections of their imminent and rapid demise as absurd nonsense. After all, they had been rulers of the continents for 150 million years and were more sophisticated and intelligent than ever before. The Cambrian Revolution, during which hard-shelled organisms suddenly appeared, was remarkably rapid. In 1901, Wilbur told Orville that man would not fly for 50 years; when Orville died in 1947, Chuck Yeager had already broken the sound barrier. Extreme, rapid, and unexpected transformations from one paradigm to another are normal in Earth's history, and humans are going to lose their status as top mind sooner or later. Probably sooner.

But back to the subject at hand. Imagine that it is a few decades from now, and you are working on your latest phylogenetic study. You are one of those stubborn humans that has refused to get a mental upgrade or undergo mind transfer. While you fiddle with a caliber, taking yet another measurement of yet another dinosaur fossil in a large collection, one of those new robopaleontologists enters the room. It is part of a global complex of cyberminds that is in the final stages of a year-long project to 3-D-image and measure every dinosaur fossil in the world. That its mind includes a cyberversion of one of your former human colleagues does not help matters. You watch as it begins to thoroughly examine and precisely remember all the bones in this collection. When you ask how long it will take, the robot cheerily replies—via its little-used audio system—that it will need only a few days before moving on to the next museum. The "mind collective" will then take all this data and crunch through it in a matter of weeks, analyzing and thinking about more information than a human can hope to mull over in a million years. There is no way any human can hope to match this performance or even understand its scope. You are as obsolete as the prosauropods you are studying.

Out in the field, things have become equally as extraordinary. Gone is the day when sun-fried, bug-swatting humans walked hither and yon in search of the superficial, limited to slowly finding what is at or very near the surface or exposed in the odd excavation, natural, or artificial. At any one time, the fossils available at the surface represent only a fraction of a percent of the remains preserved deep in the ground. Now smart robots cruise over dinosaur-bearing sediments, using ultra-sophisticated remote scanning systems to probe deep into the formations, locating each fossil, which number in the millions. If a closer examination of a particular specimen is desired, no need to do something as crude as excavate it: the cybersystems use higher-resolution scanners to focus in on the skeleton and record it in detail. If it is decided to dig something up, special roboexcavators move in. These use remote scanners to discern the shape of the still-buried bones to avoid damaging them during excavation. The bones are protected by composite materials. A robotic helicopter removes the dinosaur and carries it to where it can be loaded onto the autonomous air transport. No pilots are onboard—they make too many mistakes. The few humans left in the area are allowed to watch, to amuse themselves.

What would Cope and Marsh think?

What about traveling back in time and looking at live dinosaurs in their actual habitats? Here we push speculation to the extreme. Although current physics does not necessarily rule out time travel, it is highly questionable whether it can be achieved in principle, much less in practice. There is also the problem of time paradoxes—altering the course of time by contaminating it with the presence of things from the future—a favorite plot line of dinosaurian science fiction (time traveling dinosaur hunter steps on the wee mammal that happens to contain the genetic coding that would have led to the evolution of humans and so forth). Don't hold your breath on seeing the real Mesozoic world. If it ever becomes possible, it will probably not involve human minds and bodies, which would not be able to withstand the intense energies needed to punch through time, and would be so prone to interfer with past events that they would not be temporally transportable. Time travelers would certainly be cyber in form and mind, tough enough to survive the trip, and so stealthy that the past would never be aware of their presence.

One thing we can be sure of: twenty-first-century dinosaur research is going to be very, very interesting.

APPENDIX A:
Skeletal Gallery

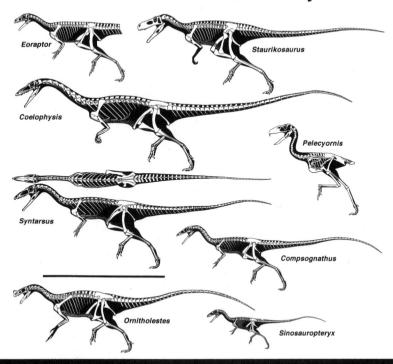

Eoraptor

Staurikosaurus

Coelophysis

Pelecyornis

Syntarsus

Compsognathus

Ornitholestes

Sinosauropteryx

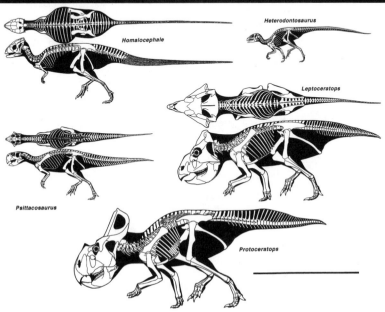

Homalocephale

Heterodontosaurus

Leptoceratops

Psittacosaurus

Protoceratops

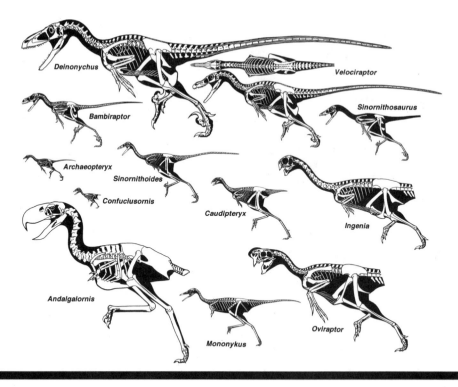

Deinonychus

Velociraptor

Bambiraptor

Sinornithosaurus

Archaeopteryx

Sinornithoides

Confuciusornis

Caudipteryx

Ingenia

Andalgalornis

Mononykus

Oviraptor

Alxasaurus

Nanshiungosaurus

Plateosaurus

Massospondylus

Aepyornis

Riojasaurus

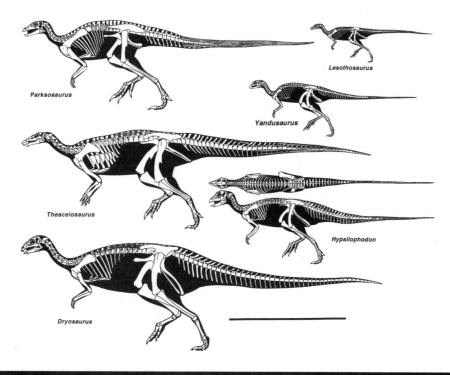

Parksosaurus

Lesothosaurus

Yandusaurus

Thescelosaurus

Hypsilophodon

Dryosaurus

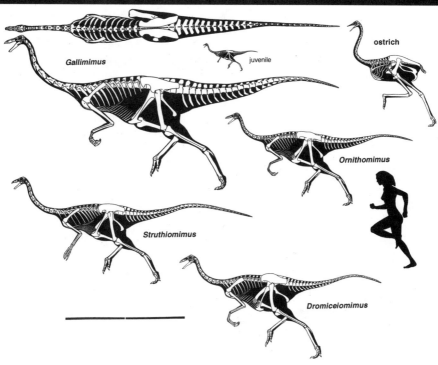

Gallimimus

juvenile

ostrich

Ornithomimus

Struthiomimus

Dromiceiomimus

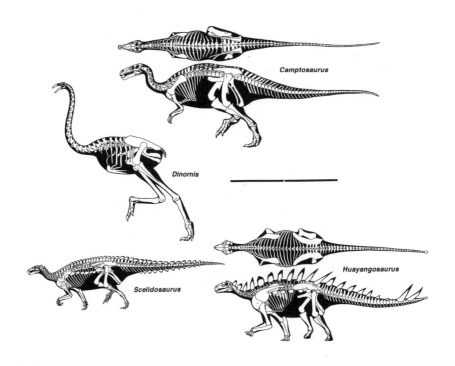

Camptosaurus

Dinornis

Scelidosaurus

Huayangosaurus

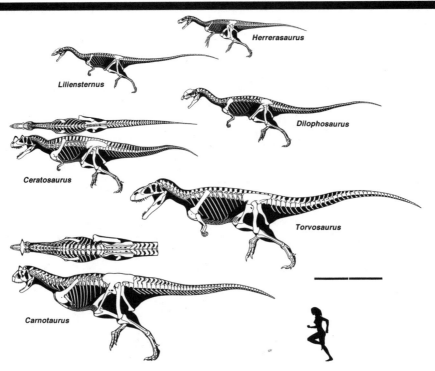

Herrerasaurus

Liliensternus

Dilophosaurus

Ceratosaurus

Torvosaurus

Carnotaurus

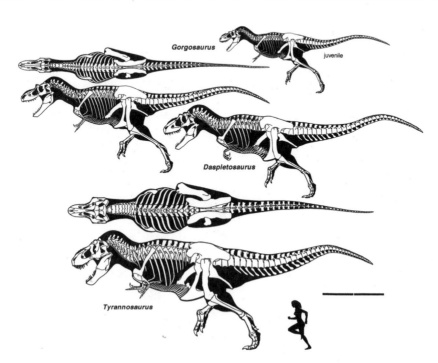

Kentrosaurus

Megalania

Stegosaurus

Tuojiangosaurus

Sauropelta

Euoplocephalus

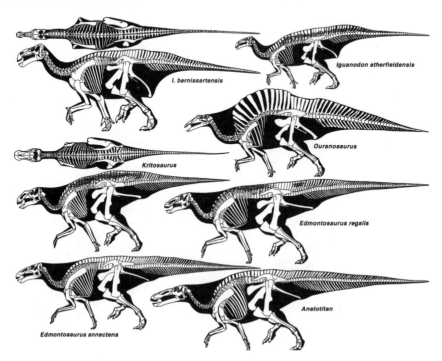

I. bernissartensis

Iguanodon atherfieldensis

Ouranosaurus

Kritosaurus

Edmontosaurus regalis

Edmontosaurus annectens

Anatotitan

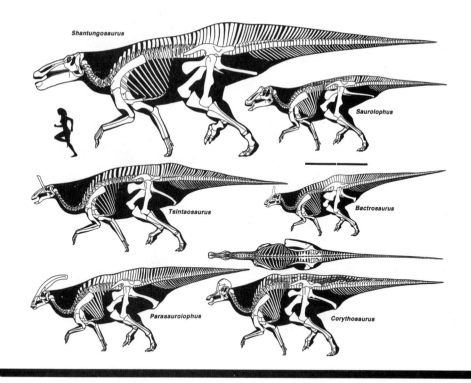

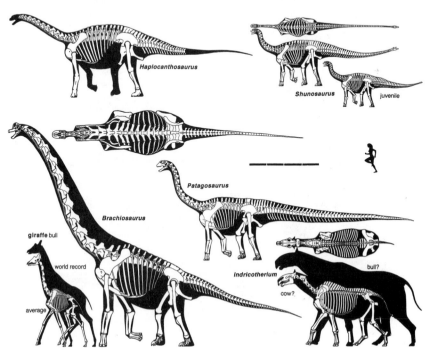

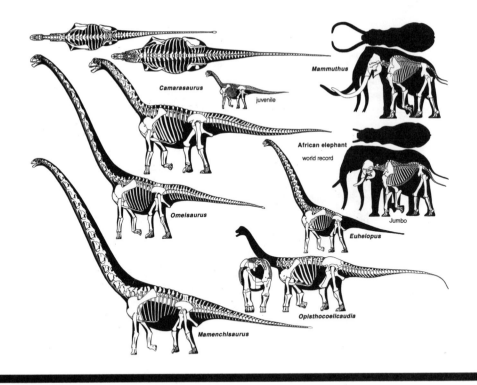

Mammuthus

Camarasaurus

juvenile

African elephant

world record

Omeisaurus

Euhelopus

Jumbo

Opisthocoelicaudia

Mamenchisaurus

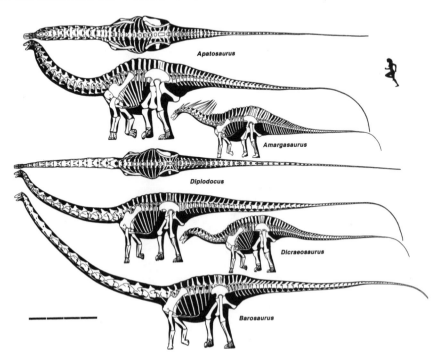

Apatosaurus

Amargasaurus

Diplodocus

Dicraeosaurus

Barosaurus

APPENDIX B:
Suggested Readings

CHAPTER ONE

Alexander, R. McN. 1989. *Dynamics of Dinosaurs and Other Extinct Giants*. New York: Columbia University Press.

Aufderheide, A. C., and C. Rodriguez-Martin, 1998. *The Cambridge Encyclopedia of Human Paleopathology*. New York: Cambridge University Press.

Chaplin, R. E. *The Study of Animal Bones from Archaeological Sites*. New York: Seminar Press.

Chure, D. J., B. B. Britt, and J. H. Madsen. "A New Specimen of Marshosaurus bicentes- imus (Theropoda) from the Morrison Formation (Late Jurassic) of Dinosaur National Monument." Paper presented at the fifty-seventh annual meeting of the Society of Vertebrate Paleontology. Paper published in the *Journal of the Society of Vertebrate Paleontology* 17, No. 3 (1997): 38A.

Christiansen, P. "Locomotion in Sauropod Dinosaurs." *GAIA 14*, (1997): 45-75.

Colbert, E. H. 1968. *Men and Dinosaurs*. New York: The Dial Press.

Currie, P.J. and K. Padian, eds. 1997. *Encyclopedia of Dinosaurs*. San Diego: Academic Press.

Czerkas, S. M and D.F. Glut. 1982. *Dinosaurs, Mammoths, and Caveman*. Dutton.

Czerkas, S. J and E.C. Olsen, eds. 1987. *Dinosaurs Past and Present*. University of Washington Press.

Damuth, J. and B.J. MacFadden, 1990. *Body Size in Mammalian Paleobiology*. Cambridge: Cambridge University Press.

Desmond, A.J. 1976. *The Hot-Blooded Dinosaurs*. New York: The Dial Press.

Edeiken, J., P. J. Hodes, and L. H. Caplan. "New Bone Production and Periosteal Reaction." *American Journal of Roentology* 97, No. 3 (1966): 708-718.

Farlow, J.O. and M.K. Brett-Surman, eds. 1997. *The Complete Dinosaur*. Bloomington: Indiana University Press.

Gallagher, W.B., T.A. Tumanova, P. Dodson, and L. Axel. "CT Scanning Asian Ankylosaurs: Paleopathology in a Tarchia Skull." Paper presented at the fifty-eighth annu- al meeting of the Society of Vertebrate Paleontology. Paper published in the *Journal of the Society of Vertebrate Paleontology* 18, No. 3 (1998): 44A-45A.

Gilmore, C. W. "Osteology of the Carnivorous Dinosauria in the United States National Museum, with Special Mention to the Genera Antrodemu (Allosaurus) and Ceratosaurus." *Bulletin of the United States National Museum*, 110 (1920): 159.

Glut, Donald F. and Don Lessem. 1993. *The Dinosaur Society's Dinosaur Encyclopedia*. Random House.

Gross, J. D., T. H Rich and Vickers-Rich, P. "Dinosaur Bone Infection." *National Geographic Research and Exploration* 9, 3 (1993): 286-293.

Hanna, R. R. (Laws), J. W. LaRock and J. R. Horner. "Pathological Brachylophosaur Bones from the Upper Cretaceous Judith River Formation, Northeastern Montana." Paper presented at the fifty-ninth annual meeting of Society of Vertebrate Paleontology. Abstracts of Papers 19, 3 (1999): 49A.

Hildebrand, M. 1974. *The Analysis of Vertebrate Structure*. New York: John Wiley and Sons.

Knight, Charles R. 1946. *Life Through the Ages*. Alfred A Knopf.

Larson, P. L. "The Black Hills Institute Tyrannosaurus—A Preliminary Report." Paper presented at the fifty-third annual meeting of the Society of Vertebrate Paleontology. Published in the *Journal of the Society of Vertebrate Paleontology* 11, No. 3 (1991): 41A-42A.

Laws, R. R. "A Specimen of *Allosaurus fragilis* from Big Horn County, Wyoming Exhibiting Several Pathologies." Paper presented at the fifty-third annual meeting Society of Vertebrate Paleontology. Published in the Journal of the Society of Vertebrae Paleontology 13, No.3 (1993): 46A.

———. "Multiple Injury and Infection in a Sub-adult Theropod Dinosaur (Allosaurus fragilis) with Comparisons to Allosaur Pathology in the Cleveland-Lloyd Collection." Journal of Vertebrate Paleontology. Forthcoming.

———. "Description and analysis of the pathologies of a sub-adult *Allosaurus fragilis* (MOR 693)." Geological Society of America Abstracts with Programs, Rocky Mountain Section 27, No. 4 (1995): 43.

———. 1996. "Paleopathological analysis of a sub-adult Allosaurus fragilis (MOR 693) from the Upper Jurassic Morrison Formation with multiple injuries and infections." Masters thesis, Montana State University, Bozeman, Montana, 61 pp.

———. "Allosaur Trauma and Infection: Paleopathological Analysis as a Tool for Lifestyle Reconstruction." Paper presented at the fifty-seventh annual meeting of the Society of Vertebrate Paleontology. Paper published in the *Journal of the Society of Vertebrate Paleontology* 17, No. 3 (1997): 59A-60A.

Madsen, J. H. "Allosaurus fragilis: A Revised Osteology." *Utah Geological and Mineral Survey, Bulletin* 109 (1976): 163.

————. "A Second New Theropod Dinosaur from the Late Jurassic of East Central Utah." *Utah Geology* 3 (1976): 51-60.

Mann, R. W., and S. P. Murphy. 1990. *Regional Atlas of Bone Disease: A Guide to Pathologic and Normal Variation in the Human Skeleton*. Springfield, Illinois: Charles C. Thomas.

McWhinney, L.A., Rothschild, B. M. and Carpenter, K. "Post-Traumatic Chronic Osteomyelitis in Stegosaurus Dermal Spikes." Paper presented at the fifty-eighth annual meeting Society of Vertebrate Paleontology. Published in the *Journal of the Society of Vertebrate Paleontology* 18, No.3 (1998): 62A.

Molnar, R. E. " Theropod Paleopathology: A Literature Survey." Edited by B. M. Rothschild and S. Shelton. *Paleopathology*. London: Archetype Press. Forthcoming.

Molnar, R. E. and J. O Farlow. 1990. "Carnosaur Paleobiology" Edited by D. B. Weishampel, P. Dodson, and H. Osmolska. *The Dinosauria*. Berkley: University of California Press. pp. 210-224.

Olshevsky, G. O. "The Archosaurian Taxa." *Mesozoic Meanderings* 1 (1978): 1-50.

Ortner, D. J. 1991 "Theoretical and Methodological Issues in Paleopathology." Edited by D. J. Ortner and A. C. Aufderheide. *Human Paleopathology: Current Syntheses and Future Options*. Washington D. C.: Smithsonian Institution Press. pp. 5-11

Ortner, D. J. and W. G. J. Putschar. 1981. *Identification of Pathological Conditions in Human Skeletal Remains*. Washington D. C.: Smithsonian Institution Press.

Petersen, K., J. I. Isakon, and J. H. Madsen, "Preliminary Study of Paleopathologies in the Cleveland-Lloyd Dinosaur Collection." *Utah Academy Proceedings* 49, No. 1 (1972): 44-47.

Rothschild, B. M. "Stress Fracture in a Ceratopsian Phalanx." *Journal of Paleontology* 62, No. 2 (1988), pp. 302-303.

————. "Radiologic Assessment of Osteoarthritis in Dinosaurs." *Annals of the Carnegie Museum* 59 (1990) : 295-301.

————. "Dinosaurian Paleopathology." Edited by J. O. Farlow and M. K. Brett-Surman. *The Complete Dinosaur*. Indianapolis: Indiana University Press, 1997. pp. 427-448

Rothschild, B. M. and L. D. Martin. 1993. *Paleopathology: Disease in the Fossil Record*. Ann Arbor: CRC Press.

Russell, D. A. "Tyrannosaurs from the Late Cretaceous of Western Canada." (National Museum of Natural Sciences) *Publications in Paleontology*, 1(1970):1-34.

Schmidt-Nielsen, K. 1984. *Scaling: Why is Animal Size so Important?* Cambridge: Cambridge University Press.

Tanke, D. H. "K/U Centrosaurine (Ornithischia: Ceratopsidae) Paleopathologies and Behavioral Implications." Paper from forty-ninth annual meeting of the Society of Vertebrate Paleontology. *Published in the Journal of the Society of Vertebrate Paleontology* 9, No. 3 (1989):41A.

————. "Paleopathologies in Late Cretaceous hadrosaurs (Reptilia: Ornithischia) from Alberta, Canada. " Papar presented at the forty-ninth annual meeting the Society of Vertebrate Paleontology. Published in the *Journal of the Society of Vertebrate Paleontology* 9, No. 3 (1989): 41A.

CHAPTER THREE

Chiappe, Luis M. "The First 85 Million Years of Avian Evolution." *Nature* 378, (November 1995): 349-355.

Currie, Philip J. and Kevin Padian, eds. 1998. *The Encyclopedia of Dinosaur.* Academic Press, 1997. (See entries "Aves" and "Bird Origins.")

Dingus L. and T. Rowe. 1998. *The Mistaken Extinction/Dinosaur Evolution and the Origin of Birds.* New York: W. H. Freeman.

Feduccia, A. 1999. *The Origin and Evolution of Birds.* New Haven and London: Yale University Press.

Gauthier, Jacques. "A Saurichian Monophyly and the Origin of Birds." *Memoirs of the California Academy of Sciences* 8, (1986):1-55.

Ostram, John H. "Archaeopteryex and the Origin of Birds." *Biological Journal of the Linnaean Society* (London) 8, No. 1(1976): 91-182.

Padian, Kevin and Luis M. Chiappe, "The Origin and Early Evolution of Birds." *Biological Reviews.* Forthcoming.

CHAPTER FOUR

Behernsmeyer, A.K. and et al. 1992.*Terrestrial Ecosystems through Time.* Chicago: The University of Chicago Press.

Padian, K., ed. 1986. *The Beginning of the Age of Dinosaurs.* Cambridge: Cambridge University Press.

Russell, D. 1989. *An Odyssey in Time: The Dinosaurs of North America.* Toronto: University of Toronto Press.

CHAPTER FIVE

Abler, William. "The Serrated Teeth of Tyrannosaurusid Dinosaurs and Biting Structures in Other Animals." *Paleobiology* 18, No.2 (1992): 161-183. (Taken from Thomas and Farlow reprint)

Abler, William. 1997. "Tooth Serrations in Carnivorous Dinosaurs." Reprint of article. *Encyclopedia of Dinosaurs*. Edited by Philip J. Currie and Kevin Padian. Academic Press.

Bakker, R. 1997. "Raptor Family Values: Allosaur Parents brought Giant Carcasses into Their Lair to Feed their Young" *Paleobiology* 18, No.2 (1992): 51-63.

Carpenter, K. 1999. *Eggs, Nests and Baby Dinosaurs*. Bloomington: Indiana University Press.

Chin, Karen, Timothy T. Tokaryk, Gregory Erickson and Lewis C. Caulk. "A King-Sized Theropod Corpolite." *Nature* 393, (June 1998): 680-682.

Chinsamy, A. "Bone Histology and Growth Trajectory of the Prosauropod Dinosaur Massospondylus carinatus Owen." *Modern Geology* 18 (1993): 319-329.

Clark, J. M., M. A., Norell, and L. M. Chiappe. "An Oviraptorid Skeleton from the Late Cretaceous of Ukhaa Tolgod, Mongolia, Preserved in an Avianlike Brooding Position Over an Oviraptorid Nest." *American Museum Novitates* 3265 (1999).

Curry, K. A. 1999. "Ontognetic Histology of Apatosaurus (Dinosauria:Sauropoda): New Insights on Growth Rates and Longvity." *Journal of Paleontology* 20:115-129.

Dodson, P. "Quantitative Aspects of Relative Growth and Sexual Dimorphism in Protoceratops." *Journal of Paleontology* 50 (1976): 929-940.

Erickson, Gregory. August 1996. "Incremental Lines of Von Ebner in Dinosaurs and the Assessment of Tooth replacement rates Using Growth Line Counts." *Proceedings of the National Academy Academy of Sciences* (United States) 382:706-708.

Erickson, Gregory, Samuel D. van Kirk, Jintung Su, Marc. E. Levenston, William E. Caler and Dennis Carter. "Bite-force Estimation for the *Tyrannosaurus Rex* from Tooth-Marked Bones." Nature 382 (August 1996):706-708

Farlow, James O. and M.K. Brett-Surnam, eds. 1997. *The Complete Dinosaur*. Bloomington: Indiana University Press.

Farlow, James O. 1987. "Lower Cretaceous Dinosaur Tracks, Paluxy River Valley, Texas." South Central G.S.A., Baylor University.

Farlow, James O. 1993. *The Dinosaurs of Dinosaur Valley State Park*. Texas Parks and Wildlife Press.

Horner, J. R. "Egg Clutches and Embryos of Two Hadrosaurian Dinosaurs." *Journal of Vertebrate Paleontology* 19 (1999):607-611.

Horner, J. R., and P.J Currie. 1994. "Embryonic and Neonatal Morphology and Ontogeny of a New Species of Hypacrosaurus (Ornithischia, Lambeosauridae) from Montana and Alberta." Edited by K. Carpenter, K. F. Hirsch, J. R. Horner. *Dinosaur Eggs and Babies*. New York: Cambridge University Press. pp. 312-336.

Horner, J. R., and D. B. Weishampel. "A Comparative Embryological Study of Two Ornithischian Dinosaurs." *Nature* 332 (1988):256-257.

Horner, J. R., A. de Ricqles, and K. Padian. "Long Bone Histology of the Hadrosaurid Dinosaur Maiasaura peeblesorum:Growth Dynamics and Physiology Based on an Ontogenetic Series of Skeletal Elements." *Journal of Vertebrate Paleontology* 20 (2000):115-129.

Horner, J.R. "Dinosaur Reproduction and Parenting." *Annual Review of Earth and Planetary Sciences* 28: 19-45. (from Abler reprint)

Larson, P. "Tyrannosaurus sex." *Paleontological Society, Special Publication* 7 (1994):139-155.

Makovicky, P. J. and G. Grellet-Tinner. 1999. "Associations Between a Specimen of Denonychus antirrhopus and Theropod Eggshell." *Annual Review of Earth and Planetary Sciences* 28:39-40. pp. 19-45. (from Abler reprint)

Manuppella, G. 1997. "Couvee, oeufs et embryons d'un Dinosaure Theropode du Jurassique superieur de Lourinha (Portugal)." C. R. Acad. Sci. Paris, Sciences de la terre et des planetes 325:71-78.

Mateus, I.H. et al. "Sauropod Dinosaur Embryos from the Late Cretaceous of Patagonia." *Nature* 396 (1998):258-261.

Norell, M. A., J. M. Clark, D. Demberelyin, et al. "A Theropod Dinosaur Embryo and the Affinities of the Flaming Cliffs Dinosaur Eggs." *Science* 266 (1994):779-782.

Raath, M.A., "Morphological Variation in Small Theropods and Its Meaning in Systematics: Evidence from Syntarsus Rhodesiensis." K. Carpenter and P. J. Currie Editors, *Dinosaur Systematics: Approaches and Perspectives*. New York: Cambridge University Press, 1990. pp. 91-106

Sampson, S. D. 1997. "Bizarre Structures and Dinosaur Evolution." *Encyclopedia of Dinosaurs*. Edited by Philip J. Currie and Kevin Padian. Academic Press pp. 39-45.

Varricchio, D. J. "Bone Microstructure of the Upper Cretaceous theropod dinosaur Troodon formosus." *Journal of Vertebrate Paleontology* 13:99-104.

Varricchio, D. J., F. Jackson, J. Borkowski, and J. R. Horner. "Nest and Egg Clutches of the Dinosaur Troodon formosus and the Evolution of Avian Reproductive Traits." *Nature* 385(1997): 247-250.

Varricchio, D. J., F. Jackson and C. N. Trueman. "A Nesting Trace with Eggs for the Cretaceous Theropod Dinosaur Troodon formosus." *Journal of Vertebrate Paleontology* 19 (1999): 91-100.

Witmer, L. M. 1995. "The Extant Phylogenetic Bracket and the Importance of Reconstructing Soft Tissues in Fossils." Edited by J. Thomason. Functional Morphology in Vertebrate Paleontology. New York: Cambridge University Press. pp. 19-33.

CHAPTER SIX

Gregory, R.T., C.B. Douthitt, et al. "Oxegen Isotopic Composition of Carbohydrate Concentrations from the Lower Cretaceeous of Victoria, Australia." *Earth and Planetary Science Letters*, 92 No.1 (February 1989): 27-42.

Parrish, J.T., R.A Spicer, J. D. Douglas, et al. "Continental Climate near the Albian South Pole and Comparison with Climate near the North Pole." *Geological Society of America, Abstracts with Programs* 23, 5 (Annual Meeting, 1991): A302.

Rich, P.V. et al. "Evidence for Tempatures and Biological Diversity in Cretaceous High Altitudes." *Science* 242 (December 1988): pp.1403-1406.

T.H., Rich, P.V. Rich et al. "Polar Dinosaurs and Biotas of the Early cretaceous of Southeastren Australia." National Geographic Research 5, No. 1 (Winter 1989): 15-53.

CHAPTER SEVEN

Alverez, W. "Toward A Theory of Impact Crisises." *Eros* 67, No.5 (September 1986): 645-658.

Muller, R.A. 1988. *Nemesis*. Weinfeld and Nicolson.

Raup, David M. 1986. *Nemesis Affair: A Story of Death of Dinosaurs and the Ways of Science*. W.W. Norton and Co.

Sharpton, V.L and P. D. Ward, eds. "Global Catastrophes in Earth History: An Interdisciplinary Conference on Impacts, Volcanism and Mass Mortality." *Geological Society of America Special Paper* 247. Forthcoming.

Silver, Leon T. and Peter H. Schultz. "Geological Implications of the Impacts of the Large Asteroids and Comets on the Earth." *Geological Society of America Special Paper* 190 (1982).

AUTHOR BIOGRAPHIES

WILLIAM L. ABLER received a doctorate in linguistics from the University of Pennsylvania in 1971. Following a postdoctoral appointment in neuropsychology at Stanford University, he joined the faculty of the linguistics department at the Illinois Institute of Technology. His interests in human origins and evolution eventually led him to contemplate animal models for human evolution and onto the study of dinosaurs, particularly their brains. The appeal of dinosaurs led him to his current position in the department of geology at the Field Museum, Chicago.

WALTER ALVAREZ and Frank Asaro are longtime collaborators on the impact hypothesis of mass extinction. It was their sensational article published in *Scientific American* that forced the paleontological community to look at and ultimately accept the catastrophic-impact theory as a main cause for the demise of the dinosaurs. Alvarez is a geologist at the University of California, Berkeley. His laboratory is the Apennine Mountains of Italy where he is an honorary citizen of the town of Piobbico.

FRANK ASARO is a nuclear chemist at the Lawrence Berkeley Laboratory. Before his work on mass extinctions with Walter Alvarez, he made chemical analysis of archeological materials. He determined that one of Berkeley's prized possessions, an inscribed brass plate attributed to Sir Francis Drake, was not authentic.

ROBERT BAKKER is dinosaur curator of the Tate Museum of Wyoming and an acknowledged rebel in the field of dinosaur paleontology. His most famous work is the book *The Dinosaur Heresies*, in which he proposed the theory that dinosaurs may have been warm-blooded, active animals rather than sluggish reptilian swamp-dwellers. This continued the revolution that began with the *Scientific American* publication of the essay "Dinosaur Renaissance." More recently, Bakker served as an unofficial consultant to the effects team that created the dinosaurs for the film *Jurassic Park*.

REESE BARRICK received her Ph.D. from the University of Southern California in the geological sciences, where she learned the utility of oxygen isotopes. She first applied this technique to marine vertebrate skeletons (dolphins, sharks, fish, and turtles) to reconstruct Miocene ocean temperatures and sea level changes due to glaciation. From this start she developed her research interests in the application of stable isotopes to paleobiological and paleoecological questions. She has used oxygen isotopes to derive heat flow distributions and body temperature variability in dinosaurs, allowing interpretations of their physiology. She is presently a visiting assistant professor at North Carolina State University.

MICHAEL BENTON's main interests include the origins and evolution of dinosaurs. At the moment, he is working on a project concerning the origins of the giant plant-eating sauropodomorph dinosaurs, focusing on remains of some of the earliest beasts from the Triassic of England and Brazil. He is currently a professor of vertebrate paleontology at the University of Bristol. He has written over 30 books about dinosaurs and paleobiology.

ANUSUYA CHINSAMY is a paleobiologist employed at the South African Museum, Cape Town, and the zoology department of the University of Cape Town, where she is an associate professor. She is a specialist on the bone microstructure of both fossil and living animals. She did her doctorate at the University of Witwatersrand, Johannesburg, South Africa. She lives near Cape Town and has two children.

PER CHRISTIANSEN has mainly worked with the problems of locomotion in large vertebrates. He has published papers on the allometry of the limb bones in extant mammals and compared the results with allometry analyses on bones from sauropods and theropods. His paper on sauropod locomotion received an award from the Danish Society of Natural History in 1997. His work on dinosaur limb anatomy has appeared in the widely acclaimed *Encyclopedia of Dinosaurs* (Philip J. Currie & Kevin Padian eds., Academic Press, 1997), as well as similar chapters in the new volume *Dinosaur Biology*, (Currie & Padian, eds.). Together with Greg Paul he has published two papers regarding ceratopsian limb postures, gaits and locomotion, demonstrating that ceratopsians did indeed walk like large mammals.

DAN CHURE has harbored a lifelong devotion to studying dinosaurs. Since 1979, he has served as park paleontologist at Dinosaur National Monument in Utah, a unit of the National Park Service established to protect the Late Jurassic Carnegie Quarry. As research scientist at Dinosaur National Monument, his interests include the terrestrial ecosystems of the Mesozoic and their paleo-biodiversity. He is particularly interested in the evolution of carnivorous dinosaurs, having just completed a worldwide revision of allosaurid theropods, as well as papers on theropods from Europe and Japan.

ANDREW CONSTANTINE is a geologist particularly interested in reconstructing ancient physical environments. He received his doctorate from Monash University with a dissertation devoted to a detailed analysis of the rocks in which dinosaur fossils from southeastern Australia are found.

VINCENT E. COURTILLOT has studied a range of geologic phenomena, including the variation of the Earth's magnetic field and polar wander. He is a professor of geophysics at the Institute of Physics of the Earth in Paris and director of the institute's Paleomagnetism and Geodynamics Laboratory. Courtillot is also director of research and graduate studies for the French university system in the Ministry of National Education.

PHILIP J. CURRIE is currently the curator of dinosaurs at the Royal Tyrrell Museum of Paleontology, an adjunct associate professor at the University of Calgary, and an adjunct professor at the University of Saskatchewan. His work on dinosaurs focuses on problems with growth and variation, the anatomy and relationships of carnivorous dinosaurs, and the origin of birds. His fieldwork, connected with his research, has been concentrated in Alberta, British Columbia, the Arctic, Argentina and China. He has appeared in more than

600 newspaper, magazine, radio, film and television articles and programs, including *Canadian Geographic, Discover, Equinox, Macleans, National Geographic Magazine, New York Times, Time, The Today Show*, the Discovery Channel's *Paleoworld*, PBS's *Nova* series, and a CBS primetime program on dinosaurs.

ANDRZEJ ELZANOWSKI is a professor and Chair of Vertebrate Zoology at the University of Wroclaw, Poland, and a longtime research associate of the National Museum of Natural History, Smithsonian Institution, Washington, D.C. He is a former senior research associate of the National Research Council and grantee of the National Geographic Society. His main areas of research are morphology and evolution of birds. As a member of the Polish-Mongolian Paleontogical Expedition to the Gobi Desert he started the ongoing revolution in avian Mesozoic paleontology by the discovery of a then-novel type of Cretaceous bird. His anatomical work contributed to our understanding of the early evolution of extant birds, especially of the ratites and tinamous. He is currently the assistant director of the Preston Health & Activity Center at Western Kentucky University.

GREGORY M. ERICKSON has studied dinosaurs since his first expedition to the Hell Creek Formations badlands of eastern Montana in 1986. He received his master's degree under Jack Horner in 1992 at Montana State University and a doctorate with Marvalee Wake in 1997 from the University of California, Berkeley. Erickson is currently conducting post-doctoral research at Stanford and Brown universities aimed at understanding the form, function, development, and evolution of the vertebrate skeleton. Tyrannosaurus rex has been one of his favorite study animals in this pursuit. He has won the Romer Prize from the society of Vertebrate Paleontology, the Stoye Award from the American Society of Ichthyologists and Herpetologists, and the Davis Award from the Society for Integrative and Comparative Biology. He will shortly become a faculty member in the department of biological science at Florida State University.

AMES O. FARLOW is a professor of geology at Indiana-Purdue University, Fort Wayne. His research is concerned with the functional morphology and paleobiology of fossil vertebrates, particularly dinosaurs. He has worked on the functional significance of tooth shape in carnivorous dinosaurs, the plates of Stegosaurus, dinosaur locomotion, and the paleobiological interpretation of dinosaur footprints.

TIM FLANNERY is director of the South Australian Museum in Adelaide. His research area involves both fossil and recent mammals. He has written a number of books on the living mammals of the southwest Pacific and New Guinea. He is best known to the Australian public for his book *The Future Eaters: An Ecological History of the Australasian Lands and Its People*. He received his Master's degree in geology from Monash University and his doctorate in biology from the University of New South Wales, Sydney.

REBECCA R. HANNA is interested in using paleopathology (the study of disease in the fossil record) to better understand dinosaur behavior and physiology, and taphonomy (the study of a fossil's life history) as a tool for site interpretation. Her thesis research at Museum

417

of the Rockies and Montana State University-Bozeman focused on interpreting injury and infection in allosaur bones. She is currently curator of paleontology at the Old Trail Museum in Choteau, Montana, where she oversees their dinosaur paleontology field school, collections, and research program.

THOMAS R. HOLTZ, Jr. is a dinosaur paleontologist with the department of geology at the University of Maryland, College Park. His primary focus of research is the evolution and ecomorphology of the theropod dinosaurs, and in tyrannosaurids in particular. His work on the phylogeny of the dinosaurs helped establish that tyrant dinosaurs were members of the advanced bird-like coelurosaurs rather than the last radiation of the carnosaurs. Biomechanical studies by Holtz have supported the hypothesis of tyrant dinosaurs, ornithomimosaurs, and some other theropods as being specialized running forms. His current work includes examining the effect of continental break-up and accretion on dinosaurian diversity and on feeding adaptations in different groups of carnivorous dinosaurs.

GUY LEAHY holds a master's degree in physical education(exercise-science emphasis) from Western Washington University. He also holds a Bachelor's degree in telecommunication and film from the University of Oregon, in addition to minors in geology and anthropology. His research has appeared in several journals, including *Nature*, *Geology* and the *Canadian Journal of Earth Sciences*. His research interests include the evolution of endothermy in vertebrates, the Cretaceous-Tertiary extinction event, and magnetobiostratigraphy. He is currently the assistant director of the Preston Health and Activity Center at Western Kentucky University.

DAVID NORMAN is the director of the Sedgwick Museum at the University of Cambridge (UK) and teaches undergraduates in the departments of earth sciences and zoology. He works primarily on ornithischian dinosaurs (the anatomy, biology, systematics and evolution) and directs a research group of vertebrate paleontologists. Dr. Norman has written a number of research articles and books on dinosaurs, and has been occasionally involved in TV and radio work associated with prehistoric life. Most recently, he consulted in the production of the BBC-TV series *Walking with Dinosaurs*. Dr. Norman accepted a research position at the Smithsonian Institution in Washington, D.C. while on sabbatical leave from his position at Cambridge.

GEORGE OLSHEVSKY is a paleontologist, mathematician, and member of the Society of Vertebrate Paleontology. His lifelong interest in dinosaurs has led him to 30 years of devoted study. He has participated in dinosaur digs in Alberta, Canada, as well as contributed to a number of booklets and periodicals about dinosaurs. Olshevsky also works as a freelance editor, indexer, writer, and publisher, and has established on of the world's largest and most complete collections of dinosaur reference libraries. Known on the Internet as "Dinogeorge," he maintains his dinosaur Web site at http://members.aol.com/Dinogeorge/index.html and may also be reached by e-mail at Dinogeorge@aol.com.

KEVIN PADIAN and **LUIS M. CHIAPPE** are frequent collaborators. Padian is a professor of integrative biology and curator in the Museum of Paleontology at the University of

418

California, Berkeley. He is also president of the National Center for Science Education. Chiappe, who has extensively studied the radiation of birds during the Cretaceous period, is Chapman Fellow and research associate at the American Museum of Natural History in New York City and adjunct professor at the City University of New York.

GREGORY S. PAUL is a freelance scientist and artist whose interests for as long as he can recall have focused on dinosaurs and evolution, past and future. He has written for *Scientific American, Bioscience, The New York Times* and numerous other technical journals and volumes. His art has appeared in *Time, Smithsonian, Science News* and many other titles. Paul is author of the popular *Predatory Dinosaurs of the World*, and the co-author of *Beyond Humanity: Cyberevolution and Future Minds*. He provided design studies for the Tyrannosaurus and "raptors" that appeared in Steven Spielberg's *Jurassic Park*, for which he received a screen credit. Paul studied informally at The Johns Hopkins University, and has on occasion helped find and dig up dinosaurs.

THOMAS H. RICH and **PATRICIA VICKERS-RICH** collaborate in the study of Mesozoic tetrapods. Rich is curator of vertebrate paleontology at Museum Victoria in Melbourne. He conducts research on the evolutionary patterns of Mesozoic vertebrates, specializing in primitive mammals and ornithischian dinosaurs. Vickers-Rich is a professor of paleontology in the earth sciences department at Monash University as well as the director of the Monash Science Center. She is interested in reconstructing ancient environments, especially those without modern analogues, and in analyzing rapid biotic change, such as mass extinctions. The Riches received undergraduate degrees in paleontology from the University of California at Berkeley and doctorates in geology from Columbia University. They live near Melbourne and have two children.

KRISTINA CURRY ROGERS is currently a Ph.D. candidate at the State University of New York at Stony Brook. She is employed by the Science Museum of Minnesota, and is adjunct faculty at Macalester College. Rodgers' work focuses on several major questions of dinosaur biology and evolution. Rodgers' doctoral dissertation is on the evolutionary history of a unique and long-lived group of sauropod dinosaurs, the Titanosauria. Crews from the State University of New York at Stony Brook and the Université d'Antanarivo discovered an amazing diversity of Mesozoic animals in the Maevarano Formation northwestern Madagascar, including two of the most complete titanosaurs ever found. Rodgers' doctoral dissertation includes description of these new animals in addition to an evolutionary study of this unique group of sauropod dinosaurs.

DALE RUSSELL currently holds a joint appointment at North Carolina State University as both a visiting professor and the senior curator for the North Carolina Museum of Natural Sciences. After earning his Ph.D. from Columbia University, he has held previous posts including Curator of Fossil Vertebrates/Dinosaurs at the Canadian Museum of Nature in Ottawa, Associate Editor for the Canadian Journal of Earth Sciences and Chief of the Paleobiology Division at the National Museum of Natural Science. He has served on numerous committees including the department of Vertebrate Paleontology at the American

Museum of Natural History, the Fernbank Museum of Natural History, and the Royal Tyrrell Museum of Paleontology. Russell has also participated in many field operations including, "The Dinosaur Project: China, Canada, Alberta" from 1986-1990. His research interest and emphasis focuses on the biogeography of southern dinosaurs, how dinosaur-dominated ecosystems differed from older and younger ecosystems, and the recognition of trends in the evolution of life on Earth that may occur on Earth-like planets elsewhere. He has published numerous titles including *An Odyssey in Time: The Dinosaurs of North America*. Dr. Russell lives with his wife in North Carolina.

DAVID A. THOMAS, a sculptor, cast the world's first life-size bronze statue of a dinosaur in 1985 for the New Mexico Museum of Natural History and Science in Albuquerque. Some of Thomas's other dinosaur creations are on display at the Anniston Museum of Natural History in Alabama, the College of Eastern Utah Museum, and the National Museum of Natural Science of Taiwan. His desire to make these statues technically accurate led him to a study of animal gaits and tracks.

DAVID J. G. VARRICCHIO's primary research focus is the unusual dinosaur Troodon. His past work has included studies of Troodon growth and reproduction and he is currently working on a detailed skeletal description of Troodon. He has geology degrees from Cornell and Georgia, and received his Ph.D. under Jack Horner at Montana State University in 1995. Since 1989, he has conducted paleontology fieldwork in the Cretaceous beds of western Montana. His research focuses on theropod dinosaurs, including birds, and on understanding the origin of fossil localities such as bonebed and nesting localities. He participated in three successful expeditions led by Paul Sereno (University of Chicago) to the Cretaceous of Argentina and Africa. He continues to work as an affiliate professor at Montana State University and as the field chief at the Museum of the Rockies Paleontology Field School.

Introduction
Sinclair Oil logo © copyright and ® registered 2000 Sinclair Oil Corporation
Bambiraptor art © 2000 Gregory S. Paul

Chapter 1
A Brief History of Dinosaur Paleontology
Iguanodon art by Waterhouse Hawkins courtesy of Gregory S. Paul
Horizontal Iguanodon art by Waterhouse Hawkins courtesy of Gregory S. Paul
Iguanodon art © 2000 Gregory S. Paul
Diplodocus art by Oliver Hays courtesy of Gregory S. Paul
Brachiosaurus art © 2000 Gregory S. Paul
Tarbosaurus vs. Therizinosaurus art © 2000 Gregory S. Paul
Troodon and Parksosaurus art © 2000 Gregory S. Paul

Chapter 2
Digging Them Up
Photographs courtesy of Gregory S. Paul

Noses, Lungs, and Guts
Allosaurus art; Diagram, nasal cavity; Diagram, Scipionyx; Diagram, lung comparison; Diagram, internal organ; Diagram, Pachycephalosaur skeleton. All art © 2000 Gregory S. Paul.

Dinosaur Biomechanics
Diagram, Kritiosaurus; Diagram, Brachiosaurus; Diagram, Diplodocus and Dicreosaurus; Diagram, Brachiosaurus; Diagram, Brachiosaurus limbs; Diagram, Triceratops skeleton; Triceratops art. All art © 2000 Gregory S. Paul.

Restoring the Life Appearances of Dinosaurs
Diagram, Tyrannosaurus rex skeleton; Tyrannosaurus rex art; Diagram, Dromaeosaur skull; Gallimimus art; Diagram, neck bones; Diagram, dinosaur trackways; Diagram, quadrupedal tracks; Diagram, restored feet; Diagram, Triceratops galloping; Muttaburrqsaurus art; Baby Muttaburrqsaurus art; Plateosaurus art; Diagram, muscle study of Brachiosaurus; Stegosaurus art; Homalocephale art; Diagram, Anchiceratops skeleton; Diagram, Chasmosaurus muscle study; Pentaceratops art; Parasaurolophus art. All art © 2000 Gregory S. Paul.

A Quick History of Dinosaur Art
Hadrosaurus by Waterhouse Hawkins courtesy of Gregory S. Paul
Allosaurus and Apatosaurus art © 2000 Gregory S. Paul
Einioisaurs sculpture by Gregory Wenzel courtesy of Gregory S. Paul
Horned Agathaumas by Charles Knight courtesy of the American Museum of Natural History

The Art of Charles R. Knight
All art courtesy of the American Museum of Natural History

Dinosaurs Got Hurt Too
Daspletosaurus tyrannosaurs © 2000 Gregory S. Paul
T. rex bite marks © 2000 Gregory Erickson

Chapter 3
Naming the Dinosaurs
Apatosaurus skeleton © 2000 Gregory S. Paul
Coelophysis skeleton © 2000 Gregory S. Paul

Classification of Evolution of the Dinosaur Groups
Cladograms courtesy of Tom Holtz

Feathered Dinosaurs
Caudipteryx skeleton © 2000 Gregory S. Paul
Caudipteryx art © 2000 Joe Tucciaroni
Sinosauropteryx photo courtesy of Philip Currie
Sinosauropteryx art © 2000 Louis Rey
Oviraptor art © 2000 Gregory S. Paul
Oviraptor art © 2000 Louis Rey

The Origin of Birds and Their Flight
All art courtesy of Scientific American Inc.

Chapter 4
The Evolution of Mesozoic Flora and Fauna
Late Triassic Forest art © 2000 Doug Henderson
Late Jurassic art © 2000 Robert Walters
Iguanodont art © 2000 Gregory S. Paul
Late Cretaceous art © 2000 Robert Walters

Color Section
Sauropod art courtesy of the American Museum of Natural History
North American dinosaurs by Charles Knight courtesy of the Field Museum of Natural History
T. rex and Triceratops courtesy of the Field Museum of Natural History
Apatosaurus art by Bill Berry
Brachiosaurus art by Bill Berry
Diplodicus art by Bill Berry
Camarasaurus head by Bill Berry
Dryosaurus and camptosaurs by Bill Berry
Camptosaurus grazing by Bill Berry
Dilophosaurus art by Bill Berry
Various dinosaurs from the Jurassic by Bill Berry
Allosaur and Iguanodon art by Bill Berry
Bill Berry art courtesy of Dinosaur National Monument

Dilophosaurus art © 2000 Gregory S. Paul
Coelophysis art © 2000 Douglas Henderson
Heterodontosaurs art © 2000 Gregory S. Paul
Gasosaurus and Shunosaurus art © 2000 Gregory S. Paul
Yangchuanosaurus © 2000 Gregory S. Paul
Omeisaurus art © 2000 Gregory S. Paul
Apatosaurus art © 2000 Gregory S. Paul
Allosaurus and Stegosaurus © 2000 Gregory S. Paul
Sinosauropteryx art © 2000 Michael Skrepnick
Caudipteryx art © 2000 Michael Skrepnick
Jobaria art © 2000 Michael Skrepnick
Suchomimus art © 2000 Michael Skrepnick
Brachiosaurus art © 2000 Gregory S. Paul
Carnotaurus art © 2000 Luis Rey
Centrosaurus art © 2000 Gregory S. Paul
Maiasaura art © 2000 Gregory S. Paul
Kritosaurus © 2000 Gregory S. Paul
Dinosaurs under the Southern Lights by Peter Trusler courtesy of Tom Rich

Chapter 5
Tracking Dinosaur Society
Anatotitan art © 2000 Gregory S. Paul
Diagram, Tetropod trackways © 2000 Gregory S. Paul
Diagram, Sauropod trackways © 2000 Gregory S. Paul
T. rex and Triceratops art © 2000 Gregory S. Paul
Acrocanthosaurus and Pleurocoelus art © 2000 Gregory S. Paul

Tracking a Dinosaur Attack
Fossil trackway © 2000 American Museum of Natural History
Diagram, Bird's trackway © 2000 Slim Films
Attack sequence © 2000 Slim Films

Feeding Adaptations in the Dinosauria
Riojasaurus art © 2000 Gregory S. Paul
T. rex, wolf and Komodo skulls © 2000 Gregory S. Paul
Struthiomimus art © Gregory S. Paul
Triceratops and other herbivore skulls © 2000 Gregory S. Paul
Ouranosaurus art © Gregory S. Paul

Breathing Life into Tyrannosaurus rex
Coprolite photograph courtesy of Karen Chin
T. rex skull photograph © 2000 Gregory Erickson
T. rex tooth photograph © 2000 Francois Gohier/Photo Researchers

Reproduction and Parenting
Cladograms courtesy of David Varricchio
Chasmosaurus art © 2000 Gregory S. Paul

Gallimimus art © 2000 Gregory S. Paul
Sauropod hatchling art © 2000 Gregory S. Paul
Nesting Oviraptor art © 2000 Gregory S. Paul

Chapter 6
Growth Rates Among the Dinosaurs
Hypacrosaurus art © 2000 Gregory S. Paul
Animal silhouettes © 2000 Gregory S. Paul
Size chart courtesy of Kristina Curry Rodgers

The Thermodynamics of Dinosaurs
Syntarsus vs. Massospondylus art © 2000 Gregory S. Paul

Australia's Polar Dinosaurs
Leaellynasaurus © 2000 Patricia Wynne
Ausktribosphenos art courtesy of Tom Rich

Dinosaur Renaissance
Therapsid art © 2000 Sarah Landry
Syntarsis art © 2000 Sarah Landry
Longisquama art © 2000 Sarah Landry
Deinonychus and Microvenator art © 2000 Sarah Landry

Chapter 7
An Extraterrestrial Impact
Iridium anomaly map © 2000 George Retseck
Iridium deposit chart © 2000 George Retseck
Basaltic spherules photo © 2000 Alessandro Montanari

A Volcanic Eruption
Volcanic hot spots map © 2000 George Retseck
Marine animal diversity chart © 2000 George Retseck

The Mass Extinction of the Late Mesozoic
K/T geological boundary photograph © 2000 Alessandro Montanari
Comparison of marine and terrestrial extinctions © 2000 Patricia Wynne

The Yucatan Impact and Related Matters
Future-dinosaur art © 2000 Gregory S. Paul
New Mexico flora and fauna art © 2000 Gregory S. Paul

Appendix A
Skeletal Gallery
All skeletal restoration art © 2000 Gregory S. Paul